Contents

INTRODUCTION IV

HOW TO USE THIS BOOK V

PART ONE — FUEL SYSTEM AND EMISSION CONTROL FUNDAMENTALS 1

Chapter 1 — Introduction to Fuel Systems and Emission Controls 2
- Air Pollution — A Perspective 2
- Major Pollutants 3
- Pollution and the Automobile 5
- Smog — Climatic Reaction With Air Pollutants 6
- Air Pollution Legislation and Regulatory Agencies 7
- Automotive Emission Controls 8
- Review Questions 9

Chapter 2 — Engine Operating Principles 10
- Air Pressure — High and Low 10
- Reciprocating Engine Components 11
- The Four-Stroke Cycle 13
- Cylinder and Piston Arrangement 14
- Valve Arrangement 16
- Engine Displacement and Compression Ratio 17
- Other Engine Types 19
- Review Questions 25

Chapter 3 — Engine Air-Fuel Requirements 26
- Airflow Requirements 26
- Air-Fuel Ratios 27
- Fuel Distribution 30
- Fuel Composition 32
- Review Questions 34

PART TWO — THE FUEL SYSTEM 35

Chapter 4 — Fuel Tanks, Lines, and Evaporative Emission Controls 36

On The Cover:
The Buick turbocharged V-6 engine, courtesy of the Buick Motor Division of the General Motors Corporation.

Tanks and Filters 36
Fuel Lines 40
Evaporative Emission Control Systems 42
American Motors EEC System 46
Chrysler EEC System 48
Ford EEC System 49
General Motors EEC System 51
Review Questions 53

Chapter 5 — Fuel Pumps and Filters 54

Pump Operation Overview 54
Pump Types 54
Mechanical Fuel Pumps 57
Electric Fuel Pumps 58
Fuel Filters 61
Review Questions 64

Chapter 6 — Air Cleaners and Filters 65

Engine Filtering Requirements 65
The Air Cleaner and Filtration 66
Air Filter Elements 67
Air Intake Ducts and Fresh Air Intakes 69
Thermostatically Controlled Air Cleaners 69
Review Questions 75

Chapter 7 — Basic Carburetion 76

Pressure Differential 76
Airflow and the Venturi Principle 78
Basic Carburetor Systems 80
Carburetor Types 88
Variable-Venturi (Constant Depression) Carburetors 90
Carburetor Linkage 93
Carburetor Circuit Variations and Assist Devices 96
Altitude-Compensating Carburetors 102
Review Questions 106

Chapter 8 — Intake and Exhaust Manifolds 107

Intake Manifold Principles 107
Basic Intake Manifold Types 109
Exhaust Manifolds 111
Intake Manifold Heat Control 113
Review Questions 116

Chapter 9 — Electronic Fuel Injection 117

Injection Advantages 117
Mechanical Fuel Injection Systems 118
Electronic Fuel Injection 118
Review Questions 125

Chapter 10 — Supercharging and Turbochargers 127

Supercharging 127
Turbochargers 128
Turbocharger Controls 130
Specific Turbochargers 131
Review Questions 133

PART THREE — SPECIFIC CARBURETORS 135

Chapter 11 — Autolite-Motorcraft Carburetors 136

Autolite 1100 One-Barrel 136
Autolite 1250 One-Barrel 136
Autolite 2100 Two-Barrel 136
Motorcraft 2150 Two-Barrel 140
Autolite 4100 Four-Barrel 140
Autolite 4300 Four-Barrel 140
Motorcraft 4350 Four-Barrel 140
Motorcraft 2700 VV Variable Venturi Two-Barrel 146
Motorcraft 740 Two-Barrel 149

Chapter 12 — Carter Carburetors 150

Carter YF and YFA One-Barrel 150
Carter RBS One-Barrel 150
Carter BBS One-Barrel 150

Contents

Carter BBD Two-Barrel 155
Carter WCD Two-Barrel 155
Carter AFB Four-Barrel 155
Carter AVS Four-Barrel 155
Carter Thermo-Quad (TQ) Four-Barrel 155

Chapter 13 — Holley Carburetors 161

Holley 1920 One-Barrel 161
Holley 1931 One-Barrel 161
Holley 1940 and 1945 Two-Barrel 161
Holley 2209 Two-Barrel 164
Holley 2210 and 2245 Two-Barrel 165
Holley 2300 Two-Barrel 165
Holley 5200 and 5210-C Two-Barrel 165
Holley 4150 and 4160 Four-Barrel 165
Holley 4165 and 4175 Four-Barrel 165
Holley Model 4360 Four-Barrel 165

Chapter 14 — Rochester Carburetors 173

Rochester B, BC, and BV One-Barrel 173
Rochester H and HV One-Barrel 173
Rochester M, MV, and 1MV One-Barrel 173
Rochester 2G, 2GV, and 2GC Two-Barrel 177
Rochester 2MC and M2MC Two-Barrel 177
Rochester 4GC Four-Barrel 180
Rochester Quadrajet Four-Barrel (Models 4M, 4MC, 4MV, M4MC, M4ME, and M4MEA) 181

PART FOUR — EMISSION CONTROL SYSTEMS AND DEVICES 183

Chapter 15 — Positive Crankcase Ventilation 184

Draft Tube Ventilation 184
Positive Crankcase Ventilation Systems 185
Original Equipment Closed PCV Systems 190
Retrofit Systems 191
Review Questions 192

Chapter 16 — Air Injection 193

Basic System Design and Operation 194
Second-Generation Air Injection Systems 197
Pulse Air Injection 200
Review Questions 201

Chapter 17 — Spark Timing Control Systems 202

Spark Timing and Combustion 202
Spark Timing and Emission Control 203
Spark Timing Emission Control Systems 203
Electronically Controlled Timing 211
Review Questions 217

Chapter 18 — Exhaust Gas Recirculation 218

NO_x Formation 218
System Components and Operating Principles 219
American Motors EGR Systems 222
Chrysler EGR Systems 224
Ford EGR System 225
General Motors EGR Systems 228
Review Questions 229

Chapter 19 — Catalytic Converters 230

Reducing Emissions 230
Converter Operating Precautions 233
Oxidation Converter Installations 234
Oxidation-Reduction Converter Installations 237
Review Questions 239

NIASE SAMPLE CERTIFICATION EXAM 240

GLOSSARY OF TECHNICAL TERMS 243

INDEX 246

ANSWERS TO REVIEW AND NIASE QUESTIONS 248

Introduction to Fuel Systems and Emission Controls

Fuel Systems and Emission Controls is part of the Harper & Row/Chek-Chart Automotive Series. The package for each course has two volumes, a *Classroom Manual* and a *Shop Manual*.

Other titles in this series include:
- Automatic Transmissions
- Automotive Electrical Systems
- Engine Performance Diagnosis and Tune-Up
- Heating and Air Conditioning (planned)
- Steering, Suspension, Alignment, Wheels, and Tires (planned).
- Brake Systems (planned)
- Engine Overhaul and Rebuilding (planned)
- Manual Transmissions, Drivelines, and Differentials (planned)

Each book is written to help the instructor teach students to become excellent professional automotive mechanics. The 2-manual texts are the core of a complete learning system that leads a student from basic theories to actual hands-on experience.

The entire series is job-oriented, especially designed for students who intend to work in the car service profession. A student will be able to use the knowledge gained from these books and from the instructor to get and keep a job. Learning the material and techniques in these volumes is a giant leap toward a satisfying, rewarding career.

The books are divided into *Classroom Manuals* and *Shop Manuals* for an improved presentation of the descriptive information and study lessons, along with the practical testing, repair, and overhaul procedures. The manuals are to be used together: the descriptive chapters in the *Classroom Manual* correspond to the application chapters in the *Shop Manual*.

Each book is divided into several parts, and each of these parts is complete by itself. Instructors will find the chapters to be complete, readable, and well thought-out. Students will benefit from the many learning aids included, as well as from the thoroughness of the presentation.

The series was researched and written by the editorial staff of Chek-Chart, and was produced by Harper & Row Publishers. For 50 years, Chek-Chart has provided car and equipment manufacturers' service specifications to the automotive service field. Chek-Chart's complete, up-to-date automotive data bank was used extensively to prepare this textbook series.

Because of the comprehensive material, the hundreds of high-quality illustrations, and the inclusion of the latest automotive technology, instructors and students alike will find that these books will keep their value over the years. In fact, they will form the core of the master mechanic's professional library.

How To Use This Book

Why Are There Two Manuals?

This two-volume text — **Fuel Systems and Emission Controls** — is not like any other textbook you've ever used before. It is actually two books, the *Classroom Manual* and the *Shop Manual*. They should be used together.

The *Classroom Manual* will teach you what you need to know about a car's fuel system, carburetion, and emission controls. The *Shop Manual* will show you how to fix and adjust those systems.

The *Classroom Manual* will be valuable in class and at home, for study and for reference. It has text and pictures that you can use for years to refresh your memory about the basics of automotive fuel systems and emission controls.

In the *Shop Manual*, you will learn about test procedures, troubleshooting, and overhauling the systems and parts you are studying in the *Classroom Manual*. Use the two manuals together to fully understand how the parts work, and how to fix them when they don't work.

What's In These Manuals?

There are several aids in the *Classroom Manual* that will help you learn more:
1. The text is broken into short bits for easier understanding and review.
2. Each chapter is fully illustrated with drawings and photographs.
3. Key words in the text are printed in **boldface type** and are defined on the same page and in a glossary at the end of the manual.
4. Review questions are included for each chapter. Use these to test your knowledge.
5. A brief summary of every chapter will help you to review for exams.
6. Every few pages you will find short blocks of "nice to know" information, in addition to the main text.
7. At the back of the *Classroom Manual* there is a sample test, similar to those given for National Institute for Automotive Service Excellence (NIASE) certification. Use it to help you study and to prepare yourself when you are ready to be certified as an expert in one of several areas of automobile mechanics.

The *Shop Manual* has detailed instructions on overhaul, test, and service procedures. These are easy to understand, and many have step-by-step, photo-illustrated explanations that guide you through the procedures. This is what you'll find in the *Shop Manual*:
1. Helpful information on using and maintaining shop tools and test equipment
2. Detailed safety precautions
3. System diagrams to help you locate troublespots while you learn to read the diagrams
4. Professional shop tips
5. A full index to help you quickly find what you need
6. Test procedures and troubleshooting hints that help you work better and faster.

Where Should I Begin?

If you already know something about a car's fuel and emission control systems and how to repair them, you may find that parts of this book are a helpful review. If you are just starting in car repair, then the subjects covered in these manuals may be all new to you.

Your instructor will design a course to take advantage of what you already know, and what facilities and equipment are available to work with. You may be asked to take certain chapters of these manuals out of order. That's fine. The important thing is to really understand each subject before you move on to the next.

Study the vocabulary words in boldface type. Use the review questions to help you understand the material. While reading in the *Classroom Manual*, refer to your *Shop Manual* to relate the descriptive text to the service procedures. And when you are working on actual car systems, look back to the *Classroom Manual* to keep the basic information fresh in your mind. Working on such a complicated piece of equipment as a modern car isn't always easy. Use the information in the *Classroom Manual*, the procedures of the *Shop Manual*, and the knowledge of your instructor to help you.

The *Shop Manual* is a good book for work, not just a good workbook. Keep it on hand while you're working on equipment. It folds flat on the workbench and under the car, and can withstand quite a bit of rough handling.

When you do test procedures and overhaul equipment, you will also need a source of accurate manufacturers' specifications. Most auto shops have either the carmaker's annual shop service manuals, which lists these specifications, or an independent guide, such as the **Chek-Chart Car Care Guide**. This unique book, which is updated every year, gives you the complete service instructions, troubleshooting tips, and tune-up information that you need to work on specific cars.

PART ONE

Fuel System and Emission Control Fundamentals

Chapter One
Introduction to Fuel Systems and Emission Controls

Chapter Two
Engine Operating Principles

Chapter Three
Engine Air-Fuel Requirements

Chapter 1

Introduction to Fuel Systems and Emission Controls

This is a book about automotive gasoline engine fuel systems. But, it is also a book about automotive emission controls. The internal combustion engine produces power by burning fuel and changing the chemical energy of that fuel into thermal (heat) energy. The thermal energy is then changed to mechanical power.

Not all of the chemical energy of the gasoline is converted to useful power. Much of it is wasted. Also, the engine's combustion process produces some harmful byproducts. These are discharged from the engine and become air pollutants. Therefore, an engine's fuel system and its emission controls are closely interrelated.

The fuel system has three important jobs. It must:
1. Store liquid fuel
2. Deliver liquid fuel to the carburetor, where it is mixed with air
3. Distribute the air-fuel mixture uniformly to each cylinder of the engine.

Emission control requirements are important factors in the design and operation of all parts of the fuel system. The ignition system, which provides the spark to begin combustion, plays an equally important role in emission control.

Because emission controls are now a vital part of the design of a modern automotive engine and its fuel and ignition systems, we will begin our study of fuel systems and emission controls by examining the harmful byproducts of combustion as a major cause of air pollution.

AIR POLLUTION — A PERSPECTIVE

Air pollution is usually defined as the introduction of a contamination into the atmosphere in an amount large enough to injure human, animal, or plant life. There are many different types and causes of air pollution, but they all fall into two general groups: natural and man-made. Natural pollution is caused by such things as the organic plant life cycle, forest fires, volcanic eruptions, and dust storms and is often beyond our control. Man-made pollution from industrial plants and automobiles *can* be controlled.

Most urban and large industrial areas suffer periodic air pollution. During the late 1940's, a unique form of air pollution was identified in the Los Angeles area. This combination of smoke and fog, which forms irritating chemical compounds, is called **photochemical smog**, figure 1-1. As smog became more of a problem, California became the first state to place controls on motor vehicles. When smog began to appear in other parts of the country, the federal government moved into the area of regulation. To

Introduction to Fuel Systems and Emission Controls

Figure 1-1. Have a nice day.

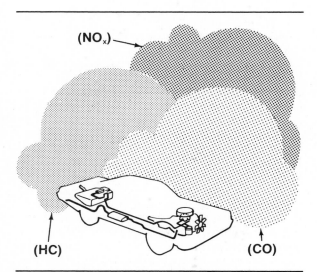

Figure 1-2. Hydrocarbons (HC), carbon monoxide (CO), and oxides of nitrogen (NO$_x$) are the three major pollutants emitted by an automobile.

understand why, we must first look at the elements produced by the automobile which form air pollution and smog.

MAJOR POLLUTANTS

An internal combustion engine emits three major gaseous pollutants into the air: **hydrocarbons** (HC), **carbon monoxide** (CO), and **oxides of nitrogen** (NO$_x$), figure 1-2. In addition, an automobile engine gives off many small liquid or solid particles, such as lead, carbon, sulfur and other **particulates**, which contribute to pollution. By themselves, all these emissions are not smog, but simply air pollutants.

Photochemical Smog: A combination of pollutants which, when acted upon by sunlight, forms chemical compounds that are harmful to human, animal, and plant life.

Hydrocarbon: A chemical compound made up of hydrogen and carbon. A major pollutant given off by an internal combustion engine. Gasoline, itself, is a hydrocarbon compound.

Carbon Monoxide: An odorless, colorless, tasteless poisonous gas. A major pollutant given off by an internal combustion engine.

Oxides of Nitrogen: Chemical compounds of nitrogen given off by an internal combustion engine. They combine with hydrocarbons to produce smog.

Particulates: Liquid or solid particles such as lead and carbon that are given off by an internal combustion engine as pollution.

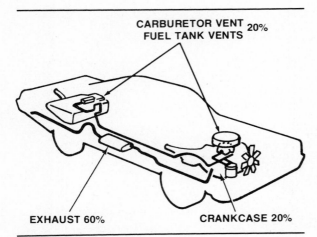

Figure 1-3. Sources of hydrocarbon emissions.

Hydrocarbons (HC)

Gasoline is a hydrocarbon material. Unburned hydrocarbons given off by an automobile are largely unburned portions of fuel. For example, every 1,000 gallons of gasoline used by a car without emission controls produces about 200 pounds of hydrocarbon emissions.

Over 200 different varieties of hydrocarbon pollutants come from automotive sources. While most come from the fuel system and the engine exhaust, others are oil and gasoline fumes from the crankcase. Even a car's tires, paint, and upholstery give off tiny amounts of hydrocarbons. The three major sources of hydrocarbon emissions in an automobile are shown in figure 1-3:
1. Fuel system evaporation — 20 percent
2. Crankcase vapors — 20 percent
3. Engine exhaust — 60 percent

Hydrocarbons are the only major automotive air pollutant that come from sources *other than* the engine's exhaust.

Hydrocarbons of all types are destroyed by combustion. If an automobile engine burned gasoline completely, there would be no hydrocarbons in the exhaust, only water and carbon dioxide. But when the vaporized and compressed air-fuel mixture is ignited, combustion occurs so fast that gasoline near the sides of the combustion chamber does not get burned. This unburned fuel then passes out with the exhaust gases. The problem is worse with engines that misfire or are not properly tuned.

Carbon Monoxide (CO)

Although not part of photochemical smog, carbon monoxide is also found in automobile exhaust in large amounts. The same 1,000 gallons of gasoline which will produce 200 pounds of hydrocarbons when burned without emission controls will also produce 2,300 pounds of carbon monoxide. A deadly poison, carbon monoxide is both odorless and colorless. In a small quantity, it causes headaches and vision difficulties; in larger quantities, it is fatal. Large quantities of carbon monoxide that collect in the air push out the oxygen, which makes breathing difficult.

Because it is a product of incomplete combustion, the amount of carbon monoxide produced depends on the way in which hydrocarbons burn. When the air-fuel mixture burns, its hydrocarbons combine with oxygen. If the air-fuel ratio has too much fuel, there is not enough oxygen for this process to happen, and so carbon monoxide is formed. To make combustion more complete, an air-fuel mixture with less fuel is used. This increases the supply of oxygen, which reduces the formation of CO by producing harmless carbon dioxide (CO_2) instead.

Oxides of Nitrogen (NO_x)

Air is about 78 percent nitrogen, 21 percent oxygen, and 1 percent other gases. When the combustion chamber temperature reaches 2,500° F or greater, the nitrogen and oxygen in the air-fuel mixture combine to form large quantities of oxides of nitrogen (NO_x). NO_x is also formed at lower temperatures, but in far smaller amounts. By itself, NO_x is of no particular concern. But when the amount of hydrocarbons in the air reaches a certain level, and the ratio of NO_x to HC is correct, the two pollutants will combine chemically to form smog.

The amount of NO_x formed can be reduced by lowering the temperature of combustion in the engine, but this causes a problem which is difficult to solve. Lowering the combustion chamber temperature to reduce NO_x results in less efficient burning of the air-fuel mixture. This automatically means an increase in hydrocarbons and carbon monoxide, both of which are formed in large quantities at lower combustion chamber temperatures. Carmakers have used various emission control systems to combat this problem, which we will study in later chapters.

Introduction to Fuel Systems and Emission Controls

Particulates

Particulates are microscopic solid particles, such as dust and soot. They are fragments of solid or liquid matter which remain in the atmosphere for long periods of time. Because of this, particulates are prime causes of secondary pollution. For example, particulates such as lead and carbon tend to collect in the atmosphere. These are all harmful substances and, in large amounts, can injure our health. Absorbed directly or indirectly into the body, particulates can damage the respiratory tract and lungs of both humans and animals.

Particulates produced by automobiles are a small percentage of the total in the atmosphere. Most come from fixed sources such as factories, which spew industrial wastes into the air. But automobiles *do* produce particulates, and studies have shown that they can be reduced considerably. This is done by eliminating additives such as lead from gasoline, and by changing other characteristics of the fuel. For this reason, the amount of additives used in gasoline has been reduced, and the types of additives are now carefully controlled.

Sulfur Oxides

The sulfur created by processing and burning gasoline and other fossil fuels (coal and oil) gets into the atmosphere in the form of **sulfur oxides**. As this material breaks down, it will combine with water in the air to form corrosive sulfuric acid, which is a secondary pollutant. Recently, such pollution has become a major problem in the northeastern United States, as well as in other parts of the world. Clouds containing sulfuric acid are carried by winds to the Northeast, where they deposit rain and snow with enough acid to kill wildlife.

POLLUTION AND THE AUTOMOBILE

While it is true that a single car gives off only a microscopic amount of these pollutants, remember that there are more than 117 million automobiles in use in the United States. Multiply each car's contribution toward air pollution by that figure, and you have the potential for a staggering amount of pollution — about 42 percent of the total problem in this country.

Great progress has been made in reducing — almost eliminating — automobile air pollution since 1966. Those HC emissions that come from the engine crankcase and fuel system — 40 percent of the total — have been almost totally eliminated. The other 60 percent of HC emissions — from the exhaust — has been lowered

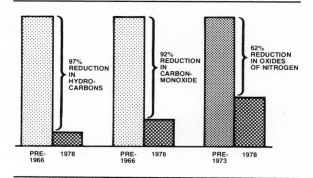

Figure 1-4. Automotive emission reductions.

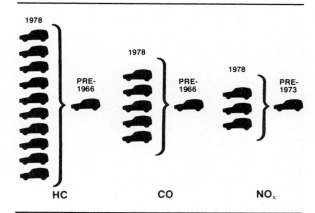

Figure 1-5. Vehicle emission comparisons.

considerably. Total hydrocarbon emissions have been reduced by about 97 percent. Total carbon monoxide emissions have been reduced by a similar amount — about 92 percent — since 1966. Since 1973, total NO_x emissions have been reduced about 62 percent. These accomplishments are shown in figure 1-4. Figure 1-5 shows us that ten 1978 cars emit less HC than one pre-1966 model; five 1978 cars produce less CO than one pre-1966 car; and three 1978 cars emit less NO_x than one pre-1973 model.

Sulfur Oxides: Chemical compounds given off by processing and burning gasoline and other fossil fuels. As they decompose, they combine with water to form sulfuric acid.

Figure 1-6. Smog engulfs the Los Angeles Civic Center. When the base of the temperature inversion is 1500 feet above ground, the inversion layer — a layer of warm air above a layer of cool air — prevents the natural dispersion of air contaminants into the upper atmosphere. (LA County)

SMOG — CLIMATIC REACTION WITH AIR POLLUTANTS

Smog and air pollution are not the same thing; smog is a form of air pollution, but air pollution is not necessarily smog. Each of the three major pollutants is a byproduct of combustion and is created in different ways. Hydrocarbons come mostly from unburned fuel; carbon monoxide is the result of air-fuel mixtures that have too much fuel; and oxides of nitrogen are created by high combustion chamber temperatures. But HC and NO_x are the two principal materials that combine in the atmosphere to form smog.

The Photochemical Reaction

Although smog can be created in a laboratory experiment, scientists still do not understand exactly what it is, or exactly how it is formed. But they do know that for smog to form in the atmosphere, three factors must be present — sunlight, relatively still air, and a high concentration of hydrocarbons and oxides of nitrogen. When these three factors are present at the same time, sunlight causes a chemical reaction between the HC and NO_x. The result is smog.

Temperature Inversion

Normally, air temperature becomes cooler at higher altitudes. Warm air near the ground rises, cooling itself by contact with the cooler air above it. When nature follows this normal pattern, smog and other pollutants are carried away. But some areas experience a natural weather pattern called a **temperature inversion**. When this happens, a layer of warm air prevents the upward movement of cooler air near the ground. This inversion acts as a "lid" over the stagnant air; the air cannot rise, so smog and pollution collect.

When the inversion layer is several thousand feet high, the smog will rise enough to provide reasonable visibility. But when the inversion layer is within a thousand feet of the ground, smog is trapped and visibility is reduced, figure 1-6. The distant landscape is impossible to see, and eye irritation, headaches, and difficulty in breathing follow. The Los Angeles area, where temperature inversion was first noted, provides a classic example of this phenomenon. Surrounding mountains form a natural basin in which temperature inversion is present to some extent for more than 300 days a year.

Introduction to Fuel Systems and Emission Controls

Year	Regulations	HC	CO	No$_x$
Before Controls		850 ppm (16.8 g/mi)*	3.4% (125.0 g/mi)*	1000 ppm (4.0 g/m)*
1966-67	Calif.	275 ppm	1.5%	none
	U.S. Federal	none	none	none
1968-69	Calif.	275 ppm	1.5%	none
	U.S. Federal	275 ppm	1.5%	none
1970	Calif. & U.S. Federal	4.6 g/mi	46.0 g/mi	none
1971	Calif.	4.6 g/mi	46.0 g/mi	4.0 g/mi
	U.S. Federal	4.6 g/mi	46.0 g/mi	none
	Canadian	2.2 g/mi**	23.0 g/mi**	none
1972	Calif.	3.2 g/mi	39.0 g/mi	3.2 g/mi
	U.S. Federal	3.4 g/mi	39.0 g/mi	none
	Canadian	3.4 g/mi	39.0 g/mi	none
1973	Calif.	3.2 g/mi	39.0 g/mi	3.0 g/mi
	U.S. Federal	3.4 g/mi	39.0 g/mi	3.0 g/mi
	Canadian	3.4 g/mi	39.0 g/mi	3.0 g/mi
1974	Calif.	3.2 g/mi	39.0 g/mi	2.0 g/mi
	U.S. Federal	3.4 g/mi	39.0 g/mi	3.0 g/mi
	Canadian	3.4 g/mi	39.0 g/mi	3.0 g/mi
1975-76	Calif.	0.9 g/mi	9.0 g/mi	2.0 g/mi
	U.S. Federal	1.5 g/mi	15.0 g/mi	3.1 g/mi
	Canadian	2.0 g/mi	25.0 g/mi	3.0 g/mi
1977-79	Calif.	0.41 g/mi	9.0 g/mi	1.5 g/mi
	U.S. Federal	1.5 g/mi	15.0 g/mi	2.0 g/mi
	Canadian	2.0 g/mi	25.0 g/mi	3.0 g/mi

ppm = parts per million
g/mi = grams per mile
* approximate estimations of uncontrolled levels
** theoretical flow rate

Figure 1-7. Exhaust Emission Limits For New Passenger Cars By Model Years

AIR POLLUTION LEGISLATION AND REGULATORY AGENCIES

Once the problem of air pollution was recognized, it became the subject of intense research and investigation. By the early 1950's it was believed that smog in Los Angeles was caused by the photochemical process, and California, as we noted earlier, became the first state to pass laws to limit automotive emissions. Standards established by California one year have usually become U.S. Federal standards the next year.

Emission Control Legislation

Beginning with the 1961-model cars, California required control over crankcase emissions. This became standard for the rest of the United States with the 1963-model cars. That year, carmakers voluntarily equipped their cars with a blowby device which almost eliminated crankcase emissions on all cars.

The first Federal air pollution research program began in 1955. In 1963, Congress passed the Clean Air Act, giving money to states for development of air pollution control programs. This law was amended in 1965, giving the Federal government authority to set emission standards for new cars. These standards were first applied nationwide to 1968 models.

The Air Quality Act of 1967 was a new approach to air pollution. Major changes followed in 1970 — changes designed to turn piecemeal programs into a unified attack on pollution of all kinds. At this time, emission standards were established for 1973-74 cars and projected for 1975 and later models, figure 1-7.

Regulatory Agencies

The Environmental Protection Agency (EPA) is the U.S. Federal agency responsible for enforcing the Air Quality Act. It is part of the Department of Health, Education, and Welfare. The EPA first set standards which required that HC

Temperature Inversion: A weather pattern in which a layer or "lid" of warm air keeps the cooler air beneath it from rising.

■ Those EPA Mileage Ratings

What do EPA mileage ratings represent? Mileage testing began in 1975 and the results have become the auto industry's yardstick. To help you understand EPA standards and estimates, here's a look at how they are figured.

EPA groups cars by model, engine size, number of cylinders, catalyst use, fuel system, and sales area (49 states or California). Because identical engine and drivetrain combinations are used in many different models, EPA does not actually test every possible combination.

The 10.2 mile fuel consumption test cycle simulates highway driving and is done on a chassis dynamometer. The speed range is 0 to 60 mph, with an average speed of 48.2, and no engine shutdown. The EPA emissions test cycle represents urban driving. Exhaust gases collected during this second test are analyzed both for pollutants and quantity of fuel used. EPA claims that exhaust analysis accurately measures fuel consumption.

Starting with the 1976 model year, EPA began providing a combined city and highway driving figure. The results of the fuel consumption and emissions test cycles are weighted to reflect a combination of 55 percent city and 45 percent highway driving.

Beginning with the 1978 model year, mandatory Corporate Average Fuel Economy (CAFE) figures are provided. These are obtained by sales weighting the combined city and highway figures of a manufacturer's entire product line. The CAFE standards begin at 18.0 mpg in 1978, and gradually increase through 1985, when the industry must meet a 27.5 mpg requirement.

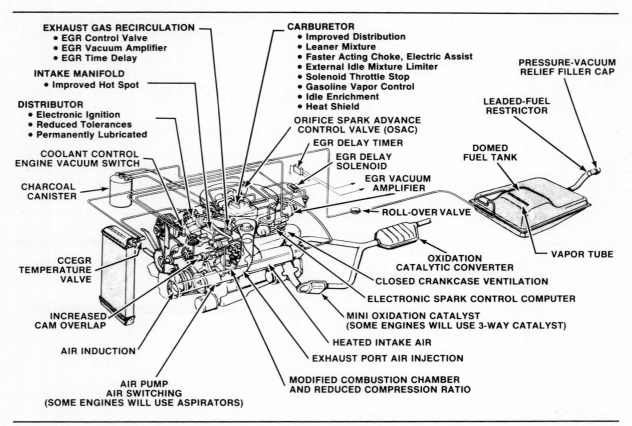

Figure 1-8. The complex emission controls on a modern automobile engine are an integral part of the fuel, ignition, and exhaust systems. (Chrysler)

and CO emissions for 1975 passenger cars be reduced 90 percent from 1970 levels, with a 90-percent reduction in NO_x by 1976. These standards were later amended to those shown in figure 1-7. In addition, the EPA has established standards in other automotive areas, such as fuel consumption and fuel additives. California established its own Air Resources Board (ARB), whose authority roughly parallels that of the Federal EPA. The California ARB operates under permission from the EPA, but its authority extends only to those vehicles sold in or brought into California.

AUTOMOTIVE EMISSION CONTROLS

Early researchers dealing with automotive pollution and smog began work with the idea that *all* pollutants were carried into the atmosphere by the car's exhaust pipe. But auto manufacturers doing their own research soon discovered that pollutants were also given off from the fuel tank and the engine crankcase. The total automotive emission system, figure 1-8, contains three different types of controls. This picture shows that the emission controls on a modern automobile are not a separate system, but are part of an engine's fuel, ignition, and exhaust systems.

In order to service a car's fuel system and emission controls, you must have a basic understanding of the internal combustion engine and how it works. Engine operating principles and air-fuel requirements are explained in Chapters 2 and 3. Emission controls are explained throughout the rest of this book as they relate to different parts of the fuel system. Chapters 15 through 19 of this book cover those major emission controls that can be studied separately from the fuel system.

Automotive emission controls can be grouped into major families, as follows:
1. Crankcase emission controls — Positive crankcase ventilation (PCV) systems control HC emissions from the engine crankcase. These are described in detail in Chapter 15.
2. Evaporative emission controls — Evaporative emission control (EEC) systems control the evaporation of HC vapors from the fuel tank, pump, and carburetor. These are described in detail in Chapter 4.
3. Exhaust emission controls — Various systems and devices are used to control HC, CO, and NO_x emissions from the engine exhaust. These controls can be subdivided into the following general groups:

Introduction to Fuel Systems and Emission Controls

a. Air injection systems — These systems inject extra air into the exhaust system to help burn up HC and CO in the exhaust. They are described in Chapter 16.

b. Engine Modifications — Various changes have been made in the designs of engines and in the operation of fuel and ignition system parts to help eliminate all three major pollutants. They are identified in appropriate sections throughout this book.

c. Spark timing controls — Carmakers have used various systems to delay or retard ignition spark timing to control HC and NO_x emissions. Most of these systems modify the distributor vacuum advance, but many late-model cars have electronic spark controls. Ignition spark timing and its role in controlling emissions are discussed in Chapter 17.

d. Exhaust gas recirculation — The most effective way to control NO_x emissions is to recirculate a small amount of exhaust gas back to the intake manifold to dilute the incoming air-fuel mixture. Exhaust gas recirculation (EGR) systems are described in Chapter 18.

e. Catalytic converters — The first catalytic converters installed in the exhaust systems of 1975 and '76 cars helped the chemical oxidation, or burning, of HC and CO in the exhaust. Later catalytic converters, which began to appear on 1977 and '78 cars, also promote the chemical reduction of NO_x emissions. Catalytic converters are described in Chapter 19.

■ How New Cars Are Emission-Certified

The availability of new cars each year depends on whether the carmakers successfully complete the emission certification process. The constant volume sampling test is performed by the Federal Environmental Protection Agency (EPA), although California's Air Resources Board tests some vehicles.

As each new model year approaches, automakers build prototype or emission data cars for EPA use. The automakers are responsible for conducting a 50,000-mile durability test of their emission control systems. The test cars are driven for 4,000 miles to stabilize the emission systems before testing.

The first step is to precondition a car. Then the car stands for 12 hours at an air temperature of 73° F to simulate a cold-start. The actual test is done on a chassis dynamometer, using a driving cycle which represents urban driving conditions. The car's exhaust is mixed with air to a constant volume and analyzed for harmful pollutants.

The entire test takes about 41 minutes. The first 23 minutes are a cold-start driving test. The next ten minutes are a waiting, or hot-soak, period. The final eight minutes are a hot-start test, representing a short trip in which the car is stopped and started several times while hot. If the emissions test results for all data cars are equal to or lower than all of the HC, CO, and NO_x standards, the manufacturer receives certification of the engine "family" and the car can be offered for sale to the public.

Review Questions

Choose the single most correct answer. Compare your answers to the correct answers on page 248.

1. Smog:
    a. Is a natural pollutant
    b. Cannot be controlled
    c. Is created by a photochemical reaction
    d. Was first identified in New York City

2. The three *major* pollutants in automobile exhaust are:
    a. Sulfates, particulates, carbon dioxide
    b. Sulfates, carbon monoxide, nitrous oxide
    c. Carbon monoxide, oxides of nitrogen, hydrocarbons
    d. Hydrocarbons, carbon dioxide, nitrous oxide

3. Fuel evaporation accounts for what percentage of total HC emissions:
    a. 10%
    b. 60%
    c. 33%
    d. 20%

4. Carbon monoxide is a result of:
    a. Incomplete combustion
    b. A lean mixture
    c. Excess oxygen
    d. Impurities in the fuel

5. Which of the following is *not* true:
    a. High temperatures increase NO_x emission
    b. High temperatures reduce HC and CO emission
    c. Low temperatures reduce HC and CO emission
    d. Low temperatures reduce NO_x emission

6. Particulates are:
    a. Microscopic particles suspended in the atmosphere
    b. Produced in emissions by additives such as lead
    c. Harmful to our lungs
    d. All of the above

7. Sulfur oxides are harmful because:
    a. They damage catalytic converters
    b. They combine with water in the air to form sulfuric acid
    c. They are primary pollutants
    d. They are visible

8. Since 1966, total HC emissions have been reduced by about:
    a. 65%
    b. 47%
    c. 82%
    d. 97%

9. Smog is created by a combination of sunlight, still air, and:
    a. High levels of CO and HC
    b. High levels of CO and NO_x
    c. High levels of HC and NO_x
    d. High levels of HC and sulfur oxides

10. A temperature inversion increases air pollution by:
    a. Pushing cool air up
    b. Forming a "lid" over stagnant air
    c. Pushing warm air down
    d. None of the above

Chapter 2
Engine Operating Principles

Engines used in automobiles are called internal combustion engines because fuel burns inside the engine. The several types of internal combustion engines used in passenger cars are covered in this chapter, and you will learn the basic principles of engine operation that relate to fuel systems and emission control.

Except for the Wankel rotary engine, all automotive engines are the **reciprocating**, or piston, type. We will begin our survey of engine types with a short look at how internal combustion engines use air pressure. Then we will discuss the common parts of all reciprocating engines.

AIR PRESSURE — HIGH AND LOW

You can think of an internal combustion engine as a big air pump. As the pistons move up and down in the cylinders, they pump in air and fuel for combustion and pump out the burned exhaust. They do this by creating a difference in air pressure. The air outside an engine has weight and exerts pressure. So does the air inside an engine.

As a piston moves down on an intake stroke with the intake valve open it creates a larger area inside the cylinder for the air to fill. This lowers the air pressure within the engine. Because the pressure inside the engine is lower than the pressure outside, air will flow in through the carburetor to try to fill the low pressure area and equalize the pressure.

We call the low pressure within the engine, **vacuum**. You can think of the vacuum as sucking air into the engine, but it is really the higher pressure on the outside that forces air into the low pressure area inside. The difference in pressure between two areas is called a **pressure differential**. The pressure differential principle has many applications in automotive fuel and emission systems.

An engine pumps exhaust out of its cylinders by creating pressure as a piston moves upward on the exhaust stroke. This creates high pressure in the cylinder, which forces the exhaust toward the lower pressure area outside the engine. Pressure differential can be applied to liquids as well as to air. Fuel pumps work on this principle. The pump creates a low-pressure area in the fuel system that allows the higher pressure of the air and fuel in the tank to force the fuel through the lines to the carburetor.

Again, you can think of vacuum or suction drawing fuel through the lines, but that is simply another way of looking at the pressure differential. We will be studying airflow and air pressure more in the next chapter and again in Chapter 7 as we study carburetors in detail.

Engine Operating Principles

Figure 2-1. The engine block contains the cylinders and water jackets for cooling. The crankshaft turns in main bearings at the bottom of the block.

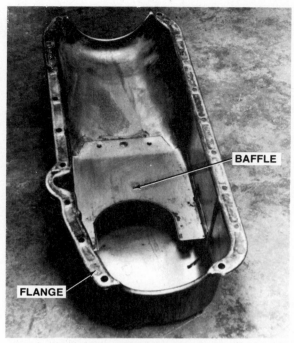

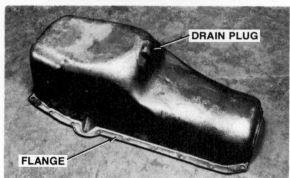

Figure 2-2. The oil pan is fastened to the bottom of the block.

RECIPROCATING ENGINE COMPONENTS

Reciprocating means "up and down" or "back and forth." It is the up and down action of a piston in a cylinder that produces power in a reciprocating engine. Almost all engines of this type are built upon a cylinder block, or engine block, figure 2-1. The block is an iron or aluminum casting that has the engine cylinders as well as passages, called water jackets, for coolant circulation. The top of the block is covered with the cylinder head, which contains additional water jackets and which forms the combustion chamber. The bottom of the block is covered with an oil pan, or oil sump, figure 2-2. An example of a major exception to this type of engine construction is the air-cooled Volkswagen engine. It is shown in figure 2-3.

Reciprocating Engine: Also called piston engine. An engine in which the pistons move up and down or back and forth, as a result of combustion of an air-fuel mixture in the top of the piston cylinder.

Vacuum: A pressure less than atmospheric pressure.

Pressure Differential: A difference in pressure between two points.

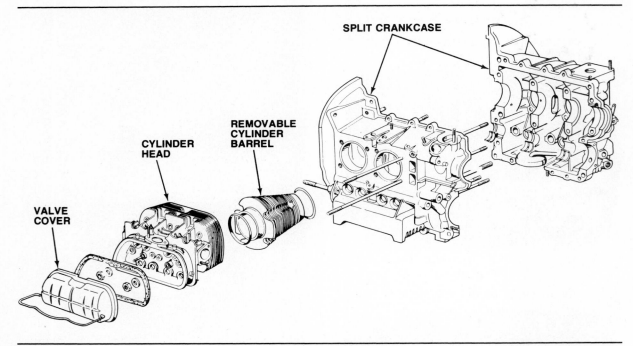

Figure 2-3. The Volkswagen horizontally opposed engine is built with a split crankcase. Individual cylinder castings bolt to the split crankcase, and cylinder heads attach to each pair of cylinders.

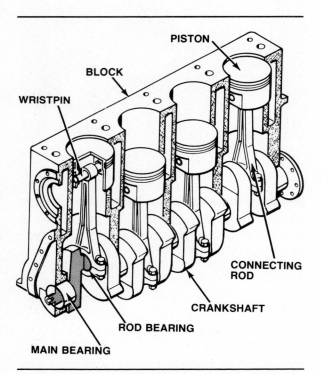

Figure 2-4. The crankshaft changes the reciprocating motion of the pistons to rotating motion.

The diameter of the cylinder is called the engine's **bore**. The distance that the piston travels from its top position to its bottom position in the cylinder is called the engine's **stroke**.

Power is produced by the reciprocating motion of a piston in a cylinder, but this up and down power must be changed to rotating power to turn the wheels of a car. The piston is attached to the top of a connecting rod by a pin, called a piston pin or a wristpin, figure 2-4. The bottom of the connecting rod is attached to the crankshaft. The connecting rod transmits the up and down motion of the piston to the crankshaft, which changes it to rotating motion. The connecting rod is mounted on the crankshaft with large bearings, called rod bearings. Similar bearings, called main bearings, are used to mount the crankshaft in the block, figure 2-4.

The combustible mixture of gasoline and air is drawn into the cylinder through valves. Automotive engines use **poppet valves**, figure 2-5. The valves can be in the cylinder head or in the block. They are opened by rotating camshaft points, called **lobes**, linkage of valve lifters and other parts. The most common arrangements of engine cylinders and valves are discussed later.

In addition to being called internal combustion, reciprocating engines, most automotive engines are classified as:
• 4-stroke or 2-stroke engines.
• Spark-ignition (gasoline) or compression-ignition (diesel) engines.

Engine Operating Principles

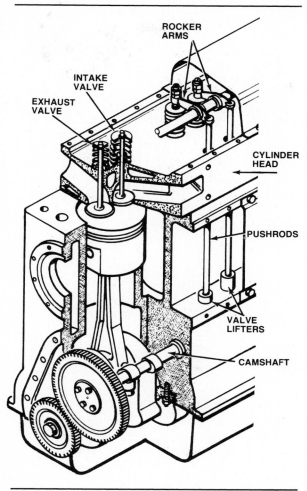

Figure 2-5. The valves in this engine are located in the cylinder head. They are operated by the camshaft, valve lifters, pushrods, and rocker arms.

Because the most common automotive engine is the 4-stroke gasoline engine, we will begin our study of operating cycles with that design.

THE FOUR-STROKE CYCLE

When gasoline is mixed with oxygen and ignited, it burns. This burning is called combustion and is a way of releasing the energy stored in the air-fuel mixture. To do any useful work in an engine, the combustion energy must be changed to mechanical energy. This is done by confining the combustion to the sealed combustion chamber, figure 2-6, where the combustion energy can work on the movable piston to produce mechanical energy. The combustion chamber must be sealed as tightly as possible for efficient engine operation. Any leakage from the combustion chamber allows part of the combustion energy to leak away without adding to the mechanical energy developed by the piston movement.

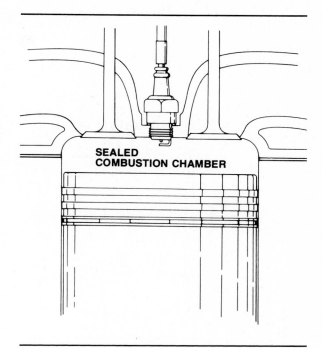

Figure 2-6. For combustion to produce power in an engine, the combustion chamber must be sealed.

The term "stroke" is used to describe the *movement* of the piston within the cylinder, as well as the *distance* of piston travel. Depending on the type of engine, the operating cycle may require either two or four strokes to complete. The **four-stroke engine** is also called the Otto cycle engine, in reference to the German engineer, Dr. Nikolaus Otto, who first applied the principle in 1876.

In the 4-stroke engine, four strokes of the piston in the cylinder are required to complete one operating cycle: two strokes up and two strokes down. Each stroke is named after the action it performs — intake, compression, power, and exhaust — in that order, figure 2-7.

Bore: The diameter of an engine cylinder.

Stroke: One complete top-to-bottom or bottom-to-top movement of an engine piston.

Poppet Valve: A valve that plugs and unplugs its opening by axial motion.

Lobes: The rounded protrusions on a camshaft that force, and govern, the opening of the intake and exhaust valves.

Four-Stroke Engine: The Otto cycle engine. An engine in which a piston must complete four strokes to make up one operating cycle. The strokes are: intake, compression, power, and exhaust.

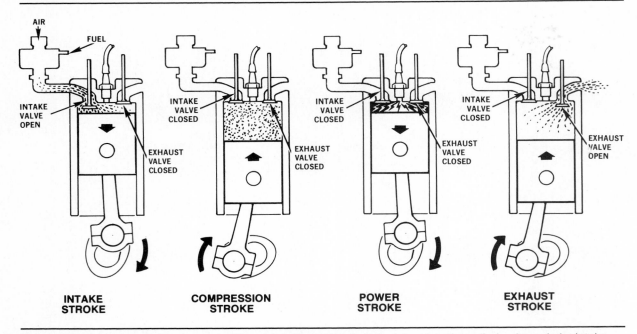

Figure 2-7. The downward movement of the piston draws the air-fuel mixture into the cylinder through the intake valve on the intake stroke. On the compression stroke, the mixture is compressed by the upward movement of the piston with both valves closed. Ignition occurs at the beginning of the power stroke, and combustion drives the piston downward to produce power. On the exhaust stroke, the upward-moving piston forces the burned gases out the open exhaust valve.

1. Intake Stroke: as the piston moves down, a vaporized mixture of fuel and air is drawn into the cylinder through the open intake valve.
2. Compression Stroke: the piston returns up, the intake valve closes, the mixture is compressed within the combustion chamber, and ignited by a spark.
3. Power Stroke: the expanding gases of combustion force the piston down in the cylinder again. The exhaust valve opens near the bottom of the stroke.
4. Exhaust Stroke: the piston returns up with the exhaust valve open, and the burned gases are pushed out to prepare for the next intake stroke. The intake valve usually opens just before the top of the exhaust stroke.

This 4-stroke cycle is continuously repeated in every cylinder as long as the engine remains running.

CYLINDER AND PISTON ARRANGEMENT

Up to this point, we have been talking about a single piston in a single cylinder. While single-cylinder engines are common in motorcycles, outboard motors, and small agricultural implements, automotive engines have more than one cylinder. Most passenger car engines have 4, 6, or 8 cylinders, although engines with 3, 5, 12, and 16 cylinders have been built. In automotive engines, all pistons are attached to a single crankshaft. The more cylinders an engine has, the more power strokes are produced for each revolution. This means that an 8-cylinder engine produces a power stroke twice as often as a 4-cylinder engine. The 8-cylinder engine runs more smoothly because the power strokes are closer together in time and in degrees of engine rotation. The cylinders of multicylinder automobile engines are arranged in one of three ways, figure 2-8.

1. Inline engines use a single bank of cylinders. Most 4-cylinder and many 6-cylinder engines are of this type.
2. V-type engines use two equal banks of cylinders usually inclined 60 degrees or 90 degrees apart. Most V-type engines are 6- or 8-cylinder, but some V-4 designs have been built.
3. Horizontally opposed, or "pancake," engines use two equal banks of cylinders 180 degrees apart. These space-saving engine designs are usually air-cooled, and are found in Corvairs, Porsches, and Volkswagens. Subaru uses a water-cooled, horizontally opposed, 4-cylinder engine.

As described earlier, each cylinder contains intake and exhaust poppet valves. These are

Engine Operating Principles

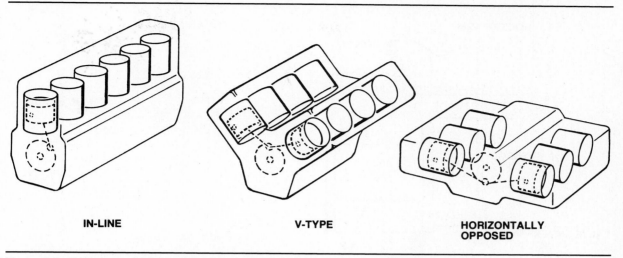

Figure 2-8. Common cylinder arrangements for automotive engines.

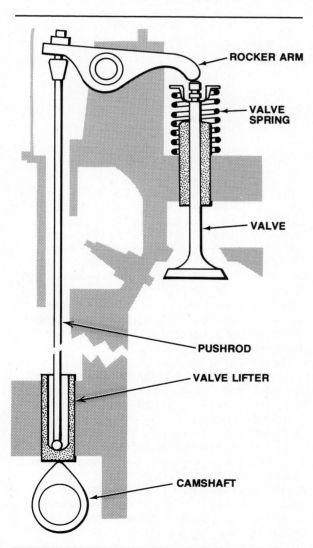

Figure 2-9. Valve mechanism for an overhead-valve engine.

opened by lobes on the camshaft and closed by valve springs, figure 2-9. To coordinate this opening and closing action with piston movement, the camshaft's rotation must be synchronized, or timed, with that of the crankshaft. This synchronization is accomplished in one of three ways: by gears, by a chain and sprockets, or by a timing belt and sprockets. Timing marks on the gears or sprockets are used to synchronize the two shafts with each other, figure 2-10.

A camshaft gear meshes with a driven gear on the ignition distributor to synchronize, or time spark plug firing with piston and valve positions at the beginning of the power stroke. Because each piston is attached to the crankshaft in a predetermined **firing order**, only one power stroke takes place at any instant, regardless of the number of cylinders in the engine. These power strokes follow each other in a rapid sequence, producing a smooth flow of continuous power.

Firing Order: The order in which combustion occurs in the cylinders of an engine.

■ Sixteen Cylinders

The chances are remote that you'll ever be called on to service a 16-cylinder 1935-37 Cadillac, but in case you do, this is the firing order: 1-8-9-14-3-6-11-2-15-10-7-4-13-12-5-16. The firing order on the 1938-40 V-16 models is: 1-4-9-12-3-16-11-8-15-14-7-6-13-2-5-10. Odd-numbered cylinders are on the left bank, even-numbered cylinders are on the right.

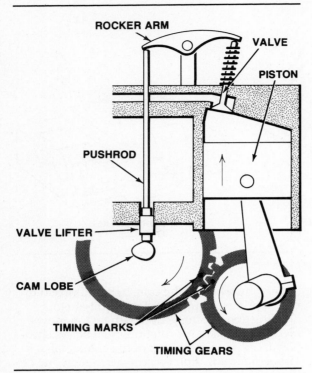

Figure 2-10. Timing marks on the timing gears synchronize valve action with piston movement.

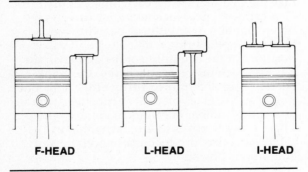

Figure 2-11. Common valve arrangements for four-stroke automotive engines.

VALVE ARRANGEMENT

Intake and exhaust valves on modern engines are located in the cylinder head but have been arranged in one of three ways, figure 2-11.
1. The L-head design has both valves located side-by-side in the engine block. Because the cylinder head is rather flat, containing only the combustion chamber, water jacket, and spark plugs, L-head engines are also called "flatheads." Once very common, this valve arrangement has not been used in a domestic engine since the early 1960's.
2. The F-head design locates the intake valve in the cylinder head and the exhaust valve in the engine block. A compromise between the

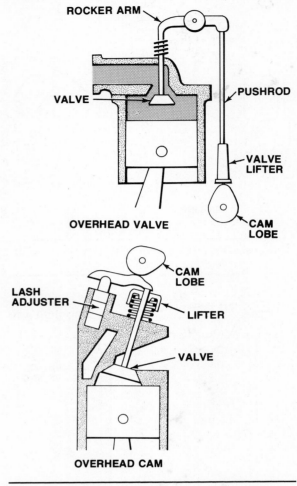

Figure 2-12. In an I-head engine, the valves may be operated by a camshaft in the block or in the head.

L-head and I-head designs, the F-head was last used in the 1971 Jeep.
3. Also known as overhead-valve and overhead-cam designs, the I-head or valve-in-head arrangement has both the intake and the exhaust valve in the cylinder head.

In the overhead-valve design, the camshaft is in the engine block, and the valves are opened by valve lifters, pushrods, and rocker arms, figure 2-12. In the overhead-cam engine, the camshaft is mounted in the head, either above or to one side of the valves, figure 2-12. This improves valve action at higher engine speeds. The valves may open directly by means of valve lifters or cam followers, or through rocker arms. The double-overhead-cam engine has two camshafts, one on each side of the valves. One camshaft operates the intake valves; the other operates the exhaust valves.

Engine Operating Principles

ENGINE DISPLACEMENT AND COMPRESSION RATIO

Two frequently used engine specifications are engine **displacement** and **compression ratio**. Displacement and compression ratio are related to each other as we will learn in the following paragraphs.

Engine Displacement

Commonly used to indicate engine size, this specification is really a measurement of cylinder volume. The number of cylinders helps determine displacement, but the arrangement of the cylinders or valves does not. Engine displacement is found by multiplying the number of cylinders in the engine by the piston displacement of one cylinder. The total engine displacement, then, is the volume displaced by all the pistons in one revolution of the crankshaft.

To calculate engine displacement, you must begin with the piston displacement of one cylinder. This is the space through which the piston's top surface moves as it travels from the bottom of its stroke (**bottom dead center**) to the top of its stroke (**top dead center**), figure 2-13. It is the volume of air displaced or pushed out, from the cylinder by the piston movement. Piston displacement can be figured out as follows:
1. Divide the bore (cylinder diameter) by two. This will give you the radius of the bore.
2. Square the radius by multiplying it by itself (2×2, 3×3, 4×4, etc.).
3. Multiply the square of the radius by 3.1416 (pi or π) to find the area of the cylinder cross section.
4. Multiply the area of the cylinder cross section by the length of the stroke.

You now know the piston displacement for one cylinder. Multiply this by the number of cylinders to determine the total engine displacement. The formula for the complete procedure reads:

$R^2 \times \pi \times$ stroke $\times$ No. of cylinders = displacement

For example, to find the displacement of a V-6 engine with a 3.80-inch bore and a 3.40-inch stroke:

$\left(\frac{3.80}{2}\right)^2 = 3.61 \times 3.1416 \times 3.40 \times 6 = 231$ cu. in.

The displacement is 231 cubic inches. Fractions of an inch are usually not included.

This procedure can be greatly simplified by using a cubic inch displacement chart, figure 2-14. You simply locate that point on the chart where the bore and stroke specifications for a given engine intersect, then multiply that figure by the total number of cylinders in the engine to find its displacement.

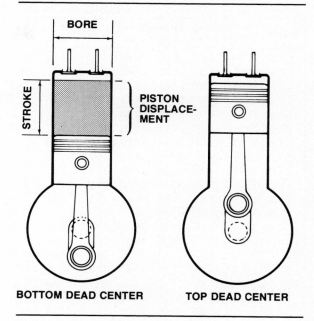

Figure 2-13. Basic engine dimensions.

The greater the engine displacement, the more air-fuel mixture the cylinders can accept, and so the greater the power output (assuming all other factors remain equal).

Metric displacement specifications
When stated in English values, displacement is given in cubic inches; the engine's cubic inch displacement is abbreviated as "cu. in." or "cid." When stated in metric values, displacement is given in cubic centimeters (cc) or in liters (one liter equals 1,000 cc). To convert engine displacement specifications from one value to another, use the following formulas:
- To change cubic centimeters to cubic inches, multiply by 0.061. (cc $\times$ 0.061 = cid)
- To change cubic inches to cubic centimeters, multiply by 16.39. (cid $\times$ 16.39 = cc)

Displacement: A measurement of the volume of air displaced by a piston as it moves from bottom to top of its stroke. Engine displacement is the piston displacement multiplied by the number of pistons in an engine.

Compression Ratio: The total volume of an engine cylinder divided by its clearance volume.

Bottom Dead Center: The exact bottom of a piston stroke. Abbreviated: bdc.

Top Dead Center: The exact top of a piston's stroke. Also a specification used when tuning an engine. Abbreviated: tdc.

CUBIC INCH DISPLACEMENT CHART

Bore → Stroke ↓	(3½) 3.50	(3⁹/₁₆) 3.56	(3⅝) 3.63	(3¹¹/₁₆) 3.69	(3¾) 3.75	(3¹³/₁₆) 3.81	(3⅞) 3.88	(3¹⁵/₁₆) 3.94	4.00	(4¹/₁₆) 4.06	(4⅛) 4.13	(4³/₁₆) 4.19	(4¼) 4.25	(4⁵/₁₆) 4.31	(4⅜) 4.38
2.50	24.05	24.91	25.80	26.69	27.61	28.53	29.48	30.44	31.41	32.40	33.41	34.43	35.46	36.51	37.58
2.60	25.01	25.90	26.83	27.75	28.71	29.67	30.66	31.65	32.67	33.70	34.74	35.80	36.88	37.97	39.08
2.70	25.97	26.90	27.86	28.82	29.82	30.81	31.84	32.87	33.92	34.99	36.08	37.18	38.30	39.43	40.58
2.80	26.93	27.90	28.89	29.89	30.92	31.95	33.02	34.09	35.18	36.29	37.41	38.56	39.72	40.89	42.09
2.87	27.61	28.59	29.62	30.64	31.69	32.75	33.84	34.94	36.06	37.20	38.35	39.52	40.71	41.92	43.14
2.94	28.28	29.29	30.34	31.38	32.47	33.55	34.67	35.79	36.94	38.10	39.29	40.48	41.70	42.94	44.19
3.00	28.86	29.89	30.96	32.03	33.13	34.23	35.37	36.53	37.69	38.88	40.09	41.31	42.55	43.81	45.09
3.10	29.82	30.89	31.99	33.09	34.23	35.38	36.55	37.74	38.95	40.18	41.42	42.69	43.97	45.28	46.60
3.20	30.78	31.89	33.02	34.16	35.34	36.52	37.73	38.96	40.21	41.47	42.76	44.07	45.39	46.74	48.10
3.25	31.26	32.38	33.54	34.69	35.89	37.09	38.32	39.57	40.84	42.12	43.43	44.75	46.10	47.47	48.85
3.30	31.74	32.88	34.05	35.23	36.44	37.66	38.91	40.18	41.46	42.77	44.40	45.44	46.81	48.20	49.60
3.38	32.51	33.68	34.88	36.08	37.33	38.57	39.86	41.15	42.47	43.81	45.17	46.54	47.94	49.39	50.81
3.40	32.71	33.88	35.09	36.30	37.55	38.80	40.09	41.40	42.72	44.07	45.43	46.82	48.23	49.66	51.11
3.44	33.09	34.27	35.50	36.72	37.99	39.26	40.56	41.88	43.22	44.58	45.97	47.37	48.80	50.24	51.71
3.50	33.67	34.87	36.12	37.36	38.65	39.94	41.27	42.61	43.98	45.36	46.77	48.20	49.65	51.12	52.61
3.56	34.25	35.47	36.74	38.00	39.31	40.62	41.98	43.34	44.73	46.14	47.57	49.02	50.50	51.99	53.51
3.60	34.63	35.87	37.15	38.43	39.76	41.08	42.45	43.83	45.23	46.66	48.11	49.57	51.07	52.58	54.11
3.62	34.82	36.07	37.36	38.64	39.98	41.31	42.69	44.07	45.49	46.92	48.37	49.85	51.35	52.87	54.41
3.64	35.02	36.27	37.56	38.86	40.20	41.54	42.92	44.32	45.74	47.18	48.64	50.13	51.63	53.16	54.72
3.66	35.21	36.47	37.77	39.07	40.42	41.77	43.16	44.56	45.99	47.44	48.91	50.40	51.92	53.46	55.02
3.68	35.40	36.67	37.97	39.29	40.64	41.99	43.39	44.81	46.24	47.70	49.17	50.68	52.20	53.75	55.32
3.69	35.50	36.77	38.08	39.39	40.75	42.11	43.51	44.93	46.37	47.83	49.31	50.81	52.34	53.89	55.47
3.70	35.59	36.87	38.18	39.50	40.86	42.22	43.63	45.05	46.49	47.95	49.44	50.95	52.48	54.04	55.62
3.75	36.07	37.36	38.70	40.03	41.41	42.79	44.22	45.66	47.12	48.60	50.11	51.64	53.19	54.77	56.37
3.78	36.36	37.66	39.01	40.35	41.74	43.14	44.57	46.02	47.50	48.99	50.51	52.05	53.62	55.21	56.82
3.80	36.56	37.86	39.21	40.57	41.96	43.36	44.81	46.27	47.75	49.25	50.78	52.33	53.90	55.50	57.12
3.87	37.23	38.56	39.94	41.31	42.74	44.16	45.63	47.12	48.63	50.16	51.71	53.29	54.90	56.52	58.17
3.90	37.52	38.86	40.25	41.63	43.07	44.51	45.99	47.48	49.00	50.55	52.11	53.71	55.32	56.96	58.62
3.94	37.90	39.26	40.66	42.06	43.51	44.96	46.46	47.97	49.51	51.07	52.65	54.26	55.89	54.54	59.23
4.00	38.48	39.86	41.28	42.70	44.17	45.65	47.17	48.70	50.26	51.84	53.45	55.08	56.74	58.42	60.13

Figure 2-14. To use this cubic inch displacement chart, find the number where the bore and the stroke dimensions of your engine intersect. Then multiply by the number of cylinders.

- To change liters to cubic inches, multiply by 61.02. (liters × 61.02 = cid)

Our 231-cid engine from the previous example is also a 3,786-cc engine (231 × 16.39 = 3,786). When expressed in liters, this figure would be rounded off to 3.8 liters.

Compression Ratio

This specification compares the total cylinder volume to the volume of only the combustion chamber, figure 2-15. Total cylinder volume may seem to be the same as piston displacement, but it is not. Total cylinder volume equals piston displacement plus combustion chamber volume. The combustion chamber volume — with the piston at top dead center — is often called the **clearance volume**.

Compression ratio is the total volume of a cylinder divided by the clearance volume. If the clearance volume is ⅛ of the total cylinder volume, the compression ratio is 8:1. The formula can be written as follows:

$$\frac{\text{Total volume}}{\text{Clearance volume}} = \text{Compression ratio}$$

To determine the compression ratio of an engine in which each piston displaces 31.12 cu. in. and which has a clearance volume of 4.15 cu. in.:

31.12 + 4.15 = 35.27 (total cylinder volume)

$$\frac{35.27}{4.15} = 8.498$$

The compression ratio is 8.498:1; this would be rounded off and expressed as a compression ratio of 8.5:1.

In theory, the higher the compression ratio, the more power an engine will develop from a given quantity of fuel. To put it another way, higher compression ratios are supposed to produce better mileage and fuel economy. The reason is that combustion takes place faster because the fuel molecules are more tightly packed and the flame of combustion passes through a smaller area.

Engine Operating Principles

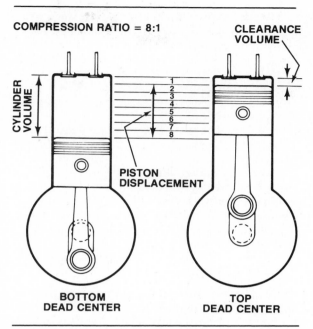

Figure 2-15. Compression ratio is the ratio of the total cylinder volume to the clearance volume.

But there are practical limits to how high a compression ratio can be. To prevent engine damage, most gasoline-burning engines are restricted to a compression ratio no greater than 11.5:1. Ratios as great as 11:1 were common in the 1960's. Ratios this high, however, make it difficult to control emissions. High-compression engines burn high-octane leaded fuel and create high combustion chamber temperatures. This in turn creates oxides of nitrogen (NO_x), a primary air pollutant. Since 1970, compression ratios have been lowered to an average of 8:1 to permit the use of lower octane, low-lead or unleaded fuel, and to reduce NO_x formation.

OTHER ENGINE TYPES

Other engine types besides the 4-stroke engine have been installed in automobiles over the years, but only four have been used with any real success — the 2-stroke, the diesel, the rotary, and the stratified charge engines.

The 2-Stroke Engine

While 4-stroke engines develop *one* power stroke for every *two* crankshaft revolutions, **two-stroke engines** produce a power stroke for *each* revolution. Poppet valves are not used in most automobile and motorcycle 2-stroke engines. The valve work is done instead by the piston, which uncovers intake and exhaust ports in the cylinder as it nears the bottom of its stroke.

Two-stroke gasoline engines used in motorcycles and some small cars suck the air-fuel mixture into the crankcase where it is partially compressed for delivery to the cylinders. It is easy to see that, as a piston moves up, pressure increases in the cylinder above it while pressure decreases in the crankcase below it. By using the crankcase to pull in the air-fuel mixture, the two-stroke engine combines the intake and compression strokes. Also, by using exhaust ports in the cylinder wall, the power and exhaust strokes are combined. The operating cycle of a two-stroke engine shown in figure 2-16, works like this:

1. Intake and Compression: As the piston moves up, a low-pressure area is created in the crankcase. A **reed valve** opens, and the air-fuel mixture is sucked into the crankcase. At the same time, the piston compresses a previous air-fuel charge in the cylinder. Ignition occurs in the combustion chamber when the piston is near top dead center.

2. Power and Exhaust: Ignited by the spark, the expanding air-fuel mixture forces the piston downward in the cylinder. As the piston travels down, it uncovers exhaust ports in one side of the cylinder. The exhaust flows through the ports and out of the engine. As the piston continues its downward movement, it compresses the air-fuel charge in the crankcase. The intake reed valve closes to hold the charge in the crankcase. After uncovering the exhaust ports, the piston uncovers similar intake ports nearer the bottom of the stroke. The compressed mixture in the crankcase moves through a transfer passage to the intake ports and flows into the cylinder. The force of the incoming mixture also helps to drive the remaining exhaust gases from the cylinder. A ridge on the top of the piston deflects the intake mixture upward in the cylinder so it will not flow directly out the exhaust ports.

Clearance Volume: The volume of a combustion chamber when the piston is at top dead center.

Two-Stroke Engine: An engine in which a piston completes two strokes to make one operating cycle.

Reed Valve: A one-way check valve. A reed, or flap, opens to admit a fluid or gas under pressure from one direction, while closing to deny movement from the opposite direction.

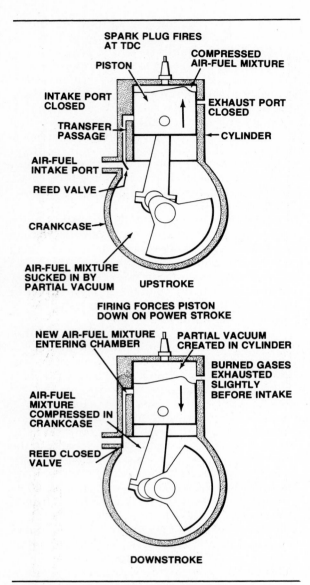

Figure 2-16. Two-stroke cycle engine operation.

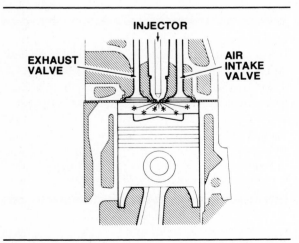

Figure 2-17. Diesel combustion occurs when fuel is injected into the hot, highly compressed air in the cylinder. (Cummins)

The piston then begins another upward compression stroke, closing off the intake and the exhaust ports, and the cycle begins again. Because the crankcase of a two-stroke gasoline engine is used for air-fuel intake, it cannot be used as a lubricating oil reservoir. Therefore, motor oil for a two-stroke gasoline engine must be mixed with fuel.

In theory, the two-stroke engine should develop twice the power of a four-stroke engine of the same size, but the two-stroke design also has its practical limitations. With intake and exhaust occurring at almost the same time, it does not breathe or take in air as efficiently. Mixing fresh fuel with unburned fuel preheats the mixture. Because this increases volume, it reduces efficiency.

The Diesel Engine

In 1892, a German engineer named Rudolf Diesel perfected the compression-ignition engine named after him. The diesel engine uses heat created by compression to ignite the fuel, so it requires no spark ignition system.

The diesel engine requires compression ratios of 16:1 and higher. Incoming air is compressed until its temperature reaches about 1,000° F. As the piston reaches the top of its compression stroke, fuel is injected into the cylinder, where it is ignited by the hot air, figure 2-17. As the fuel burns, it expands and produces power.

Diesel engines differ from gasoline-burning engines in other ways. Instead of a carburetor to mix the fuel with air, a diesel uses a precision pump called a **fuel injector**. This measures the fuel exactly and sprays it into the combustion chamber at the precise time required for efficient combustion, figure 2-18. The diesel fuel injection system performs the fuel-delivery job of the carburetor and the ignition-timing job of the distributor in a gasoline engine.

The air-fuel mixture of a gasoline engine remains nearly constant — changing only within a narrow range — regardless of engine load or speed. But in a diesel engine *air* remains constant and the amount of *fuel* injected is varied to control power and speed. The air-fuel mixture of a diesel can vary from as little as 85:1 at idle, to as rich as 20:1 at full load. This higher air-fuel ratio and the increased compression pressures make the diesel more efficient in terms of fuel consumption than a gasoline engine.

Like gasoline engines, diesel engines are built in both 2-stroke and 4-stroke versions. The most common 2-stroke diesels are the truck and industrial engines made by the Detroit Diesel Allison division of General Motors, figure 2-19. In these engines, air intake is through ports in the cylinder walls, and exhaust is through

Engine Operating Principles

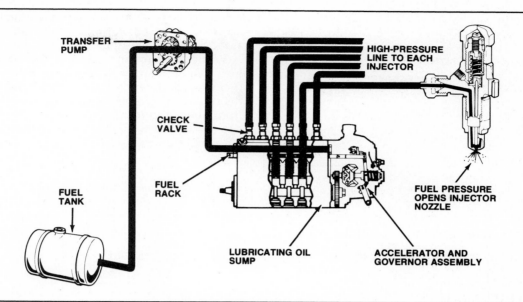

Figure 2-18. Typical automotive diesel fuel injection system.

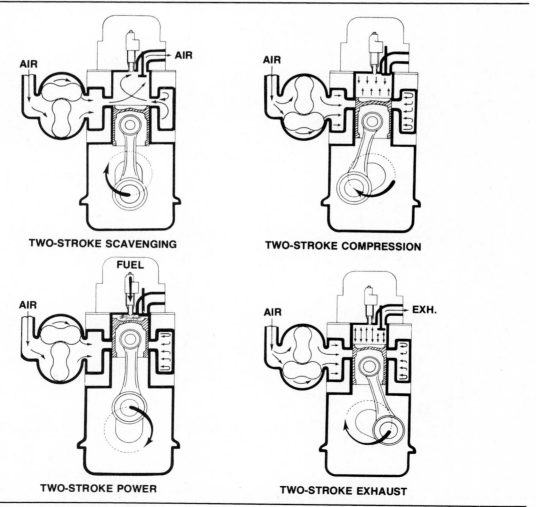

Figure 2-19. These four pictures show the operating cycle of a two-stroke diesel. Crankcase compression is not used as it is in a gasoline two-stroke engine.

valves in the head. Crankcase fuel induction is not used in two-stroke diesels, and air intake is aided by supercharging.

While diesel engines are primarily used in heavy equipment, Mercedes-Benz has used them in passenger cars since 1936. Now, with the recent appearance of diesel-powered cars from Volkswagen and Oldsmobile, the trend appears to be toward greater use of diesel engines in automobiles.

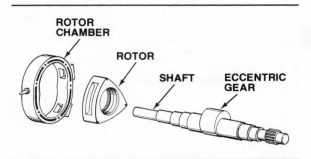

Figure 2-20. The main parts of a Wankel rotary engine are the rotor chamber, the three-sided rotor, and the shaft with an eccentric gear.

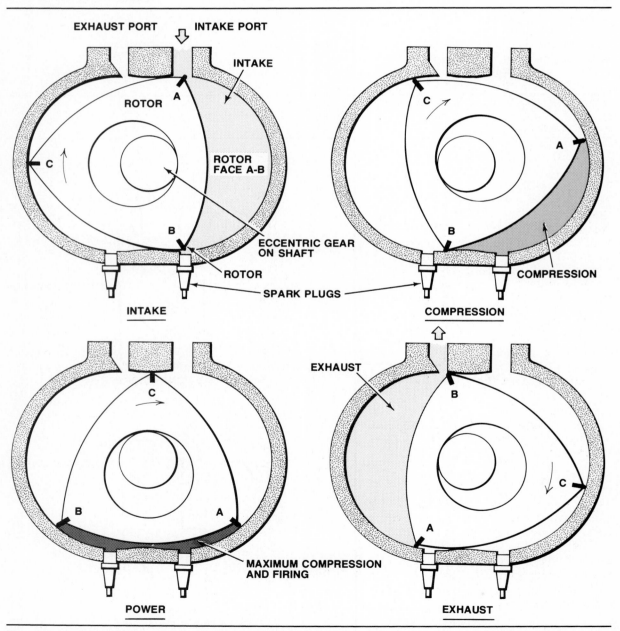

Figure 2-21. This illustration shows the four stages of rotary engine operation. They correspond to the intake, compression, power, and exhaust strokes of a four-stroke reciprocating engine. The sequence is shown for only one rotor face, but each face of the rotor goes through all four stages during each rotor revolution.

Engine Operating Principles

The Rotary (Wankel) Engine

The reciprocating motion of a piston engine is both complicated and inefficient. For these reasons, engine designers for years have attempted to devise engines in which the working parts would all rotate on an axis. The major problem with this rotary concept has been the sealing of the combustion chamber. Of the various solutions proposed, only the rotary design of Felix Wankel (as later adapted by NSU, Curtiss-Wright, and Toyo Kogyo-Mazda), has proven practical.

Although the same sequence of events takes place in both a rotary and a reciprocating engine, the rotary is quite different in design and operation. A curved triangular rotor moves on an **eccentric**, or off-center, geared portion of a shaft within a long chamber, figure 2-20. As it turns, the rotor's corners follow the housing shape. The rotor thus forms separate chambers whose size and shape change constantly during rotation. The intake, compression, power, and exhaust functions occur within these chambers as shown in figure 2-21. Wankel engines can be built with more than one rotor. Mazda engines, for example are two-rotor engines.

One revolution of the rotor produces three power strokes or pulses, one for each face of the rotor. In fact, each rotor face can be considered the same as one piston. Each pulse lasts for about three-quarters of a rotor revolution. The combination of rotary motion and longer power pulses which overlap results in a smooth-running engine. While the rotary overcomes many of the disadvantages of the piston engine, it has its *own* disadvantages.

About equivalent in power output to that of a 6-cylinder piston engine, a two-rotor engine is only one-third to one-half the size and weight. With no pistons, rods, valves, lifters, and other reciprocating parts, the rotary engine has 40 percent fewer parts than a piston engine. But it is also basically a very "dirty" engine. In other words, it gives off a high level of emissions, and so it requires additional external devices to clean up the exhaust.

The Stratified Charge Engine

Like the rotary design, the concept of a **stratified charge engine** has been around in many forms for many years. Honda, however, was the first carmaker to produce and use one successfully. Honda's Compound Vortex Controlled Combustion (CVCC) design was the first stratified charge gasoline engine used in a mass-produced car.

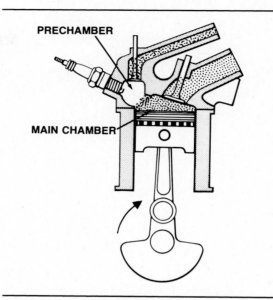

Figure 2-22. Honda CVCC cylinder head, showing main combustion chamber and precombustion chamber.

The CVCC engine has a separate small precombustion chamber located above the main combustion chamber and containing a tiny extra valve, figure 2-22. Except for this feature, the CVCC is a conventional 4-stroke piston engine. However, it uses a two-stage combustion process. Figure 2-23 shows the stages in the operating cycle.

Fuel Injector: A nozzle that meters, atomizes, and injects fuel at high pressure into a combustion chamber.

Eccentric: Off center. A shaft lobe which has a center different from that of the shaft.

Stratified Charge Engine: An engine that uses 2-stage combustion: first is combustion of a rich air-fuel mixture in a precombustion chamber, then combustion of a lean air-fuel mixture occurs in the main combustion chamber.

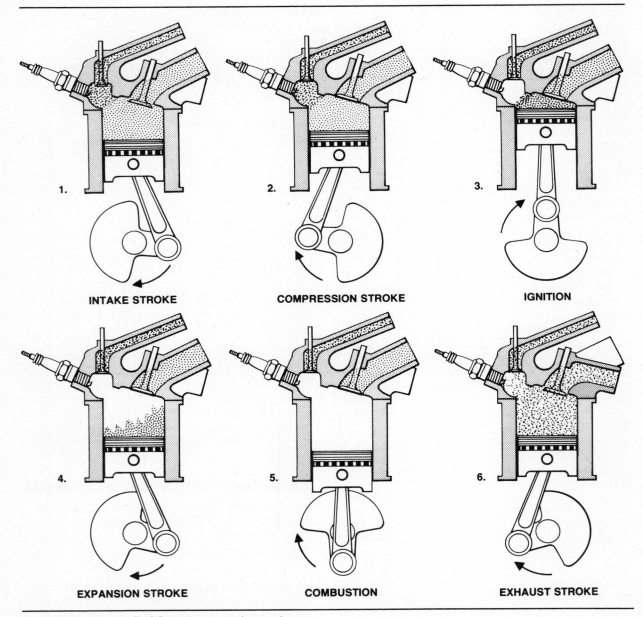

Figure 2-23. Honda CVCC engine operating cycle.

The first stage is one of precombustion, in which the air-fuel mixture is ignited in the precombustion chamber. In the second stage, the flame front created moves down into the main combustion chamber to ignite a mixture with less fuel in it. The stratified charge engine takes its name from this layering, or stratification, of the air-fuel mixture just before combustion. At that time, there is a rich mixture (with lots of fuel) near the spark plug, a moderate mixture in the auxiliary combustion chamber, and a lean mixture (with little fuel) in the main chamber. The result is a more complete combustion of the air-fuel mixture, which keeps unburned fuel and emissions to a minimum.

The stratified charge principle is a *method* of controlling the combustion process. It does not represent a *type* of engine construction, such as the reciprocating or rotary engine types. In fact, charge stratification has been applied both to reciprocating diesel engines and to rotary gasoline engines. Most passenger car diesel engines have a precombustion chamber into which the fuel is injected, figure 2-24. This allows the combustion to occur in two stages: in the precombustion chamber and in the main chamber. This improves combustion efficiency and reduces engine noise and vibration.

Engine Operating Principles

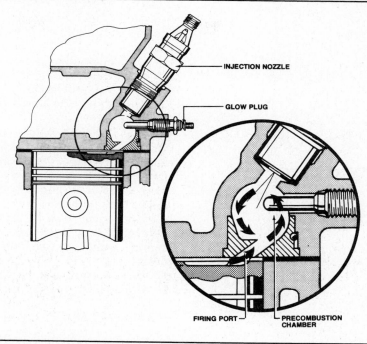

Figure 2-24. Volkswagen's passenger car diesel engine uses a precombustion swirl chamber.

Review Questions

Choose the single most correct answer.
Compare your answers to the correct answers on page 248.

1. The combustion chamber is usually contained in the:
   a. Engine block
   b. Piston
   c. Cylinder head
   d. Water jacket

2. The bore is the diameter of the:
   a. Connecting rod
   b. Cylinder
   c. Crankshaft
   d. Combustion chamber

3. The four-stroke engine is also called the:
   a. Otto cycle engine
   b. Diesel engine
   c. Rotary engine
   d. None of the above

4. The four-stroke cycle operates in which order:
   a. Intake, exhaust, power, compression
   b. Intake, power, exhaust, compression
   c. Compression, power, intake, exhaust
   d. Intake, compression, power, exhaust

5. Which of these engines is most often air cooled:
   a. Rotary
   b. Horizontally opposed
   c. V-type
   d. Incline

6. Valves are opened by:
   a. Camshaft lobes
   b. Connecting rods
   c. The crankshaft
   d. Valve springs

7. Synchronization of camshaft and crankshaft rotations is accomplished by:
   a. Gears
   b. A chain and sprockets
   c. A timing belt and sprockets
   d. Any of the above

8. In an eight-cylinder engine, the number of power strokes at a given instant is:
   a. 8
   b. 2
   c. 4
   d. 1

9. In an I-head engine:
   a. The intake valves are in the block; the exhaust valves are in the head
   b. All the valves are in the block
   c. All the valves are in the head
   d. None of the above

10. Which of the following is *not* used in calculating engine displacement:
    a. Stroke
    b. Bore
    c. Number of cylinders
    d. Valve arrangement

11. To change cubic centimeters to cubic inches, multiply by:
    a. 0.061
    b. 16.39
    c. 61.02
    d. 1000

12. Compression ratio is:
    a. Piston displacement plus clearance volume
    b. Total volume times number of cylinders
    c. Total volume divided by clearance volume
    d. Stroke divided by bore

13. A 2-stroke engine will:
    a. Produce a power stroke for each crankshaft revolution
    b. Have intake and exhaust ports
    c. Use the crankcase for fuel induction
    d. All of the above

14. Diesel engines:
    a. Have no valves
    b. Produce ignition by high heat
    c. Have low compression
    d. Use special carburetors

15. Because of its fuel injection system, a diesel engine:
    a. Needs no carburetor or distributor
    b. Has a constant fuel mixture
    c. Is inefficient
    d. Operates only in a two-stroke configuration

Chapter 3
Engine Air-Fuel Requirements

Automobile engines run on a mixture of gasoline and air. Gasoline has several advantages as a fuel:
1. Vaporization, or evaporation, occurs easily.
2. It burns quickly, but under control, when mixed with air and ignited.
3. It has a high heat value and produces a large amount of heat energy.
4. It is easy to store, handle, and transport.

Gasoline also has certain disadvantages. The chief disadvantage is that combustion produces air pollutants that are given off into the atmosphere through the engine's exhaust.

As a fuel, however, nothing better than gasoline is now available as a substitute. To better understand how the fuel system works in an engine, we must understand the engine's air-fuel requirements. This chapter discusses those requirements and describes how the fuel gets from the fuel tank to the combustion chamber.

AIRFLOW REQUIREMENTS

All gasoline automobile engines share certain air-fuel requirements. For example, a 4-stroke engine can take in only a certain amount of air and fuel at any one time. How much it consumes depends upon how much air the engine can take in; this, in turn, depends upon four major factors:
1. Engine displacement
2. Maximum engine revolutions per minute (rpm)
3. Carburetor airflow capacity
4. Volumetric efficiency.

The first two factors can be used to figure the engine's airflow requirement, which the carburetor must provide. This is measured in cubic feet per minute (cfm). To do this, we assume that the engine has 100-percent **volumetric efficiency** (see below), or what is often called "perfect breathing."

Determining Airflow Requirements

To determine airflow in cubic feet per minute:
1. Divide the engine displacement by 2.
2. Divide the maximum revolutions per minute by 1728.
3. Multiply the results of the two previous steps.
4. Multiply again by volumetric efficiency. Assuming it to be 100 percent, use 1 as the multiplier.

The mathematical formula for this procedure reads:

$$\frac{cid}{2} \times \frac{rpm}{1728} \times \text{volumetric efficiency} = cfm$$

Engine Air-Fuel Requirements

For example, to determine the maximum airflow capability of a 300-cid engine with a maximum engine speed of 3600 rpm:

$$\frac{300 \text{ cid}}{2} \times \frac{3600 \text{ rpm}}{1728} \times 1 = 312 \text{ cfm}$$

Volumetric Efficiency

Volumetric efficiency is a term used to describe the airflow volume actually entering an engine, compared to the engine displacement, which is the maximum volume that it *could* take in. Volumetric efficiency is expressed as a percentage, and it changes with engine speed. For example, an engine might have 75 percent volumetric efficiency at 2,000 rpm. The same engine might be rated at 85 percent at 1,000 rpm and 60 percent at 3,000 rpm.

If the airflow volume is taken in slowly, a cylinder might be filled to capacity. A definite amount of time is required for the airflow to pass through all the curves of the intake manifold and valve port. So, manifold and port design directly relate to the engine's breathing, or volumetric efficiency. Cam timing and exhaust tuning are also important.

If the engine is running fast, the intake valve is not open long enough for a full volume to enter the cylinder. At 1,000 rpm, the intake valve might be open for 1/10 of a second. As engine speed increases, this time is greatly reduced to a point where only a small airflow volume can enter the cylinder. Therefore, volumetric efficiency decreases as engine speed increases. At high speed, it may drop to as low as 50 percent.

To find volumetric efficiency, the airflow volume must be measured at a specified temperature and pressure. This is because the airflow volume will increase as pressure increases and as temperature decreases. Standard pressure for measuring volumetric efficiency is **atmospheric pressure** at sea level, which is 14.7 pounds per square inch (psi) or 760 millimeters of mercury (mm Hg). Standard temperature is 0° C or 32° F.

Measuring volumetric efficiency is a laboratory exercise, but the idea is valuable for understanding airflow requirements and breathing ability of an engine. Through supercharging — explained in Chapter 10 — and other methods, it is possible to build an engine with more than 100-percent volumetric efficiency. In actual practice, the concept of volumetric efficiency must often be changed to have any practical use.

Although we calculated an engine's cubic feet per minute of airflow at a volumetric efficiency of 100 percent, this figure is seldom if ever reached by a stock engine. For this reason, the 300-cid engine in our example will *not* flow 312 cfm, but quite a bit less. With a stock engine, you can expect a volumetric efficiency of about 75 percent at maximum speed, or 80 percent at the highest torque, or turning force. A high-performance engine will be about 5 percent more efficient, and a racing engine will add another 10 percent to that. Returning to our example: a stock 300-cid engine will actually flow about three-quarters of 312 cfm, or 234 cfm.

AIR-FUEL RATIOS

Because liquid fuel will not burn, it must first be changed into a vapor and mixed with air before it is ignited in the cylinders. For most engines, this is done by the carburetor. For fuel-injected engines, the fuel vaporization and mixing with air is done in the intake manifold or in the combustion chamber before ignition.

In either case, there is a direct relationship between an engine's airflow and its fuel requirements. This relationship is called the **air-fuel ratio**.

The air-fuel ratio is the proportions by weight of air and gasoline mixed by the carburetor as required for combustion by the engine. This ratio is important, since there are limits to how rich (with more fuel) or how lean (with less fuel) it can be, and still remain fully combustible for efficient firing. The mixtures with which an engine can operate efficiently range from 8:1 to 18.5:1, figure 3-1. These ratios are usually stated this way: eight pounds of air combined with one pound of gasoline (8:1) is the richest mixture which an engine can tolerate and still fire regularly; 18.5 pounds of air mixed with one pound of gasoline (18.5:1) is the leanest. Richer or leaner air-fuel ratios will cause the engine to misfire, or simply refuse to run at all.

To get the best engine efficiency and economy, about 9,000 gallons of air are needed to burn one gallon of gasoline. When expressing this is terms of volume, we find the air-fuel ratio to be 9,000:1. Not only are ratios of this size

---

Volumetric Efficiency: The comparison of the *actual* volume of air-fuel mixture drawn into an engine to the *theoretical maximum* volume that could be drawn in. Written as a percentage.

Atmospheric Pressure: The pressure on the earth's surface caused by the weight of air in the atmosphere. At sea level, this pressure is 14.7 psi at 32° F (0° C).

Air-Fuel Ratio: The ratio of air to gasoline in the air-fuel mixture drawn into an engine.

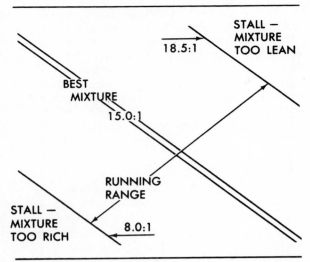

Figure 3-1. Air-fuel ratio limits for a four-stroke gasoline engine. (Chevrolet)

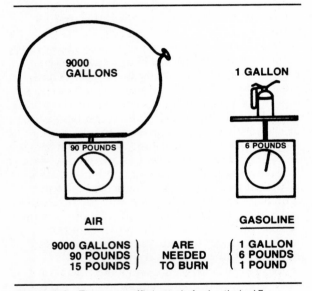

Figure 3-2. The most efficient air-fuel ratio is 15 pounds of air to 1 pound of gasoline (15:1). (Chevrolet)

hard to understand, but they are also hard for engineers to use in their designs and experiments. Therefore, *weight* rather than *volume* is used to figure out air-fuel ratios, since it is easier to work with pounds than with gallons.

For example, to convert a volume ratio of 9,000:1 into a more useful weight ratio:
1. 100 gallons of air = 1 pound
2. 9,000 gallons of air = 9000 ÷ 100 = 90 pounds
3. 1 gallon of gasoline = 6 pounds (approximate).

This means that it takes approximately 15 pounds of air to burn one pound of gasoline, so our air-fuel ratio is 15:1, figure 3-2.

Stoichiometric Air-Fuel Ratio

The ideal mixture or ratio at which all the fuel will blend with all of the oxygen in the air and be *completely* burned is called the **stoichiometric ratio** — a chemically perfect combination. In theory, an air-fuel ratio of about 14.7:1 will produce this ratio, but the *exact* ratio at which perfect mixture and complete combustion take place depends on the molecular structure of gasoline, which can vary somewhat.

This relationship between the amounts of air and fuel flow in an engine is sometimes called the fuel-air ratio. Because an engine uses far more air than fuel, the fuel-air ratio is always a number less than one, such as 0.0625. The fuel-air ratio is just a different way of expressing the more familiar air-fuel ratio.

Engine Air-Fuel Requirements

An automobile engine will work with the air-fuel mixture ranging from 8:1 to 18.5:1. But the ideal ratio would be one that provides both the most power *and* the most economy, while producing the least emissions. But such a ratio does not exist because the fuel requirements of an engine vary widely depending upon temperature, load, and speed conditions.

Research has proved that the best fuel economy is obtained with a 15:1 to 16:1 ratio, while maximum power output is achieved with a 12.5:1 to 13.5:1 ratio. A rich mixture is required for idle, heavy load, and high speed conditions; a leaner mixture is required for normal cruising and light load conditions. As you can see, no single air-fuel ratio provides the best fuel economy *and* the maximum power output at the same time, figure 3-3.

Just as outside conditions such as speed, load, temperature, and atmospheric pressure change the engine's fuel requirements, other forces at work inside the engine cause additional variations. Here are three examples:
1. Exhaust gases remain inside the cylinders and dilute the incoming air-fuel mixture, especially during idle.
2. The mixture is imperfect because complete vaporization of the fuel may not take place.
3. Mixture distribution by the intake manifold to each cylinder is not exactly equal; some cylinders may get a richer or leaner mixture than others.

If an engine is to run well under such a wide variety of outside and inside conditions, the carburetor must be able to vary the air-fuel ratio quickly, and to give the best mixture possible for the engine's requirements at a given moment.

Engine Air-Fuel Requirements

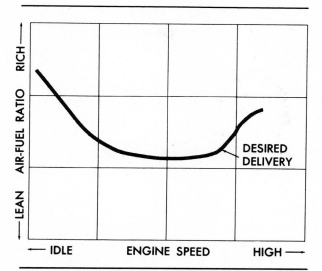

Figure 3-3. The desired air-fuel ratio changes as engine operating conditions change. (Chevrolet)

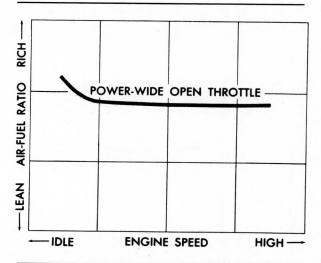

Figure 3-4. The air-fuel ratio needed for maximum power is relatively constant, except at low speed, where it must be slightly richer. (Chevrolet)

The best air-fuel ratio for one engine may not be the best ratio for another, even when the two engines are of the same size and design. Engines are mass produced, and will have slight variations in manifolding, combustion chambers, valve timing, and ignition timing. To accurately determine the best mixture, the engine should be run on a **dynamometer** to measure speed, load, and power requirements for all types of driving conditions.

Power Vs. Economy

If the goal is to get the most power out of an engine, all of the oxygen in the mixture must be burned, because the power output of any engine is limited by the amount of air it can pull in. To be sure that the oxygen combines properly with the available fuel, extra fuel must be provided. This increases the air-fuel ratio (makes it richer in fuel), resulting in some fuel which remains unburned.

This is also true at idle because of the exhaust gases left over in the cylinders. These tend to dilute the incoming mixture since some of the fuel combines with the exhaust. To make certain that the mixture is properly combustible during idle, more fuel must be delivered to make up for the fuel that combines with the exhaust gases. This makes it more difficult to equally distribute the mixture to the cylinders and creates waste material in the form of carbon monoxide, which is emitted into the atmosphere as a pollutant. To get the best fuel economy and the least emissions, the gasoline must be burned as completely as possible in the combustion chamber. This means that the greatest amount of energy (economy) will be produced with the least amount of leftover waste material (emissions). If enough oxygen is to be available to combine with the gasoline, then more air must be provided. This results in a leaner air-fuel mixture (less gasoline) than the ideal ratio.

The air-fuel ratio required to provide maximum power will change very little, except at low speeds, figure 3-4. Reducing speed reduces the airflow into the engine. The result is a poorer mixing of the air and fuel, and less efficiency in its distribution to the cylinders. Thus, at low speeds, a slight enrichment of the mixture is needed to make up for this.

The same is true for maximum fuel economy — the leaner air-fuel ratio used will remain virtually the same throughout most of the operating range, figure 3-5. But enrichment will be required during idle and low speeds, as well as during higher speeds and under load — two conditions which require more power.

Stoichiometric Ratio: An ideal air-fuel mixture for combustion in which all oxygen and all fuel will be completely burned.

Dynamometer: A device used to measure the power of an engine or motor.

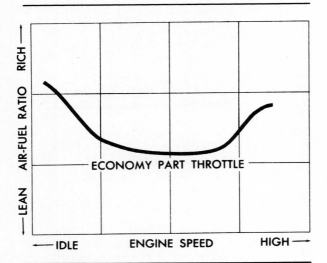

Figure 3-5. The air-fuel ratio for best economy is lean in the middle of the speed range but needs enrichment at high and low speeds. (Chevrolet)

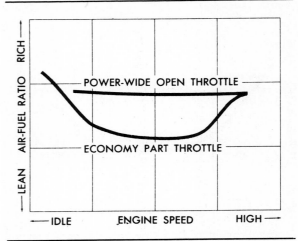

Figure 3-6. The engine must receive lean air-fuel ratios for best economy or rich ratios for maximum power at any given speed. (Chevrolet)

Enrichment can also occur when it is not required or wanted, as in the case of high altitude driving. As altitude increases, atmospheric pressure drops and the air becomes thinner than it is at sea level; the same amount of air actually weighs less and contains less oxygen at higher altitudes. This means that an engine will take in fewer pounds of air and less oxygen. The result is a richer air-fuel ratio, which must be corrected for efficient high-altitude engine operation. Altitude-compensating carburetors correct this problem.

For these reasons, the carburetor must be able to deliver fuel so that the best mileage is provided during normal cruising, with maximum power available whenever the engine is under load, acceleration, or high-speed, figure 3-6.

FUEL DISTRIBUTION

Before gasoline can do its job as a fuel, it must be metered, atomized, and distributed to each cylinder in the form of a burnable mixture. To do this, a metering device — a carburetor, figure 3-7 — mixes the gasoline with air in the correct ratio and distributes the mixture as required by engine load, speed, throttle plate position, and operating temperature.

Proper fuel distribution depends on six factors:
1. Correct fuel **volatility**
2. Proper fuel **atomization**
3. Complete fuel **vaporization**
4. Intake manifold passage design
5. Carburetor throttle plate angle
6. Carburetor location on the intake manifold.

Fuel Atomization and Vaporization

Several things are involved in changing gasoline from a liquid into a combustible form. First, the liquid fuel enters the carburetor, where it is sprayed into the incoming air and atomized (reduced to a mist), figure 3-8. The resulting air-fuel mixture then moves into the intake manifold, where the mist is changed into a vapor.

Vaporization will take place only when the fuel is hot enough to boil. The boiling point is related to pressure: the higher the pressure, the higher the boiling point; the lower the pressure, the lower the boiling point. Because intake manifold pressure is quite a bit less than atmospheric pressure, the boiling point of gasoline drops when it enters the manifold.

Heat from the intake manifold floor combines with heat absorbed from air particles surrounding the fuel particles to begin vaporization. It is helped by raising the temperature of the intake manifold, since the higher the temperature, the more complete the vaporization will be. This heated area in the intake manifold is called a "hot spot."

Sometimes, there will be poor vaporization. This can be caused by several things:
1. A mixture velocity that is too low
2. A cold manifold or low manifold vacuum
3. Cold incoming air
4. Insufficient fuel volatility
5. Poor manifold design
6. Low carburetor flow capacity.

When there is poor vaporization, too much liquid fuel reaches the cylinders. Some of this

Engine Air-Fuel Requirements

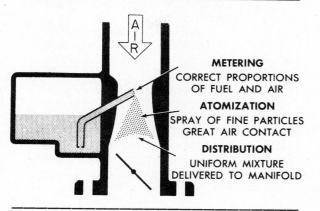

Figure 3-7. The carburetor does the basic job of fuel metering, atomization, and distribution. (Chevrolet)

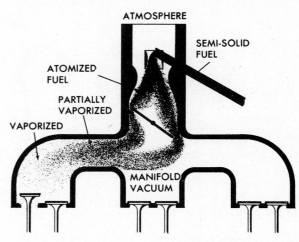

Figure 3-8. Changing liquid fuel to a combustible material is a two-stage process. First, it is atomized to a mist and mixed with air. The air-fuel mixture then must be vaporized. (Chevrolet)

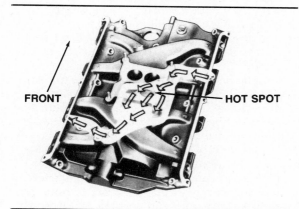

Figure 3-9. Exhaust gases are routed from ports in the cylinder heads through separate passages in the intake manifold to form the manifold hot spot. (Ford)

extra fuel is given off as unburned hydrocarbons, and some will wash oil from the cylinder walls and cause engine wear. The rest will be carried past the piston rings as blowby gases.

Intake Manifold Design

The design of an intake manifold has a direct bearing on mixture distribution and volumetric efficiency over the speed range of an engine. The location, size, and surface area of the hot spot on the manifold floor affects vaporization. The hot spot is normally just under the carburetor, figure 3-9. Although the hot spot is usually heated by exhaust, engine coolant is sometimes circulated through passages between the carburetor base and manifold, figure 3-10.

Both velocity and heating are also affected by the size of the manifold passages through which the mixture must travel. If the passages are large, the mixture will travel slowly, allowing fuel particles to cling to the manifold walls and avoid vaporization. Small passages create a

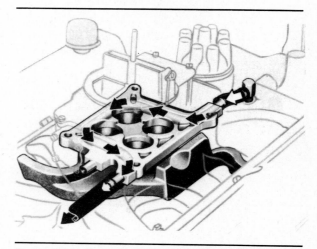

Figure 3-10. This fuel-vaporizing hot spot is created by engine coolant flowing through a spacer under the carburetor. (Ford)

higher velocity but restrict the travel and distribution of the mixture. The angles at which internal manifold passages turn can be critical, too, figure 3-11. When they are too sharp, fuel tends to separate out of the mixture.

Volatility: The ability of a liquid to change from a liquid to a vapor.

Atomization: Breaking down into small particles or a fine mist.

Vaporization: Changing a liquid, such as gasoline, into a gaseous state.

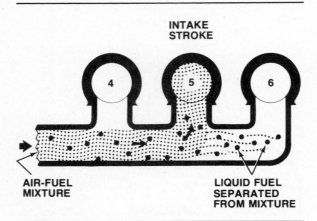

Figure 3-11. An intake manifold with large passages and sharp angles will cause liquid fuel to separate out of the air-fuel mixture. (Chevrolet)

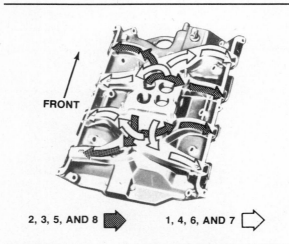

Figure 3-12. Typical V-8 intake manifold fuel passages. (Ford)

■ **Poor Driving Means Poor Mileage**

Carmakers these days are scrambling to produce cars that use less fuel and are more efficient. Still, the driving patterns of individual drivers can make a big difference in how much fuel the car will use.

Here are some interesting points to remember:
• Stopping and restarting a car engine consumes less fuel than running it at idle for one minute.
• Driving at 55 mph rather than 65 mph reduces fuel consumption by about 12 percent. Cutting the speed to 40 mph is even more economical.
• A minor tune-up will increase mileage by about 10 percent.
• The failure of one spark plug in an 8-cylinder engine can cut 12 percent from the car's mpg. It also can increase HC emissions by as much as 300 percent.
• Turning off the air conditioner can improve gas mileage by 2 or 3 mpg.

The air-fuel mixture should be distributed as evenly as possible among the cylinders. Figure 3-12 shows a typical V-8 intake manifold designed for good distribution to all cylinders. If one or more cylinders receives an overly lean mixture, an increase in the overall mixture will be necessary for that cylinder to fire properly. This will cause the other cylinders to receive a mixture that is too rich. Overly lean combustion produces oxides of nitrogen, while overly rich combustion creates unburned hydrocarbons and carbon monoxide. Neither condition is desirable, since they raise emissions and lower fuel economy.

Carburetor Size and Placement

Carburetor airflow must be matched to the airflow requirements of the engine. A carburetor that will provide more air and fuel than the engine needs will produce a rich mixture. This can reduce *both* fuel economy *and* power. A carburetor that provides less fuel and air than the engine needs will cause the engine to work harder to provide the power for any speed and load condition. Again, this means that the engine will not be providing the best combination of economy and power.

The location of the carburetor on the intake manifold is important. Incorrect placement in relation to the manifold passages can interfere with proper fuel distribution. If the carburetor is located closer to one cylinder than to the others, improper vaporization and distribution to the cylinders may result.

FUEL COMPOSITION

Gasoline is a clear, colorless liquid — a complex blend of various basic hydrocarbons (hydrogen and carbon). As a fuel, it has good vaporization qualities and is capable of producing tremendous power when combined with oxygen and ignited. Yet it is impossible to accurately predict how a certain blend of gasoline will perform in a particular engine, since no two engines are identical. Remember, mass-produced engines are subject to individual variations in production, which can affect fuel efficiency.

In laboratory tests, oil refiners calculate and measure the characteristics most important in producing gasolines suitable for specific jobs. It is blended to meet particular temperature and altitude conditions. The gasoline you use during summer months is not the same blend available in the winter, nor is the gasoline sold in Denver the same as that sold in Death Valley. In addition to temperature and altitude, refiners must consider several other things during the blending process: volatility, chemical impurities, octane rating, and additives.

Engine Air-Fuel Requirements

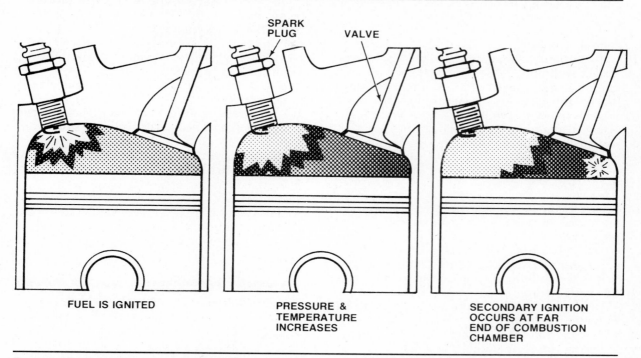

Figure 3-13. Detonation is a secondary ignition of the air-fuel mixture, caused by high cylinder temperatures. It is commonly called "pinging," or "knocking."

Volatility

Volatility is a measure of gasoline's ability to change from a liquid to a vapor and is related to temperature and altitude. The more volatile it is, the more efficiently the gasoline will vaporize. As we've seen, efficient vaporization is needed for even fuel distribution to all of the engine's cylinders and for complete combustion.

Volatility is controlled by blending different hydrocarbons that have different boiling points. In this way, it is possible to produce a fuel with a high boiling point for use in warm weather, and one with a lower boiling point for cold-weather driving. Such blending involves some guesswork about weather conditions, so severe and unexpected temperature changes can cause a number of temperature-related problems ranging from hard starting to vapor lock.

Chemical Impurities

Gasoline is refined from crude oil and contains a number of impurities which can harm engines and carburetors. For example, if the sulfur content is too high, some of it may reach the engine crankcase, where it will combine with water to form sulfuric acid. This substance will corrode engine parts, but proper crankcase ventilation helps keep damage to a minimum. Another impurity, gum, tends to form sticky deposits, which will eventually clog carburetor passages and cause piston rings and valves to stick.

To a large extent, the amount of chemical impurities present in gasoline depends upon the type of crude oil used, the refining process, and the oil refiner's desire to keep his production costs low. The more expensive process of **catalytic cracking** produces a gasoline with a lower sulfur content than the **thermal cracking** method, which is more common and cheaper.

Octane Rating

As engine compression pressure reaches a certain level, a great deal of heat is generated as the air-fuel mixture is compressed. Unless gasoline is formulated to hold up under such high pressures and temperatures, there will often be a secondary explosion called **detonation**, figure 3-13. This is popularly called "knocking" or "pinging." Detonation causes a loss of power

Catalytic Cracking: A process of refining oil so that the gasoline which is produced has a low sulfur content.

Thermal Cracking: A process commonly used to refine oil. The gasoline which is produced has a higher sulfur content than gasoline produced by catalytic cracking.

Detonation: Also called knocking, pinging. An unwanted explosion of an air-fuel mixture caused by high heat and compression.

and overheating of valves, pistons, and spark plugs. The overheating in turn causes more detonation, and may eventually damage the engine.

To prevent detonation, gasoline must have a certain **antiknock value**. This characteristic derives from the type of crude oil and the refining processes used to extract the gasoline. It is measured by an octane rating. Gasolines with a high **octane rating** resist detonation during combustion, while those with a low octane value do not.

Additives

Chemicals not normally present in gasoline are added during refining to improve its performance.
- Anti-icers are specially treated alcohols which act as antifreeze in the gasoline to prevent moisture in the air from causing carburetor icing at low temperatures.
- **Antioxidant inhibitors** are used to prevent the formation of gum.
- Phosphorus compounds prevent spark plug misfiring and preignition.
- Metal deactivators prevent gasoline from reacting chemically with metal storage containers in which it is stored and transported.
- **Tetraethyl lead** (TEL) is used to prevent detonation. Gasoline containing TEL is colored for identification.

Tetraethyl lead aids the combustion process by raising fuel octane which reduces detonation. This allows engines to be built with high compression ratios, which increase power and fuel economy. Lead also prolongs valve life by lubricating valve stems and seats.

Unfortunately, lead has some undesirable effects, as well:
1. It is poisonous and is given off in the exhaust as a particulate. Lead particulate emissions from a single car are quite small, but they can be a health problem if a large quanitity from many cars collects in a small geographic area.
2. Lead reduces the efficiency of catalytic converters.
3. Lead deposits in the engine will dilute crankcase oil, shorten spark plug life, and increase the wear of moving parts in the lower part of the engine.

Antiknock Value: The characteristic of gasoline that helps prevent detonation or "knocking."

Octane Rating: The measurement of the antiknock value of a gasoline.

Antioxidant Inhibitors: A gasoline additive used to prevent the formation of gum.

Tetraethyl Lead: A gasoline additive used to help prevent detonation.

Review Questions
Choose the single most correct answer. Compare your answers to the correct answers on page 248.

1. A disadvantage of gasoline is that it:
   a. Vaporizes easily
   b. Burns quickly
   c. Produces pollutants upon combustion
   d. Has a high heat value

2. Which is *not* a factor in determining airflow requirement:
   a. Engine displacement
   b. Maximum rpm
   c. Carburetor size
   d. Volumetric efficiency

3. Volumetric efficiency:
   a. Is the ratio of air entering the engine to engine displacement
   b. Decreases as engine speed increases
   c. Is expressed as a percentage
   d. All of the above

4. At maximum speed, the volumetric efficiency of a stock engine is approximately:
   a. 75%
   b. 50%
   c. 10%
   d. None of the above

5. The richest air-fuel ratio that an internal combustion engine can tolerate is about:
   a. 4:1
   b. 2.5:1
   c. 8:1
   d. 18.5:1

6. To burn one pound of gasoline with maximum efficiency requires about:
   a. 8 pounds of air
   b. 15 pounds of air
   c. 18.5 pounds of air
   d. None of the above

7. A rich air-fuel mixture is needed for:
   a. Idle
   b. Heavy load
   c. Acceleration
   d. All of the above

8. An internal engine condition affecting fuel requirements is:
   a. Engine load
   b. Mixture distribution
   c. Atmospheric pressure
   d. Engine speed

9. Obtaining maximum power results in:
   a. No change in air-fuel ratios
   b. Leaner mixtures
   c. Unburned oxygen
   d. Excess unburned fuel

10. Maximum fuel economy requires:
    a. Less air
    b. Leaner air-fuel mixtures
    c. Richer air-fuel mixtures
    d. Higher temperatures

11. For maximum power, the air-fuel ratio:
    a. Becomes leaner at low speeds
    b. Becomes richer at low speeds
    c. Becomes richer at high speeds
    d. None of the above

12. For maximum fuel economy, the air-fuel ratio:
    a. Is lean for middle speeds
    b. Becomes richer at high speeds
    c. Becomes richer at low speeds
    d. All of the above

PART TWO

The Fuel System

Chapter Four
Fuel Tanks, Lines, and Evaporative Emission Controls

Chapter Five
Fuel Pumps and Filters

Chapter Six
Air Cleaners and Filters

Chapter Seven
Basic Carburetion

Chapter Eight
Intake and Exhaust Manifolds

Chapter Nine
Electronic Fuel Injection

Chapter Ten
Supercharging and Turbochargers

Chapter 4
Fuel Tanks, Lines, and Evaporative Emission Controls

Part I introduced you to the basic parts, principles, and problems involved in the automotive fuel system and emission controls. Part II covers the major components in the fuel system in more detail. This will help you to develop a working knowledge of the relationships between the fuel system and emission controls. In this chapter, you will learn how the fuel system works. We will also take a detailed look at how each carmaker has tackled the problems of evaporative emission controls.

TANKS AND FILLERS

The automobile fuel tank, figure 4-1, is made of two corrosion-resistant steel halves, which are ribbed for additional strength and welded together. Exposed sections of the tank may be made of heavier steel for protection from road damage and corrosion.

Some cars and light trucks may have an optional fuel tank. Some of these auxiliary tanks — particularly on Ford products — are made of polyethylene plastic, and greater use of plastics in fuel tank construction is likely in the future.

Tank design and capacity are a compromise between available space, filler location, fuel expansion room, and fuel movement. Some late model tanks deliberately limit tank capacity by extending the filler tube neck into the tank low enough to prevent complete filling, figure 4-1. A vertical **baffle** in this same tank limits fuel sloshing as the car moves.

Regardless of its size and shape, a fuel tank must have the following:
- An inlet or filler tube, through which fuel can enter the tank, and a filler cap.
- An outlet to the fuel line leading to the fuel pump.
- A vent system.

Tank Location and Mounting

Most domestic sedans and coupes generally use a horizontally mounted fuel tank. It is usually suspended below the rear of the floor pan, figure 4-2, between the frame rails, and behind the rear axle. Many station wagons use a vertically positioned tank, located on one side of the car between the outer and inner rear fender panels. To prevent squeaks, some GM and Chrysler cars have felt insulator strips cemented on the top or sides of the tank where it touches the underbody.

Location of the fuel inlet depends on the tank design and filler tube placement. It is usually behind a filler cap or a hinged door in the center of the rear panel or in the outer side of either rear fender panel. Imported cars often have their fuel inlet in other positions. The Capri's

Fuel Tanks, Lines, and Evaporative Emission Controls

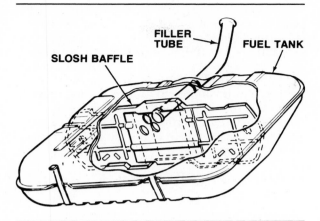

Figure 4-1. The filler tube is located in this tank so that the tank cannot be filled completely. The air space at the top of the tank allows room for fuel expansion. (Oldsmobile)

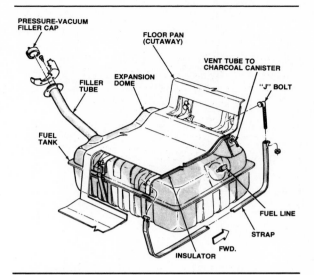

Figure 4-2. Typical fuel tank installation. (Chrysler)

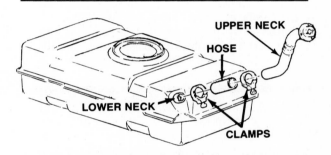

Figure 4-3. Three-piece filler tube assembly. (Oldsmobile)

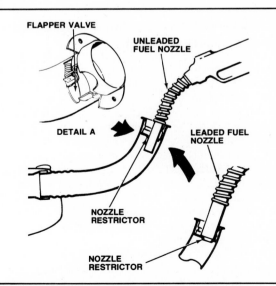

Figure 4-4. Cars that require unleaded fuel have restrictors in the filler tubes to allow only the entry of the smaller unleaded fuel pump nozzles. Restrictors may be spring-loaded flapper valves or simple ring-shaped pieces inside the tube.

rear pillar location, and the Type 1 Volkswagen's placement under the front deck or in the front body panel are common examples. On those 1975 and later cars requiring unleaded fuel, a decal reading "Unleaded Fuel Only" is located beside the filler cap.

Fuel tanks are generally mounted by a pair of metal retaining straps. The strap ends are bolted to underbody brackets or support panels. The free ends are drawn underneath the tank to hold it in place, and then bolted to other support brackets or to a frame member on the opposite side of the tank. The retaining straps used to hold station wagon tanks are often fastened between the inner wheel well and quarter panel.

Filler Tubes

Two types of filler pipes are used: a rigid, 1-piece tube, soldered to the tank, and a 3-piece unit, figure 4-3. The 3-piece unit has a lower neck soldered to the tank and an upper neck fastened to the inside of the body sheet metal panel. The two metal necks are connected by a length of hose, clamped at each end.

All cars that require unleaded fuel (cars with catalytic converters and a few without) have a special filler tube, figure 4-4. This filler has a restriction in it so that the only nozzles that can be inserted are the smaller-diameter nozzles of pumps dispensing unleaded fuel.

Tank Venting Requirements

Fuel tanks must be vented, or a **vacuum lock** will prevent fuel delivery. Before 1970, a direct vent to the atmosphere was provided at the fuel tank by means of a vent line, or through the

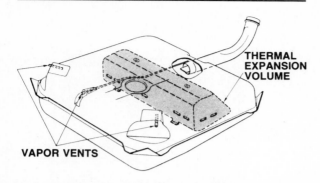

Figure 4-5. This fuel tank has an internal expansion tank to allow for changes in fuel volume due to changes in temperature.

filler cap. But these systems both added to air pollution by passing fuel vapors into the air. To reduce evaporative hydrocarbon (HC) emissions, controls have been installed on all 1971 and later cars (1970 in California). These systems will be covered at length later in this chapter.

Because fuel tanks are no longer vented directly to the atmosphere, the tank must be designed to allow for fuel expansion, contraction, and overflow that results from changes in temperature. One way is to use a separate expansion tank within the main fuel tank, figure 4-5. Another way is to provide a dome in the top of the tank, figure 4-2. As we mentioned earlier, some tanks are limited in capacity by the angle of the fuel filler tube, figure 4-1. This design usually includes a vertical slosh baffle, and reserves about 12 percent of the tank's total capacity for fuel expansion.

Rollover Leakage Protection

All 1976 and later cars have one or more devices to prevent fuel leaks in case of vehicle rollover. Carmakers have met this requirement in different ways.

American Motors

American Motors uses several different rollover leak protection devices, figure 4-6. All of them work on the **check valve** principle. An inline check valve is in the fuel return line at the carburetor. Normally, it protects the carburetor from fuel liquid or vapor feedback through the return line. It also protects against fuel leakage from the return line in case of a rollover accident.

A rollover check valve is installed in the fuel vapor vent line, from the tank to the vapor storage canister. Some AMC cars also have a liquid check valve in the fuel tank outlet to the vapor vent line. The fuel tank filler caps of 1976 and later AMC cars also have a built-in check valve to prevent leakage in case of a rollover.

Baffle: A plate or obstruction that restricts the flow of air or liquids. The baffle in a fuel tank keeps the fuel from sloshing as the car moves.

Vacuum Lock: A stoppage of fuel flow caused by insufficient air intake to the fuel tank.

Check Valve: A valve that permits flow in only one direction.

■ Fuel System Development

The first automobiles used a gravity-feed system to provide fuel to the engine. The fuel tank was mounted in the cowl, higher than the engine, and gravity drew fuel out of the tank and carried it to the engine. But front-mounted fuel tanks were limited in capacity and were dangerous.

Moving the fuel tank to the rear of the car solved the problems of safety and storage capacity, but required the use of a vacuum tank. This was a small fuel tank, still positioned above the engine in the cowl, but connected to the rear tank as well. Suction created by engine vacuum provided fuel for the vacuum tank from the larger rear-mounted tank.

If the car was not driven for long periods of time, the gasoline in the vacuum tank would eventually evaporate. In this case, it was necessary to prime the engine in order to start it and create vacuum which would move fuel through the system. With the appearance of the mechanical fuel pump after World War I, the vacuum tank was retired.

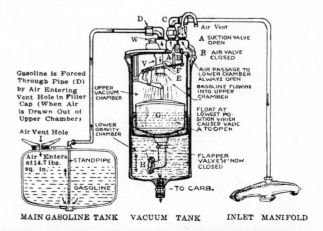

Fuel Tanks, Lines, and Evaporative Emission Controls

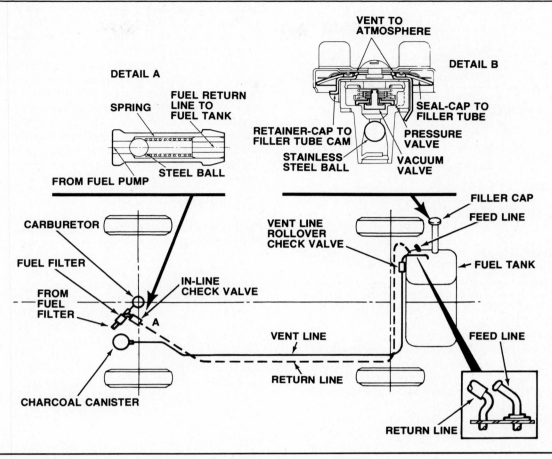

Figure 4-6. American Motors rollover leakage protection devices. (AMC)

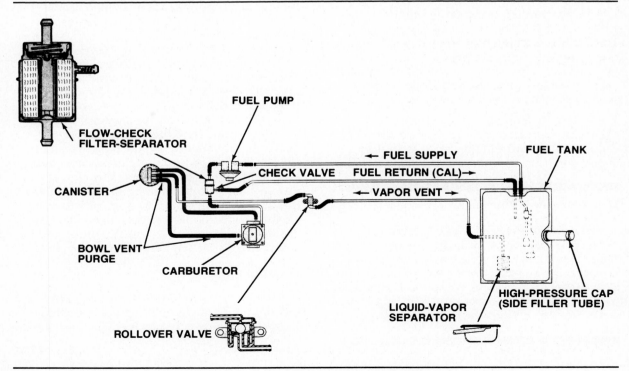

Figure 4-7. Chrysler Corporation rollover leakage protection devices. (Chrysler)

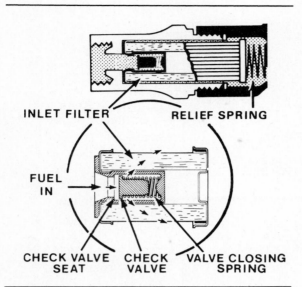

Figure 4-8. The rollover protection check valve is built into the fuel filter on late-model GM cars. (Chevrolet)

Chrysler Corporation

Chrysler uses a rollover valve that is similar to the basic AMC check valve. This valve replaced the overfill limiting valve in the vapor vent line between the vapor canister and fuel tank, figure 4-7. Cars sold in California have a return line from the fuel filter to the tank. These have a special filter that contains a one-way check valve, figure 4-7.

General Motors

General Motors uses a spring-loaded plastic check valve located in the carburetor inlet fuel filter, figure 4-8. The check valve is pushed off its seat by fuel pump pressure when the engine is running and returns to its seat when the engine is off. This valve is built into all replacement filters for Rochester and Holley carburetors used on 1976 and later GM cars.

Ford Motor Company

For rollover protection, Ford uses a spring-operated **float valve** in the vapor separator, figure 4-9. It closes whenever the car is at a 90-degree angle or more. A redesigned fuel pump is also used on 1976 and later cars to reduce fuel spillage during an accident. The filler cap on 1976 and later Ford products looks identical to that used on 1975 cars, but the opening pressure of the relief valve has been increased to prevent gasoline pressure from opening the valve during a rollover accident. The new caps can be identified by a silver-colored valve assembly and the number "50" stamped into the cap handle, or body. While the new caps are interchangeable with earlier ones, the proper cap should always be installed.

FUEL LINES

The parts of the fuel system are connected by fuel and vapor lines and hoses. These are used to supply fuel to the carburetor, to return excess fuel to the tank, and to carry fuel vapors.

Fuel lines must be located to remain as cool as possible. If any part of the line is close to too much heat, the gasoline passing through it will vaporize more rapidly than the fuel pump can create suction. **Vapor lock** will then occur. When it does, the fuel pump will pump only vapor, which passes into the carburetor and out the bowl vent without the engine receiving any gasoline. Depending on their function, fuel and vapor lines may be rigid or flexible.

Rigid Lines

All fuel lines that are fastened to the body, frame, or engine are made of seamless steel tubing. Steel springs may be wound around the tubing at certain points to protect against damage.

When fuel line replacement is necessary, only steel tubing should be used. *Copper and aluminum tubing should never be substituted for steel tubing.* These materials will not withstand normal vehicle vibration, and they could combine with gasoline to form a chemical reaction.

In some cars, rigid fuel lines are secured along the car's frame from the tank to a point close to the fuel pump. The gap between frame and pump is then bridged by a short flexible hose, which absorbs engine vibrations. Other cars run a rigid line directly from tank to pump. To absorb vibrations, the line crosses 30 to 36 inches of open space between the pump and its first point of attachment to the frame.

Flexible Lines

In most fuel systems, synthetic rubber hose sections are used where flexibility is needed. Connections between steel fuel lines and other

---

■ **Is There Life In Your Tank?**

Occasionally you may find a gel-like substance forming in car gasoline tanks, clogging the fuel lines and filters. This is caused by bacteria growth in the fuel tank. This bacteria will grow if there is water present, such as from condensation. To get rid of it, remove the gas tank, clean it with hot water or steam, and air dry it and rinse it with alcohol.

Fuel Tanks, Lines, and Evaporative Emission Controls

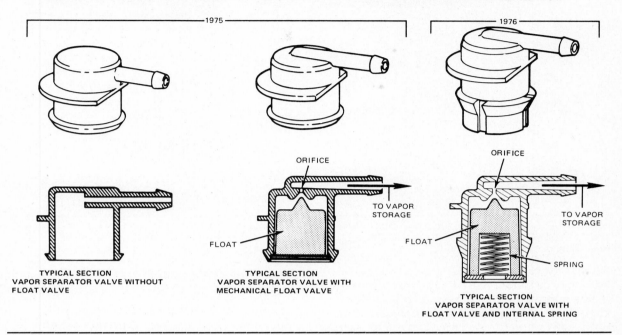

Figure 4-9. Orifice-type vapor separators on some 1974 and later Ford products have a mechanical check valve that also provides rollover leakage protection. (Ford)

system components are often made with short hose sections. The inside diameter of fuel delivery hose is generally larger (5/16 to 3/8 of an inch) than that of fuel return hose (1/4 of an inch).

Fuel system hoses must be made of special, fuel-resistant material. Ordinary rubber hose, such as that used for vacuum lines, would deteriorate when exposed to gasoline. Only hoses made for fuel systems should be used for replacement. Similarly, vapor vent lines must be made of material that will resist attack by fuel vapors. Replacement vent hoses are usually marked with the designation EVAP to indicate their intended use.

A metal or plastic restrictor is often used in vent lines to control the rate of vapor flow. These may be located either in the end of the vent pipe, or in the vapor-vent hose itself, figure 4-10. When used in the hose instead of the vent pipe, the restrictor must be removed from the old hose and installed in the new one whenever hose replacement is necessary.

Fuel Line Mounting

Fuel supply lines from the tank to the carburetor are routed to follow the frame along the under-

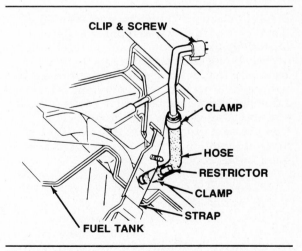

Figure 4-10. Many fuel vent lines have restrictors to control the rate of vapor flow. (Chevrolet)

Float Valve: A valve that is controlled by a hollow ball floating in a liquid, such as in the fuel bowl of a carburetor.

Vapor Lock: A condition in which bubbles are formed in a car's fuel system when the fuel gets hot enough to boil. Flow is stopped or restricted as a result.

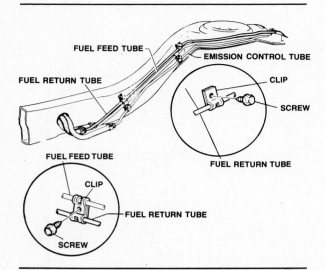

Figure 4-11. Fuel lines are routed along the car frame and secured with clips. Hoses are fastened to steel lines with hose clamps. (Buick)

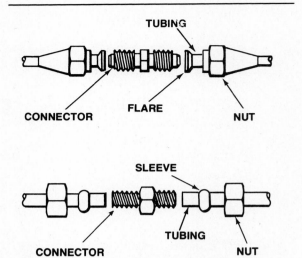

Figure 4-12. Fuel line fittings are either the flare type (top) or the compression type (bottom).

body of the car, figure 4-11. Vapor and return lines may be routed with the fuel supply lines, but usually are on the frame rail opposite the supply line. All rigid lines are fastened to the frame rail or underbody with screws and clamps or clips.

Fittings and Clamps

Brass fittings used in fuel lines are either the flared type or the compression type, figure 4-12; the flared fitting is more common. Its inverted, or SAE 45-degree, flares slip snugly over the connector to prevent leakage when the nuts are tightened. When replacement tubing is installed a double flare should be used to ensure a good seal and to prevent the flare from cracking. Compression fittings use a separate sleeve, a tapered sleeve, or a half sleeve nut to make a good connection.

Various types of clamps, are used to secure fuel hoses. Only the screw-type (aircraft) clamp should be reused when hoses are changed. Keystone, Corbin, and other spring-type clamps will not hold securely when reused and should be replaced with new ones if they are removed.

EVAPORATIVE EMISSION CONTROL SYSTEMS

California's stringent emission laws brought forth the first **evaporative emission controls** (EEC) on 1970 cars sold in that state. Use of EEC systems was extended to all 1971 cars, regardless of where they were sold.

Common Components

The fuel tank filler caps used on cars with EEC systems differ from those used on non-EEC cars. Most EEC caps have pressure-vacuum relief, figure 4-13, built into them. If a sealed cap is used on an EEC system that requires a pressure-vacuum relief design, a vacuum lock may develop in the fuel system, or the fuel tank may be damaged by fuel expansion or contraction. Fuel tanks are protected in various ways against fuel expansion and overflow caused by heat. An overfill limiter, or temperature expansion tank, was used on many 1970-73 EEC systems to prevent total filling of the tank. This is attached to the inside of the fuel tank and contains small holes which open it to the fuel area. When the fuel tank appears to be completely full — it will hold no more and the fuel gauges reads full — the expansion tank remains virtually empty. This provides enough space for fuel expansion and vapor collection if the vehicle is parked in the hot sun after filling the tank.

The dome design of the upper fuel tank section used in some later-model cars, or the overfill limiting valve contained within the vapor-liquid separator, eliminates the need for the overfill limiter tank used in earlier systems.

Some Ford-built cars have a **combination valve** which does three things:
1. It isolates the fuel tank from engine pressures and allows vapor to escape from the vapor separator tank to the vapor storage canister.
2. It vents excess fuel tank pressure to the atmosphere in case of a block in the vapor delivery line.

Fuel Tanks, Lines, and Evaporative Emission Controls

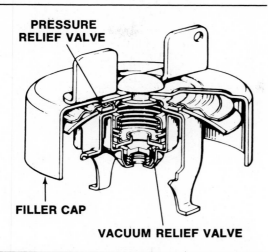

Figure 4-13. Fuel tank caps for EEC systems have vacuum and pressure relief valves.

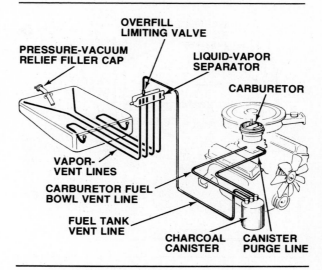

Figure 4-14. This EEC system has a liquid-vapor separator mounted separately from the tank.

3. It allows fresh air to be drawn into the fuel tank to replace the gasoline as it is used.

All EEC systems use some form of **liquid-vapor separator** to prevent liquid fuel from reaching the engine crankcase or vapor storage canister. Some liquid-vapor separators are built into the tank and use a single vapor vent line from the tank to the vapor canister. When the separator is not built in, figure 4-14, it is usually mounted on the outside of the tank or on the frame near it. In this case, vent lines run from the tank to the separator and are arranged to vent the tank, regardless of whether the car is level or not. Liquid fuel entering the separator will return to the tank through the shortest line.

Carburetor Venting

Carburetors must be vented to keep atmospheric pressure in the fuel bowl, which provides the pressure differential necessary for precise fuel metering. Two types of vents are used: internal and external.

Internal vents

Carburetors may be vented internally through the balance tubes that connect the fuel bowl to the airhorn, figure 4-15. The main purpose of this tube is to keep atmospheric pressure pushing down on the fuel in the bowl. This causes the fuel to flow from the bowl, through the circuits and jets, to the lower-pressure area created by the carburetor venturi. The internal vent, or balance tube, also helps to compensate for an air pressure drop caused by a dirty air cleaner. This helps prevent overly rich air-fuel mixtures.

We will learn more about this in Chapter 7. This balance tube also allows vapors from the fuel bowl to collect in the air cleaner when the engine is off. Collecting fuel vapors in this way controls evaporative emissions.

Evaporative Emission Control (EEC): A way of controlling HC emissions by collecting fuel vapors from the fuel tank and carburetor fuel bowl vents and directing them through an engine's intake system.

■ Tools For Making Fuel Lines

With few exceptions, replacement fuel lines cannot be bought preformed. Tubing is stocked in large rolls and must be shaped and formed however the mechanic wants it. Ordinary hand tools cannot be used to properly make a replacement fuel line. When tubing is cut with a hacksaw, its shape is frequently distorted, and the cut edge will be jagged instead of clean and smooth.

Four special tools are required: a tube cutter, a tube reamer, a tube bender, and a tube flaring device. The tube cutter uses sharpened metal discs to make a smooth, distortion-free cut. After cutting, a tapered reamer is necessary to remove any burrs which could make a bad seal. The tube bender shapes the tubing without kinking or bending it. Flaring tools are available to make either single or double flares. It is essential to use them to properly shape the connecting ends of any new fuel line.

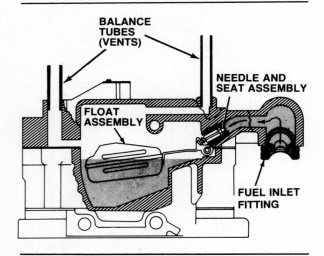

Figure 4-15. Internal carburetor vents. (Chrysler)

External vents
Many carburetors have external vents for the fuel bowl. On older cars without EEC systems, these vents opened directly to the atmosphere. They released vapors from the fuel bowl to prevent the buildup of vapor pressure, which could cause **percolation**. On 1970 and later cars with EEC systems and external carburetor vents, the vents are connected to the vapor storage canister by a rubber hose. External vents are often opened by carburetor linkage, figure 4-16, so that they are closed when the throttle is open and open when the engine is shut off or idling.

Combination Valve: A valve on the fuel tanks of some Ford cars that allows fuel vapors to escape to the vapor storage canister, relieves fuel tank pressure, and lets fresh air into the tank as fuel is withdrawn. Similar to a liquid-vapor separator valve.

Liquid-Vapor Separator Valve: A valve in some EEC fuel systems that separates liquid fuel from fuel vapors.

Percolation: The bubbling and expansion of a liquid. Similar to boiling.

Positive Crankcase Ventilation (PCV): A way of controlling engine emissions by directing crankcase vapors (blowby) back through an engine's intake system.

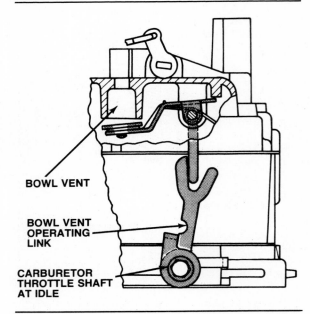

Figure 4-16. External carburetor vent operated by a link from the carburetor throttle shaft. (Chrysler)

Vapor Storage
The EEC system traps gasoline vapors from the fuel tank and carburetor. These trapped vapors are fed into the engine intake system when it is running, or stored until the engine is started. Vapors are stored in one of two places: the engine crankcase, or a charcoal-granule-filled canister.

Engine crankcase storage
The 1970-71 Chrysler, AMC, and some 1970 Ford-built models sold in California, used the crankcase as a vapor storage area. When the engine is started, the stored vapors are drawn from the crankcase through the **Positive Crankcase Ventilation (PCV)** system and into the engine, where they are burned.

Vapor canister storage
This method of fuel vapor storage appeared on all 1972 domestic cars and has been used on most cars ever since. The canister is located under the hood, figure 4-14, and is filled with activated charcoal granules which will hold up to one-third their own weight in fuel vapors. A vent line connects the fuel tank to the canister. Carburetors with external bowl vents are also vented to the canister.

Activated charcoal is used as a vapor trap because of its great surface area. Each gram of activated charcoal has a surface area of 1,100

Fuel Tanks, Lines, and Evaporative Emission Controls

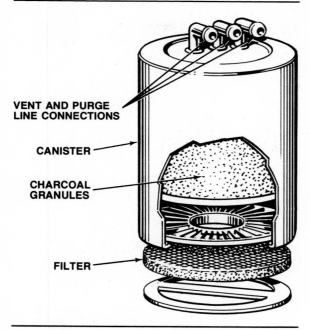

Figure 4-17. A typical vapor storage canister contains 300 or 625 grams of activated charcoal to trap and store fuel vapors.

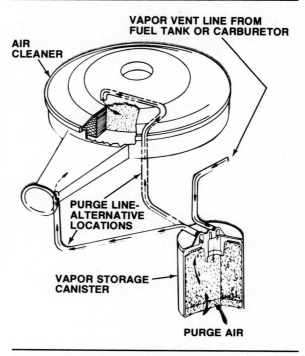

Figure 4-18. Purging the vapor storage canister through the air cleaner results in a variable purge rate. (Ford)

square meters, or more than a quarter acre. Typical canisters hold either 300 or 625 grams of charcoal, figure 4-17, with a surface area equivalent to that of 80 or 165 football fields. Fuel vapor molecules are attached to the carbon surface by **adsorption**. Since this attaching force is not strong, the molecules can be removed quite simply by flowing fresh air through the charcoal.

Vapor Purging

During engine operation, the stored vapors are drawn from the canister to the engine through a hose connected to either the carburetor base or the air cleaner, figure 4-18. This process is called "purging" the canister and can be accomplished as follows:

Constant purge
The purge air rate remains fixed, regardless of engine air consumption. By "teeing" into the PCV line at the carburetor, intake manifold vacuum is used to draw vapor from the canister. Even though manifold vacuum fluctuates, an **orifice** in the purge line provides a constant flow rate. Figure 4-19 shows both a constant purge connection and a variable purge connection.

Variable purge
The amount of purge air drawn through the canister is "proportional" to the amount of fresh air drawn into the engine. In other words, the more air the engine sucks in, the more purge air is sucked through the canister. Two ways are used to draw the purge air through the canister: either by using a **pressure drop** across the air filter, or by using the velocity of the air moving through the air cleaner snorkel. Both ways are shown in figure 4-19.

Adsorption: A chemical action by which liquids or vapors are gathered on the surface of a material. In a vapor storage canister, fuel vapors are attached (adsorbed) to the surface of charcoal granules.

Orifice: A small opening in a tube, pipe, or valve.

Pressure Drop: A loss of pressure between two points.

Purge Valve: A vacuum-operated valve used to draw fuel vapors from a vapor storage canister.

Ported Vacuum: Vacuum immediately above the throttle valve in a carburetor.

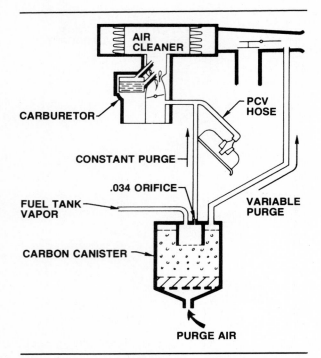

Figure 4-19. In this EEC system, a variable-purge hose runs from the canister to the air cleaner, and a constant-purge hose runs to the intake manifold. (Ford)

Two-stage purge

If the air-cleaner purge flow is not enough, a **purge valve** may be necessary, figure 4-20. This valve is operated by a **ported vacuum** signal and opens a second passage from the canister to the intake manifold.

A carburetor purge port also may be used with the constant purge system. This port is above the high side of the carburetor throttle plate so that there is no purge flow at idle but the flow increases as the throttle opens.

The flow rate and purge method are determined by two factors:
1. They must reactivate the charcoal.
2. They must have little effect on the air-fuel ratio and driveability.

AMERICAN MOTORS EEC SYSTEM

The AMC EEC system, figure 4-21, has remained almost unchanged since its introduction. Some 1970-71 tanks contain a built-in expansion tank. Other tanks, including most of those currently in service, use the filler tube location in the tank to limit the fuel level, and provide space for fuel expansion.

AMC filler caps have a **pressure-vacuum relief valve**, which remains closed under normal pressure conditions. The valve opens whenever a pressure of 0.5 to 1.0 psi, or a vacuum of 0.25

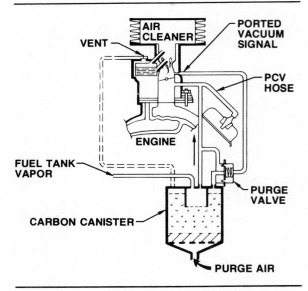

Figure 4-20. The two-stage purge arrangement in this EEC system uses a vacuum-operated valve to open a second purge line from the canister to the manifold. (Ford)

to 0.5 inches of mercury (0.5 to 1.0 on 1976 and later models) occurs in the tank. Once the pressure or vacuum has been relieved, the valve returns to its closed position.

Vent lines extend from the top of the tank to the engine compartment. All 1970 California cars and some 1971 models route the vapors to the engine crankcase. Other 1971 and 1972-73 models route vapors to a canister and then to the PCV system on 6-cylinder engines or to the air cleaner on 8-cylinder engines. From 1974 on, all vapors are routed to the canister and purged through the air cleaner.

A liquid check valve is used on most AMC cars. There are two different designs, but both are float-and-needle types and operate in the same way. One type was discussed under roll-over leakage protection and shown in figure 4-6. The other type of liquid check valve is mounted in the top of the fuel tank to pick up vapors directly from the tank. A single vent line carries vapors to the canister. Fuel tanks with a large air displacement space at the top do not have a liquid check valve because the displacement space prevents liquid fuel from entering the vent lines.

The charcoal canister has a replaceable filter pad that is removed from the bottom of the canister. Two variations are used, figure 4-22: one with two hose connections and one with three. Both use a variable purge that depends on air velocity through the air cleaner snorkel. Until the middle of the 1973 model year,

Fuel Tanks, Lines, and Evaporative Emission Controls

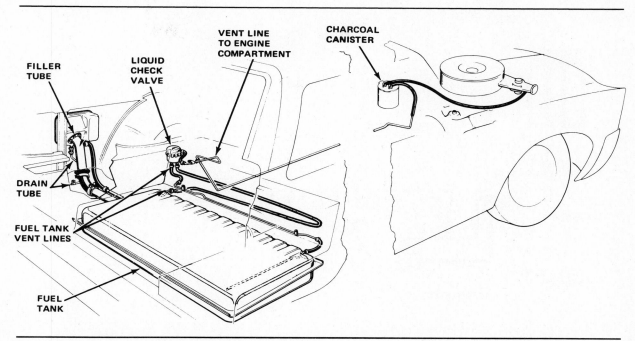

Figure 4-21. Typical American Motors EEC system. (AMC)

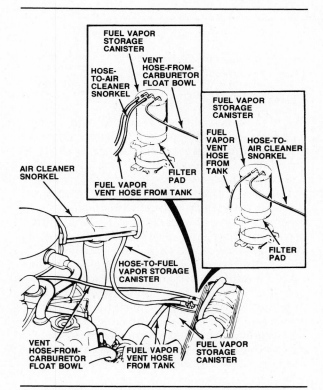

Figure 4-22. Late-model AMC systems use canisters with two or three hose connections. (AMC)

■ Why Vapor Lock?

When gasoline vapors form in the fuel system, vapor lock occurs. This is the partial or complete stoppage of fuel flow to the carburetor. Partial vapor lock will lean the air-fuel mixture and reduce both the top speed and the power of an engine. Complete vapor lock will cause the engine to stall, and make restarting impossible until the fuel system has cooled off.

Four factors usually cause vapor lock:
1. High gasoline temperature and pressure in the fuel system
2. Vapor-forming characteristics of a particular gasoline
3. The fuel system's inability to minimize vapors
4. Poor engine operating conditions, such as overheating.

Vapor may form anywhere in the fuel system, but the critical temperature point is the fuel pump.

Engineers have improved fuel pumps and fuel systems to make today's cars less likely to have vapor lock. Oil companies have succeeded in reducing the vapor-locking tendencies of gasoline by adjusting its volatility according to weather requirements. But vapor lock may still occur during long periods of idle (such as bumper-to-bumper driving), or when the car's fuel system is not properly maintained. Periodically inspect the fuel system and correct all air leaks and defects to prevent vapor lock with today's cars.

Chapter Four

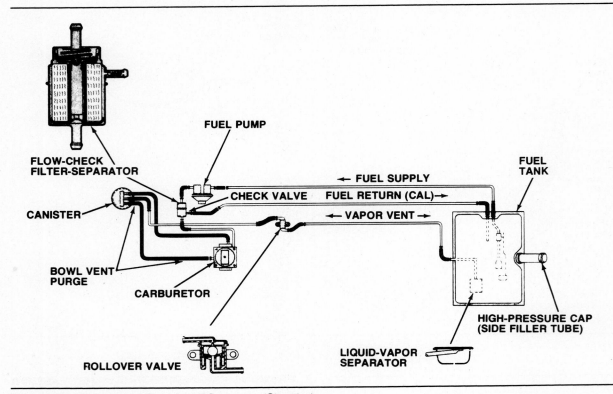

Figure 4-23. Late-model Chrysler EEC system. (Chrysler)

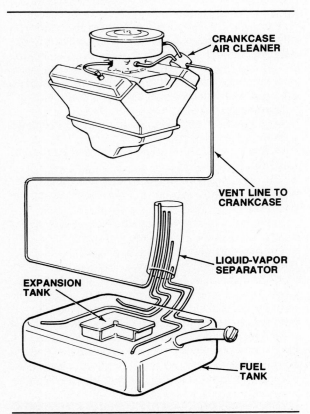

Figure 4-24. Original 1971 Chrysler EEC system. (Chrysler)

6-cylinder systems also used a three-connection canister with a purge valve.

CHRYSLER EEC SYSTEM

Originally called the vapor saver system, Chrysler's approach to evaporative controls, figure 4-23, has undergone several changes since its introduction. The original Chrysler EEC system is shown in figure 4-24.

Fuel tank designs vary. Some use a temperature expansion tank within the main tank, others have an expansion dome, and still others use a vent at each of the tank's top four corners. When tank design requires more overfill protection than is provided by the angle of the filler tube, a limiting valve is used. It is located in the vent line, forward of the liquid-vapor separator or to the rear of the vapor canister. This valve opens for vapor flow from the tank to the canister whenever fuel tank or vent line pressure rises above 0.5 psi. The limiting valve was not used on 1970-71 models.

Pressure-Vacuum Relief Valve: A valve that will correct a pressure differential between two points, such as the valve used in the filler cap of some fuel tanks.

Fuel Tanks, Lines, and Evaporative Emission Controls

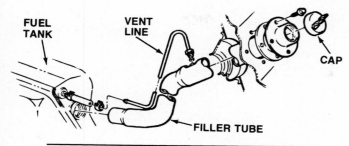

Figure 4-25. Ford EEC fill control vent system. (Ford)

Chrysler filler caps use a safety relief valve, which is closed under normal conditions. Whenever pressure or vacuum is more than the calibration of the cap, the valve opens. Once the pressure or vacuum has been relieved, the valve closes.

Vent lines extend from the tank to the vapor canister in the engine compartment. Both 1970 California cars and all 1971 models vent directly from the tank to the crankcase air cleaner, figure 4-24. The 1972 models vent to a canister, which is purged through the PCV system. On 1973 and later models, the canister vents directly to the carburetor base.

Some tank designs do not require a liquid-vapor separator. On those that do, an external standpipe design, figure 4-24, is used through the 1973 model year. This incorporates vent lines from each corner of the tank. The lines differ in height inside the separator so they can vent the tank regardless of whether or not the car is level. Beginning with 1974 models, the separator designs differ, and the unit may be located inside the dome of fuel tanks, figure 4-23, or attached to the outside top of the tank.

The charcoal canister is a fuel-resistant unit with a replaceable fiber glass filter. The canister on all 1972 and high-performance 1973 vehicles is a two-stage unit with a built-in valve and four hose connections. This was replaced by a three-connector canister on other 1973 and all later cars. The three-connector canister is purged by carburetor ported vacuum, using the throttle plates as a purge valve. This allows the engine to run better at hot idle by eliminating canister purging. If used with an internally vented carburetor, the bowl vent port on this canister is plugged.

A roll-over leakage protection valve replaces the overfill limiting valve in the vent line between the fuel tank and vapor canister on 1976 and later cars.

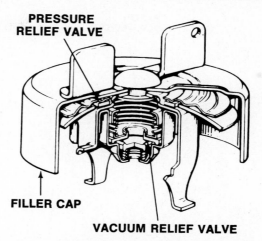

Figure 4-26. Ford uses a pressure-vacuum relief filler cap on 1971 and later EEC systems.

FORD EEC SYSTEM

Ford's EEC system has been simplified since it was first used nationwide on all Ford cars and light trucks in 1971. The numerous variations in the system are detailed below.

While fuel tank designs vary all tanks provide positive control of fuel height when filled. This is accomplished by filler tube design and by the use of vent lines within the filler tube or tank, figure 4-25. Approximately 10 percent of the tank remains empty when the tank is filled with fuel. This space at the top of the tank permits the fuel to expand during hot weather.

The 1970 Ford EEC system on California cars uses a nonvented, sealed filler cap. A three-way control (combination) valve in the vapor line vents directly to the atmosphere when vacuum or pressure begins to build in the main fuel tank. This was replaced on 1971 and later cars by a filler cap with a built-in pressure-vacuum relief valve, figure 4-26. The cap calibration is higher on 1976 and later caps, and the correct cap should be installed when replacement is necessary. Higher calibration caps can be identified by the number "50" stamped on them, and by their silver-colored valve and camcup.

Vent lines carry vapors from the fuel tank to the charcoal canister. Some 1970 California cars purged the canister through the PCV system, but from 1971 on, canisters have been purged through the air cleaner.

Two variations of the vapor vent system are used; one on vertically mounted tanks, and one on horizontally mounted tanks. Both use a vapor separator. Vertical tanks have the separator in the center of the tank top. Horizontal tanks have the separator on a raised section, also located in the center of the tank top. This raised

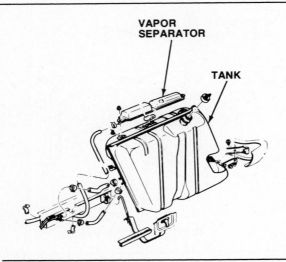

Figure 4-27. An auxiliary "wing" tank is used as a vapor separator on some 1970 Ford EEC systems. (Ford)

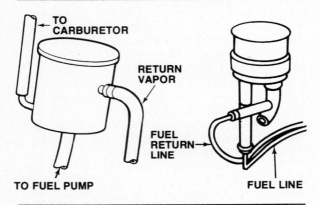

Figure 4-29. Fuel return line vapor separators used in some Ford EEC systems. (Ford)

■ The Good Old Days?

Well, yes, prices have increased in the automotive industry. Just take a look at this 1926 price list for doing repairs on the hugely popular Ford Model T.

These are the flat-labor rate charges recommended by Ford to all its dealers and mechanics:

Overhaul motor and transmission	$25.00
Overhaul motor only (includes R&R, rebabbiting, reboring)	$20.00
Overhaul transmission only	$14.00
Grind valves, clean carbon	$ 3.75
Replace head gasket	$ 1.00
Overhaul carburetor	$ 1.50
Adjust transmission bands	$ 0.40

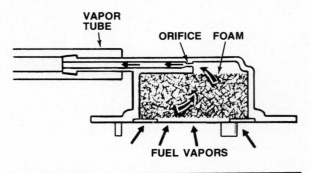

Figure 4-28. An orifice-type vapor separator is installed in the top of fuel tanks on 1971 and later Ford EEC systems. (Ford)

section is necessary to provide fill control venting, because the space for fuel expansion at the top of the horizontal tank is not as great as it is at the top of the vertical tank.

A small expansion tank is used as a vapor separator on 1970 EEC versions. This is located above the main tank, figure 4-27, and traps any vapor from fuel rising in the vents. A vent line connects this "wing" tank design with the three-way control valve.

An orifice vapor separator is installed on all 1971 and later Ford vehicles. Mounted directly on the fuel tank, it is filled with an open-cell foam. This acts as a baffle to separate liquid fuel and vapor and keeps the liquid fuel out of the vapor line. The open bottom of the orifice vapor separator, figure 4-28, draws fuel vapors inside. The orifice at the top allows the vapors to enter the vent line, but restricts the entry of liquid fuel.

A different design is used on some 1974 models, with a mechanical float valve, figure 4-9. For 1976, changes permit the vapor separator to act as a rollover check valve. The auxiliary fuel tank on some 1975 and later Ford products has its own vapor separator.

To reduce the amount of fuel vapors reaching the carburetor, a small vapor separator, figure 4-29, is installed in the supply line between the fuel pump and the carburetor on certain 1973 and later engines. A vapor return line connects this separator with the fuel tank. Vapors collected in the separator are routed back to the tank to recondense or they travel through the regular vent line to the canister. By continuously venting these vapors back to the fuel tank instead of allowing free travel to the carburetor, engine surging from fuel overenrichment is prevented.

Fuel vapors are stored in the charcoal canister and purged as shown in figure 4-30. Canister

Fuel Tanks, Lines, and Evaporative Emission Controls

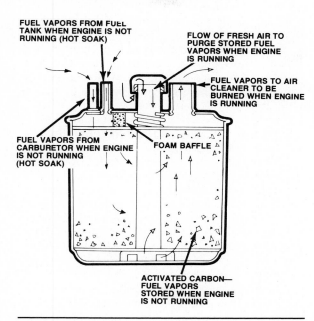

Figure 4-30. Typical Ford EEC canister. (Ford)

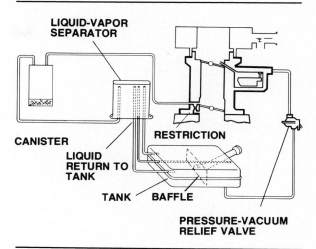

Figure 4-31. General Motors EEC system with liquid-vapor separator and constant-purge canister. (Pontiac)

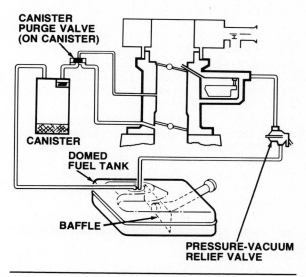

Figure 4-32. General Motors EEC system with domed fuel tank and purge valve on canister. (Pontiac)

size and shape vary from year to year. A larger canister was used with 1973-74 vehicles for greater vapor storage capacity. Those used with 1975 and later cars are a smaller 300-gram unit. The circular canister design was changed to a rectangular unit for 1977 and later applications.

GENERAL MOTORS EEC SYSTEM

The GM approach to controlling evaporative emissions from the fuel tank and carburetor is similar to that used by AMC, Chrysler, and Ford, but several different parts are used. Two variations of the GM EEC system are shown in figures 4-31 and 4-32.

Early GM fuel tanks (1970-72) have an expansion cell inside the main tank for temperature expansion of the fuel. Attached to the upper half of the main tank, the cell is open at the bottom. This offers an expansion capacity of about 12 percent when the filled tank is level. It also helps to control fuel slosh.

Most 1973 and later fuel tanks, except those used in station wagons, rely on a baffle in the tank and the angle of the filler tube for overfill protection and for controlling fuel slosh. Since station wagon fuel tanks are vertically mounted, the expansion volume is determined by the level at which the filler tube enters the tank.

Pressure and vacuum relief is provided in different ways. Some vehicles use a pressure-vacuum relief valve in the filler cap. Others use a nonvented cap with a relief valve placed in the line between the vapor separator and the canister. A pressure relief valve located in the fuel tank vent system, as shown in figures 4-31 and 4-32, is another method. These all work the same way, opening whenever pressure or vacuum is more than their calibration, and closing when the pressure or vacuum has been relieved.

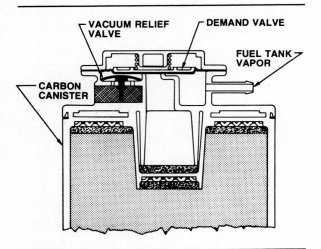

Figure 4-33. A demand valve is used in some GM vapor storage canisters.

A variation called a demand valve, figure 4-33, is used with some charcoal canisters. This is a pressure relief device that requires pressure be present before vapor is allowed to pass from the fuel tank into the canister. A rubber umbrella-type valve, next to the demand valve, is used to relieve vacuum.

A three-point vent system is used with some GM fuel tanks. Vent pipes located at the inside top of the tank are connected to a liquid-vapor separator. Placement of the rear vent above the filler pipe flange, figure 4-31, ensures that at least one vapor vent will always be open.

A single vent system combined with a domed fuel tank, figure 4-32, eliminates the need for a vapor separator. The single vent is high enough to keep liquid fuel out of the line.

Diaphragm: A thin, flexible wall to separate two cavities, such as the diaphragm in a mechanical fuel pump.

■ **Try This Tube Tool**

Next time you have a piece of steel or copper tubing break off inside an engine block or other casting, try this simple home-made tool to pull it out. Use a wrench socket or bushing, a washer, and a self-tapping screw. Place the socket or bushing, with an inside diameter larger than the broken tubing, over the hole in the block. Then place a large flatwasher on the self-tapping screw and insert it through the bushing or socket into the broken tubing. As you tighten the screw, it will pull the tubing out of the hole.

Two types of liquid-vapor separators are used: a metal standpipe, figure 4-31, and a nylon unit with a needle valve operated by a float ball. Both are used with three-point vent systems.

The standpipe has unequal-length vertical pipes in the separator chamber. When the car is pointing downhill, liquid fuel, which enters the separator from the front vent, returns to the tank through the shorter rear vent pipe. When the car is pointed uphill, the rear vent will be submerged, but no liquid fuel can enter the standpipe in this position.

The needle valve separator is mounted on the underbody behind the rear axle so that liquid fuel must travel upward from the front vents to reach it. When liquid fuel *does* enter its volume chamber (during severe braking, for example) the needle valve closes off the front vents. This prevents more fuel from traveling up the vent pipes to the separator, while allowing the tank to breathe through the rear vent.

General Motors vapor storage canisters may have two, three, or four hose connections, depending on the purge method used. The three- and four-hose canisters have a built-in purge valve. This valve consists of a spring-loaded **diaphragm**, a diaphragm cover, and metered purge restrictions. The purge valve limits vapor flow to the carburetor at idle. As carburetor airflow increases, ported vacuum unseats the spring-loaded diaphragm to provide maximum purge.

SUMMARY

Automobile fuel tanks can be mounted either vertically or horizontally, depending on how much room there is. Filler tubes, besides allowing the tanks to be filled, also are used as fill limiters, leaving room for fuel to expand. Cars requiring unleaded gasoline will not accept a leaded gas nozzle in the tube. Tanks must be vented, but evaporative emission control (EEC) requirements state that the fuel vapors must not be vented to the atmosphere. Carmakers have devised many ways to make sure all the vapors stay within the fuel system. They have provided rollover leak protection, pressure-vacuum relief valves, liquid-vapor separators, and positive crankcase ventilation. The EEC systems all use vapor storage canisters. Vapors stored in the canisters are purged into the engine; each carmaker has a slightly different way to do this. The most important thing is that all of these EEC systems prevent vapors containing unburned hydrocarbons from reaching the atmosphere, where they pollute the air.

Fuel Tanks, Lines, and Evaporative Emission Controls

Review Questions

Choose the single most correct answer.
Compare your answers to the correct answers on page 248.

1. All fuel tanks must:
   a. Be vertically mounted
   b. Be horizontally mounted
   c. Have a vent system
   d. Contain a vertical baffle

2. Filler necks with restricted openings:
   a. Provide better venting
   b. Prevent entry of leaded fuel dispensers
   c. Reduce emissions
   d. Act as rollover check valves

3. Fuel lines fastened to the frame, body, or engine are made of:
   a. Steel
   b. Aluminum
   c. Copper
   d. Rubber

4. Ordinary rubber hose can be used for:
   a. Fuel supply lines
   b. Vapor vent lines
   c. Vacuum lines
   d. Fuel return lines

5. Fuel line fittings:
   a. Are made of copper
   b. Crack easily
   c. Are used to secure rigid lines to the car frame
   d. Are either the flared or compression type

6. Clamps that can be reused when hoses are changed are:
   a. Keystone
   b. Corbin
   c. Screw type
   d. Flat spring

7. EEC systems were used on all domestic passenger cars beginning in:
   a. 1971
   b. 1972
   c. 1973
   d. 1975

8. Liquid vapor separators:
   a. Act as pressure-vacuum relief devices
   b. Are overfill limiters
   c. Prevent vapor lock
   d. Prevent liquid fuel from reaching the crankcase or the vapor storage canister

9. Carburetors are vented:
   a. To allow fuel to return to the fuel tank
   b. To prevent overfill
   c. To maintain atmospheric pressure in the float bowl
   d. All of the above

10. Crankcase vapor storage was not used after:
    a. 1969
    b. 1970
    c. 1971
    d. 1972

11. Activated charcoal is used as a vapor trapping agent because:
    a. It has the area of a football field
    b. It has a great surface area — 1100 square meters per gram
    c. There are 625 grams of charcoal in a canister
    d. Charcoal is a light material

12. Variable purge:
    a. Takes place only when the engine is off
    b. Is proportional to the air drawn in by the engine
    c. Is controlled by intake manifold vacuum
    d. Requires a graduated orifice

13. In American Motors EEC systems, which of the following is false:
    a. All vapors are routed to the canister in 1971 models
    b. A liquid check valve is used on most AMC cars
    c. Filler caps have pressure relief valves
    d. Most AMC cars use filler tube location to limit fuel level

14. 1973 and later Chrysler models:
    a. Vent directly from the tank to the crankcase air cleaner
    b. Vent to a canister which is purged through the PCV system
    c. Have the canister vented directly to the carburetor base
    d. All of the above

15. In Ford fuel tanks, empty space amounts to about:
    a. 5%
    b. 18%
    c. 12%
    d. 10%

16. Vapor venting on 1971 and later model Fords:
    a. Is through the crankcase
    b. Purges through the air cleaner
    c. Purges through the PCV system
    d. Both b and c

17. The device in the diagram is:
    a. A PCV valve
    b. A charcoal canister
    c. An orifice-type vapor separator
    d. An air cleaner

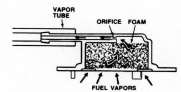

18. A demand valve is used on which canister:
    a. GM
    b. Ford
    c. Chrysler
    d. AMC

19. An integral purge valve is used on:
    a. The carburetor in Fords
    b. AMC canisters
    c. GM vent lines
    d. Three- and four-hose GM canisters

20. All EEC systems eliminate vapors containing:
    a. NO_x
    b. CO
    c. SO_2
    d. Unburned hydrocarbons

Chapter 5
Fuel Pumps and Filters

In Chapter 4, we talked about the fuel lines and the fuel system. But we still have to get the fuel from the tank, through the carburetor, to the engine. And once it gets there, it must be clean, or the engine will not run properly. In this chapter, then, we will discuss the fuel pumps that move the fuel from the tank to the engine, and the filters that keep it clean.

PUMP OPERATION OVERVIEW

The fuel pump and the fuel lines, figure 5-1, deliver gasoline from the tank to the carburetor. The pump moves the fuel with a mechanical action that creates a low-pressure, or suction, area at the pump inlet. This causes the higher atmospheric pressure in the fuel tank to force fuel to the pump. The pump spring also exerts a force on the fuel within the pump and delivers it under pressure to the carburetor. All pumps, except electric turbine pumps, develop this mechanical action through a reciprocating, "push-pull," motion. Various kinds of fuel pumps are described in detail in the following paragraphs.

PUMP TYPES

While all pumps deliver fuel through mechanical action, they are generally divided into two groups:
1. Mechanically driven by the car engine
2. Electrically driven by an electric motor or vibrating **armature**.

MECHANICAL FUEL PUMPS

The most common type of fuel pump used by domestic and foreign carmakers is the single-action, diaphragm-type mechanical pump, figure 5-2. The rocker arm is driven by an eccentric lobe on the camshaft. The pump makes one stroke with each revolution of the camshaft. The eccentric lobe (often called simply, "the eccentric") may be part of the camshaft. On some engines, the eccentric is a pressed steel lobe that is bolted to the front of the camshaft, along with the drive gear, figure 5-3.

Sometimes the rocker arm is driven directly by the eccentric, figure 5-2. Other engines have a pushrod between the eccentric and the rocker arm, figure 5-4. The most common example of this arrangement is the small-block Chevrolet V-8.

Some older cars have double-action diaphragm pumps. These pumps have two diaphragms driven by a single rocker arm. The second diaphragm is a vacuum pump that adds to intake manifold vacuum to run vacuum-operated windshield wipers. This type of windshield wiper and double-action fuel pump gen-

Fuel Pumps and Filters

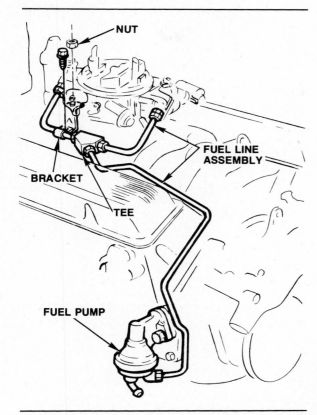

Figure 5-1. Typical fuel pump and line installation on a V-8 engine. (Chevrolet)

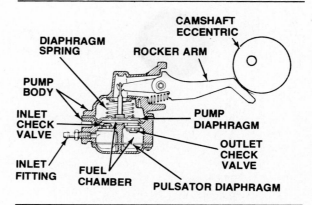

Figure 5-2. Typical diaphragm-type mechanical fuel pump.

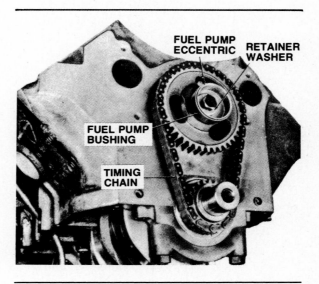

Figure 5-3. This fuel pump eccentric is bolted to the front of the camshaft. (Pontiac)

erally disappeared in the early 1960's. American Motors, however, did keep the double-action pump on some cars through 1971, figure 5-5. A sediment bowl and a pleated paper fuel filter are contained within the fuel section of the AMC two-stage pump.

Mechanical Pump Operation

The fuel intake stroke begins when the rotating camshaft eccentric pushes down on one end of the pump rocker arm. This raises the other end, which pulls the diaphragm up, figure 5-2, and tightens up the diaphragm spring. Pulling the diaphragm up creates a vacuum, or low pressure area, in the fuel chamber. Since there is a constant high pressure in the fuel lines, the inlet check valve in the pump is forced open and fuel enters the fuel chamber.

As the camshaft eccentric continues to turn, it allows the outside end of the rocker arm to "rock" back up. This, along with the push given by the diaphragm spring, allows the diaphragm to relax back down. This is the start of the fuel output stroke. As the diaphragm relaxes, it causes a pressure buildup in the fuel chamber.

This pressure closes the inlet check valve and opens the outlet check valve. The fuel flows out of the fuel chamber and into the fuel line on the way to the carburetor. The outlet check valve keeps a constant pressure in the outlet line and prevents fuel from flowing back into the pump.

We measure a fuel pump by the *pressure* and the *volume* of the fuel it delivers. Delivery pressure is controlled by the diaphragm spring. Delivery rate, or volume, is controlled by the carburetor.

The fuel pump delivery rate is proportional to the fuel required by the carburetor. When the carburetor inlet needle valve is open, fuel will flow from the pump, through the lines, and into the carburetor. When the needle valve is closed (due to the fuel bowl in the carburetor being full), then no fuel flows through the lines.

Chapter Five

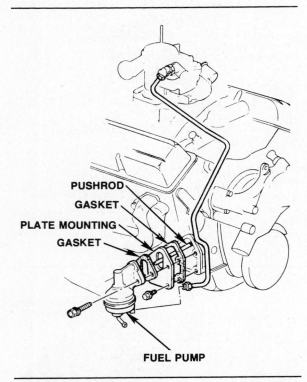

Figure 5-4. A pushrod is installed between the camshaft eccentric and the fuel pump rocker arm on this engine. (Chevrolet)

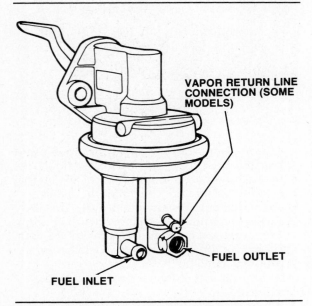

Figure 5-6. This fuel pump has a vapor return line to the fuel tank. (Ford)

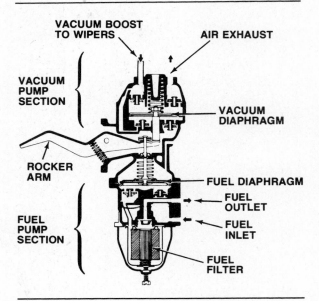

Figure 5-5. This double-action pump has a diaphragm fuel pump and a vacuum booster pump for vacuum-operated windshield wipers. (AMC)

With the needle valve closed and the pressure in the fuel line increasing, the fuel pump diaphragm stays up, even though the rocker arm continues to go up and down in a "freewheeling" motion. No fuel is pumped until the fuel level in the carburetor bowl drops enough for the inlet needle valve to open again.

During different operating conditions, the fuel level in the carburetor bowl varies. Therefore, the position of the inlet needle valve varies between fully open and fully closed. The opening of the needle valve and the rate of fuel flow into the carburetor is always controlled by, and proportional to, the rate of fuel flow out of the carburetor.

A rocker arm link, used in some fuel pumps, figure 5-2, also responds to fuel outlet line pressure and permits a partial stroke of the fuel pump.

When an engine with a mechanical fuel pump is shut off, pressure in the fuel line to the carburetor is maintained by the pump diaphragm spring. If engine compartment heat expands the gasoline in the fuel line, the fuel will push the carburetor inlet needle valve open and pass through. The result is too much fuel in the carburetor, and the engine will not restart easily. This is known as **flooding** the carburetor. Also, since fuel expands when it's hot, it may turn from a liquid into a vapor. This causes a vapor lock in the pump and lines. Four methods, described in the following paragraphs, are used to maintain fuel pressure and to prevent flooding and vapor lock.

Fuel Pumps and Filters

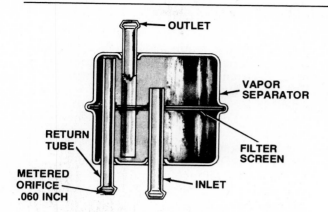

Figure 5-7. This vapor separator is installed between the pump and the carburetor to relieve pressure in the fuel line. (Chrysler)

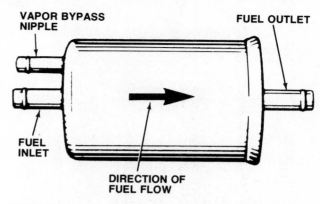

Figure 5-8. A vapor bypass filter combines the fuel filter and vapor relief functions in one unit.

Some older pumps used an air chamber on the outlet side of the pump to separate and recirculate vaporized and heated fuel to the fuel tank through a vapor return line, figure 5-6.

Pumps without the air chamber may use a vapor separator, figure 5-7, in the fuel line between the pump and the carburetor. Fuel from the pump fills the vapor separator. The outlet tube in the separator picks up fuel from the bottom of the unit and passes it into the fuel line to the carburetor. Vapor which has gathered rises to the top of the separator, where it is forced through a tube and into a return line to the fuel tank.

The vapor bypass filter, figure 5-8, has been used mostly on cars with factory-installed air conditioning. It combines the fuel filter and vapor separator into a single unit. Like a vapor separator, a bypass filter uses a restricted nipple and fuel tank return line to relieve vapor pressure buildup. Both the vapor separator and the bypass filter are directional and must be properly installed, or they will not pass fuel to the carburetor.

Many pump designs now have a bleed-down system, in which tiny holes are drilled through each check valve. This permits pressure buildup in the fuel outlet line to bleed back to the fuel inlet line.

Mechanical Fuel Pump Applications

The mechanical fuel pumps used by all domestic cars are manufactured by Carter, AC, or Airtex, all long-time industry suppliers. As a general rule, AMC and Ford use Carter and AC pumps, Chrysler uses Carter and Airtex, and General Motors uses AC pumps.

The diaphragm-type fuel pump is a simple device. Most of them operate the same, and the main differences are usually how they look on the outside. The outside design depends on what engine the pump will be used on, and how much room there is in the engine compartment.

Pumps are so similar in some cases that a production run of the same engine block may use pumps from two different manufacturers. However, replacement pumps must be identical in every respect. Installing a pump which *looks* like the one removed can result in a broken camshaft as soon as the engine is started.

Mechanical fuel pumps are quite dependable. If they break down, it is usually due to one of the following problems:
1. A diaphragm that leaks.
2. Worn-out inlet or outlet check valves.
3. A worn or broken pushrod.
4. Worn linkage, which reduces the pump stroke.

Occasionally, the camshaft eccentric may wear enough to reduce the pump stroke, or a bolt-on eccentric may come loose from the camshaft. In these cases, the camshaft or the eccentric must be replaced. It is also possible to install an electric fuel pump to bypass a defective mechanical pump.

Armature: The movable part in a relay. The revolving part in a generator or motor.

Flooding: A condition caused by heat expanding the fuel in a fuel line. The fuel pushes the carburetor inlet needle valve open and fills up the fuel bowl even when more fuel is not needed. Also, the presence of too much fuel in the intake manifold.

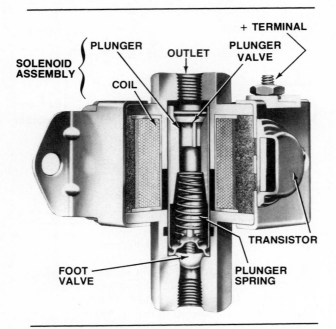

Figure 5-9. The operation of this plunger-type, electric fuel pump is controlled by a transistor. (Facet Enterprises, Inc.)

Figure 5-10. This AC diaphragm-type electric pump can be used to supplement a mechanical pump or as the only pump in the fuel system. (AC-Delco)

Many pre-1966 vehicles used fuel pumps which could be serviced by installation of repair kits. These pumps have been replaced by sealed, non-serviceable designs that reduce fuel leakage and manufacturing costs.

ELECTRIC FUEL PUMPS

There are four basic kinds of electric fuel pumps: plunger, diaphragm, bellows, and **impeller** (turbine). The first three kinds are driven by an electromagnet and vibrating armature. The impeller, or turbine, pump is driven by a small electric motor. None of the first three types is now used as original equipment on standard automobiles.

The plunger type pump, figure 5-9, has a stainless steel plunger in a brass cylinder. Valves at the bottom of the pump and above the plunger work like the check valves in a diaphragm pump. No rings or seals are used in this design. The fuel leaks constantly between the plunger and its cylinder wall to provide lubrication. The pump maintains pressure in the fuel line even when the engine is not running.

A **solenoid** controls the plunger. A magnet senses plunger position through the brass cylinder wall and makes and breaks the electrical contact, causing the plunger to move up and down inside the solenoid. Wear of the piston and cylinder wall surfaces may cause the pump to operate faster. A sediment chamber, a strainer, and removable valves make cleaning the pump easy.

A diaphragm-type electric fuel pump differs from a mechanical diaphragm pump by using an electromagnet and vibrating armature to operate the diaphragm. Pumping action is the same as in a mechanical pump. AC makes the only diaphragm electric pump currently on the market, figure 5-10.

A bellows-type electric pump, figure 5-11, is similar in design and operation to a diaphragm pump. Like the electric plunger and diaphragm pumps described above, a bellows-type pump can be used as a helper unit or as the only pump in a fuel system.

∎Solid State Fuel Pumps

As the electric fuel pump slowly replaces the mechanical pump, manufacturers are turning to electronics to produce a fuel pump with no electrical contacts and fewer moving parts. This solid state design is more reliable and lasts a lot longer.

Solid state pumps have a transistor that turns on when voltage is applied. A rapidly increasing series current passes through the transistor and pump solenoid. When energized, the solenoid activates a plunger, which cocks the pumping spring. Once the plunger has reached the length of its stroke, the transistor shuts off immediately. The pumping spring returns the plunger and the fuel is delivered. This occurs rapidly, pushing the fuel in a steady stream.

Fuel Pumps and Filters

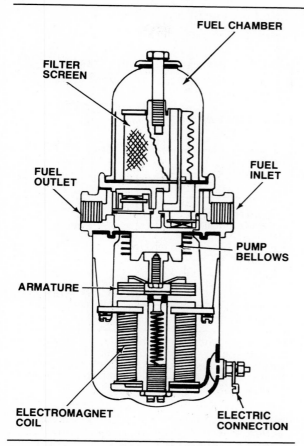

Figure 5-11. This electric fuel pump uses a bellows driven by a vibrating armature to deliver fuel.

The impeller-type pump, figure 5-12, is sometimes called a turbine or a rotary vane pump. It sucks fuel into the pump, then pushes it out through the fuel line to the carburetor. Since this type of pump uses no valves, the fuel is moved in a steady flow rather than the **pulsating** motion of all other electric and mechanical pumps.

Electric Fuel Pump Location

The electric fuel pump is principally a pusher unit: it pushes the fuel through the supply line. Because it does not rely on the engine camshaft for power, an electric pump can be mounted in the fuel line anywhere on the vehicle — including inside the fuel tank.

Pusher pumps are most efficient when they are mounted as near as possible to the fuel tank and at or below its level. This allows the pump to use gravity to transfer fuel from the tank to the pump. It also eliminates the problem of vapor lock under all but the most severe conditions. With the pump mounted at the tank, the entire fuel supply line to the carburetor is

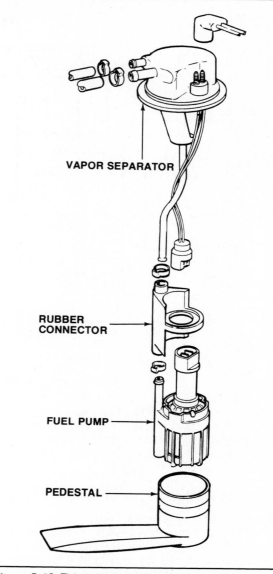

Figure 5-12. This impeller-type electric pump assembly is installed inside the fuel tank. (Ford)

Impeller: A rotor or rotor blade used to force a gas or liquid in a certain direction under pressure.

Solenoid: An iron core with a wire coil surrounding it. The core moves when electrical current is applied to the coil. Used to convert electrical energy to mechanical energy.

Pulsating: To expand and contract rhythmically and regularly.

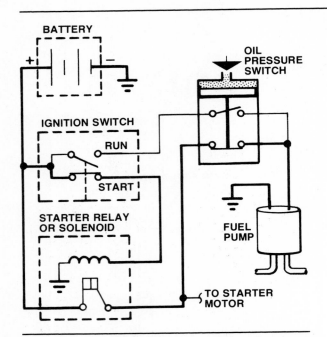

Figure 5-13. During cranking, the fuel pump receives current through the starter relay (solenoid) and the normally closed contacts of the oil pressure switch.

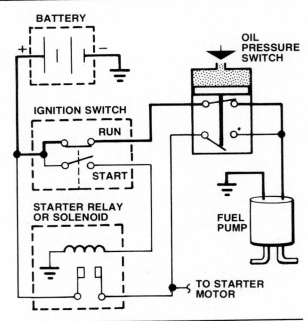

Figure 5-14. When the engine starts, the oil pressure switch opens one set of contacts and closes another. Current then flows through the ignition switch, through the oil pressure switch, and to the fuel pump as long as the oil pressure is above a minimum level.

pressurized. Regardless of how hot the fuel line gets, it is unlikely that vapor bubbles will form to interfere with fuel flow. Having the pump close to the tank also allows the pump to stay cooler because it is away from engine heat. It is therefore less likely to overheat during hot weather.

Electric Fuel Pump Operation

Most original equipment electric fuel pumps are controlled by a pressure switch in the engine oil system. This switch opens the electric circuit to the pump motor when the engine is off, and controls the operation of the pump when the engine is started and while it is running.

The pressure switch has two sets of contact points. One set is normally closed, and allows current to flow from the battery, through the starter solenoid or relay, to the fuel pump. The other set is normally open. When closed, it allows current to flow from the battery, through the ignition switch, to the fuel pump.

Turning the ignition key to Start energizes the pump by providing current through the normally closed contact points, figure 5-13. Once the engine is running, the pump receives current through the normally open contacts (which have been closed by the oil pressure), figure 5-14.

Engine oil pressure opens the normally closed contacts and closes the normally open contacts to keep the pump energized. When the ignition switch is turned off, the pump circuit is deenergized. If oil pressure drops below the specified level for any reason (usually 2 psi), contact is broken at the pressure switch and the fuel pump stops operation immediately.

In-Tank Pumps

GM compacts of the Astre, Monza, Skyhawk, and Vega class use an impeller-type electric fuel pump located in the fuel tank. This is part of the fuel pickup unit and is similar to the in-tank pumps used on 1969-70 Buick Rivieras and 1972-74 Fords with 460 police engines, figure 5-12.

The Cadillac Seville and all 1975 and later full-size Cadillacs with fuel injection use two electric fuel pumps: an in-tank impeller pump like that on the GM compacts, and a high-pressure, chassis-mounted impeller pump. The in-tank pump provides fuel to the high pressure pump to prevent vapor lock on the suction side of the fuel system. Cadillac calls the in-tank pump a booster pump.

Cadillac fuel tanks fitted with the in-tank pump contain a special "bathtub" reservoir below the pump to make sure the pump gets enough fuel, regardless of tank level or vehicle maneuvering. Excess fuel returned to the tank by the fuel pressure regulator is also fed directly into this reservoir to ensure that the pump intake remains below the fuel level.

Fuel Pumps and Filters

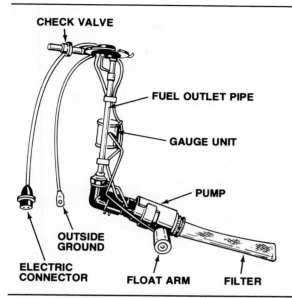

Figure 5-15. Almost all fuel systems have a filter or strainer attached to the pickup tube inside the tank. This one is part of an electric pump and gauge sender assembly.

FUEL FILTERS

Despite all the care taken in refining, storing and delivering gasoline, some impurities get into the automotive fuel system. Fuel filters are designed to remove dirt, rust, water, and other harmful materials from the gasoline before it can reach the carburetor. The useful life of all these filters is limited. The filters will become clogged and restrict fuel flow if they are not cleaned or replaced according to the maker's recommendations.

Several different types of fuel filters are used, and some systems may contain two or more. Filters can be located in several places within the fuel system.

Fuel Tank Filters and Strainers

A sleeve-type filter of woven Saran is usually fitted to the end of the fuel pickup tube inside the fuel tank, figure 5-15. This filter prevents sediment, which has settled at the bottom of the tank, from entering the fuel line. It also protects against water contamination by plugging itself up. If enough water somehow enters the fuel tank, it will accumulate on the outside of the filter and form a jelly-like mass. If this happens the filter must be replaced; except for this, no maintenance is required.

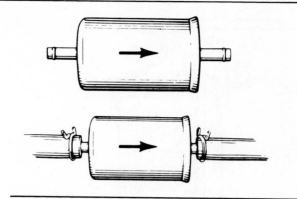

Figure 5-16. Inline fuel filters must be installed so that gasoline flows in the direction indicated by the arrow.

Inline Filters

Used primarily by AMC and Chrysler, the inline filter, figure 5-16, is in the line between the fuel pump and carburetor. This protects the carburetor from contamination but does not protect the fuel pump. The inline filter is usually a throw-

---

■ **No One Misses The Good Old Fuel Pump**

Today's fuel pump may seem to be a simple device, but pump manufacturers have worked hard to make it so. Pump designs, capacities, pressures, and performance requirements make the modern pump a rather sophisticated device. This is especially true when you consider that a fuel pump is expected to transport large amounts of gasoline for thousands of miles without failure.

Back in the thirties, fuel pump breakdown and replacement was a common occurence every few thousand miles. Remember the '36 Ford V-8? Its fuel pump operated from a pushrod. As the pushrod wore, the pump stroke lessened. Most mechanics and a lot of owners kept the fuel pump operating with a wad of chewing gum or tinfoil stuffed into the push rod cup to compensate for wear. Rather simple, but it worked.

The vacuum booster fuel pump was the first big change in pump design. Since no one could keep their windshield wipers running at a constant speed, pump designers provided additional vacuum with a dual pump design. But super highways, higher horsepower, and emission controls brought new approaches to pump design. Windshield wipers went electric, and the vacuum booster fuel pump disappeared. Intake electric pumps are gradually supplementing and even replacing the traditional mechanical pump design. Automakers now build modern fuel pumps to supply at least 30,000 trouble-free miles.

Figure 5-17. Sediment bowls may contain a paper, fiber, ceramic, or metal filter element.

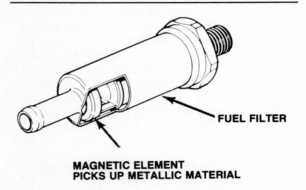

Figure 5-19. This late-model Ford filter has a magnetic element to remove metallic contamination from the fuel. (Ford)

away plastic or metal container with a pleated paper element sealed inside.

Inline filters must be installed so that gasoline flows through them in the direction shown by the arrow. If an inline filter is installed backwards, it can restrict fuel delivery to the carburetor.

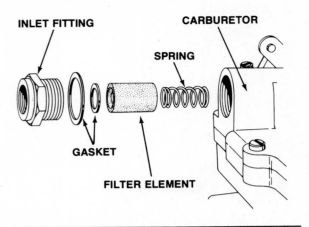

Figure 5-18. Ford and GM vehicles use carburetor inlet filters like this one.

Some inline filters have a built-in vapor bypass system. These filters have a third nipple, figure 5-8, that connects a fuel return line back to the fuel tank.

Some imported cars and older domestic cars have a sediment bowl between the fuel pump and carburetor. The bowl contains a pleated paper, ceramic, fiber, or metal filter element. The filter element works much like an inline filter. The bowl cover is held in place by a wire bail and clamp screw. It can be removed for filter cleaning or replacement, figure 5-17. Ceramic and metal elements can be cleaned and reused if necessary. Paper and fiber filter elements must be replaced when they are dirty.

Carburetor Inlet Filters

Ford and General Motors equip most of their cars with carburetor inlet filters, figure 5-18. The Ford filter is a 1-piece throw-away metal unit. It has a filter screen and magnetic washer, figure 5-19, to trap dirt and metal particles. The filter screws into the carburetor fuel inlet and clamps to the inlet hose at the other end.

Rochester carburetors on GM vehicles use a throw-away pleated paper element, which is installed as shown in figure 5-21. Older cars had a bronze filter element, but reusing was not recommended. Filters on 1976 and later Rochester carburetors must contain the new rollover check valve as described in Chapter 4. A similar paper filter is used on 1977 and later Motorcraft 5200 and 2700 VV carburetors.

Fuel Pumps and Filters

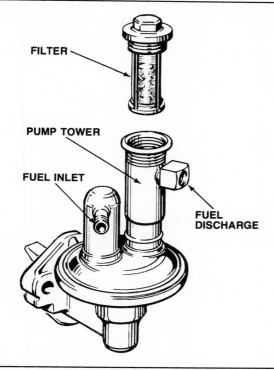

Figure 5-20. Some 6-cylinder Chrysler engines have a filter in the fuel pump outlet tower.

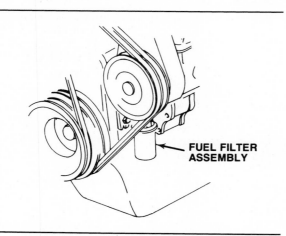

Figure 5-22. Disposable element filters may be mounted on the engine or near the fuel tank.

Pump Outlet Filters

Some cars have fuel filters in the outlet side of the fuel pump. Those pumps used on Chrysler 6-cylinder engines during the early seventies, figure 5-20, contain a throw-away filter element in the fuel outlet tower. Cadillacs through 1974 use a fuel pump outlet filter, located on the bottom of the pump, figure 5-21.

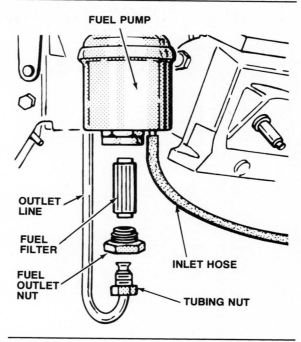

Figure 5-21. Some Cadillacs have a filter in the outlet side of the fuel pump.

Disposable Element Filters

Screw-on, throw-away element filters, figure 5-22, look much like replaceable oil filters. Ford has used this type on the fuel pump of some V-8 engines. The Cadillac Seville uses a disposable filter mounted to the frame near the left rear wheel. Other fuel-injected Cadillacs have the filter mounted to a bracket on the lower left front of the engine.

SUMMARY

Fuel pumps move the fuel from the tank to the carburetor. All pumps do this through a mechanical action that creates a low-pressure area into which the fuel will flow. With check valves and high pressure, the fuel is then forced out of the pump and into the carburetor. There are two types of fuel pumps: mechanical and electrical. Mechanical pumps use the engine camshaft eccentric for power. There are four types of electrical pumps (plunger, diaphragm, bellows, and impeller) but only the impeller type is now used in most cars. Electric pumps push the fuel, rather than pull it, so they are frequently installed in the fuel tank. Many types of filters are used in the fuel system, and they must be replaced or cleaned as directed by the manufacturer. Filters are used as in-line filters, or at the fuel tank or the carburetor.

Review Questions

Choose the single most correct answer. Compare your answers to the correct answers on page 248.

1. The most common type of fuel pump is:
   a. The double-action diaphragm type
   b. The single rocker arm type
   c. The single-action diaphragm type
   d. The pushrod-rocker arm type

2. The intake stroke in the fuel pump:
   a. Exerts pressure in the fuel tank line
   b. Creates a vacuum in the fuel chamber
   c. Opens the outlet valve
   d. All of the above

3. The output stroke of the fuel pump:
   a. Increases pressure on the diaphragm spring
   b. Opens the inlet valve
   c. Increases pressure in the pump chamber and opens the outlet valve
   d. Draws fuel into the fuel tank line

4. Fuel pump pressure is controlled by:
   a. The carburetor inlet needle valve
   b. The strength of the diaphragm spring
   c. The carburetor float
   d. None of the above

5. Vapor lock in the pump and lines can be prevented if:
   a. An external air chamber with a vapor return line is used
   b. A vapor separator is installed in the line between the carburetor and pump
   c. A bleed-down system is used
   d. Any of the above

6. When the fuel filter in the diagram figure 5-8 is installed, the arrow must point toward the:
   a. Carburetor
   b. Fuel tank
   c. Canister
   d. Fuel pump

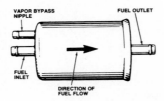

7. Which is *not* a manufacturer of domestic fuel pumps:
   a. Carter
   b. AC
   c. Airtex
   d. GMAC

8. Mechanical fuel pumps:
   a. Are not reliable
   b. Are easily repaired
   c. Can no longer be disassembled
   d. Are all exactly alike

9. Which of the following electric fuel pumps are *not* driven by an electromagnet and a vibrating armature:
   a. Plunger
   b. Diaphragm
   c. Impeller
   d. Bellows

10. Which of the following is *not* true:
    a. Diaphragm-type electric pumps have a vibrating armature to operate the diaphragm
    b. The plunger-, diaphragm-, and bellows-type electric pumps are not used as original equipment
    c. Impeller pumps have inlet and outlet valves
    d. An impeller is also called a rotary vane pump

11. Electric fuel pumps are most efficient when located near:
    a. The engine
    b. The fuel tank
    c. The carburetor
    d. The camshaft

12. Electric fuel pumps are controlled by:
    a. An oil pressure switch
    b. The starter relay
    c. The ignition switch
    d. Fuel line pressure

13. All 1975 and later fuel-injection Cadillacs used:
    a. A mechanical pump and an in-tank impeller pump
    b. Two mechanical pumps
    c. Two plunger-type pumps
    d. An in-tank impeller and a high-pressure, chassis-mounted impeller pump

14. Fuel tank filters eliminate:
    a. SO_2
    b. NO_x
    c. Sediment and water
    d. HC

15. Inline filters:
    a. Are used mostly by AMC and Chrysler
    b. Are located in the fuel line between the pump and carburetor
    c. Protect the carburetor but not the pump
    d. All of the above

Chapter 6
Air Cleaners and Filters

Gasoline must be mixed with air to form a combustible mixture. Like gasoline, air contains dirt and other materials which cannot be allowed to reach the engine. Just as fuel filters are used to clean the gasoline of impurities, an air cleaner and a filter, figure 6-1, are used to remove contamination from the air.

The air cleaner and the filter have three primary functions:
1. They clean the air before it is mixed with fuel.
2. They silence intake noise.
3. They act as a flame arrester in case of a backfire.

In 1957, engineers at the Lincoln-Mercury Division of Ford Motor Company were searching for ways to improve driveability under different weather conditions. They discovered that by using the air cleaner to provide warm air to the carburetor at low temperatures, driveability is improved. It also allows more efficient carburetor adjustments which:
- Reduces exhaust emissions without reducing engine performance
- Permits leaner air-fuel ratios
- Gives better fuel economy
- Reduces **carburetor icing** in cold weather.

Since then, the air cleaner has become a separate emission control system for intake air temperature control. It also has become a part of other emission controls, such as the PCV system.

In this chapter, you will learn the filtering requirements of the automobile engine and how the various air cleaner and filter designs are used to meet them. You will also learn about the temperature-controlled air cleaner's role in emission control and how carmakers put it to work in the fight against pollution.

ENGINE FILTERING REQUIREMENTS

As explained in Chapter 3, the automotive engine burns about 9,000 gallons of air for every gallon of gasoline at an air-fuel ratio of 14.7:1. With many of today's engines operating on even leaner ratios, the quantity of air consumed per gallon of fuel is closer to 10,000 gallons, or 200,000 gallons of air with every 20-gallon tank of fuel.

That is enough air to fill a good-sized swimming pool, and just like the pool water, the air is filled with particles of dust and dirt which must be removed. Without proper filtering of the air before intake, these particles can affect the operation of the carburetor and upset the air-fuel ratio. Given enough time, they will seriously damage engine parts and shorten engine life.

While abrasive particles cause wear at any place inside the engine where two surfaces move against each other, they first attack piston

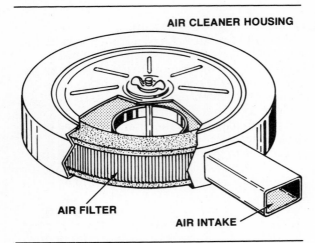

Figure 6-1. A simple air cleaner and filter.

Figure 6-3. This cam and these valve lifters were destroyed by abrasive particles drawn into an engine operating without an air cleaner.

Figure 6-4. Air cleaner with removable top housing. (AC Delco)

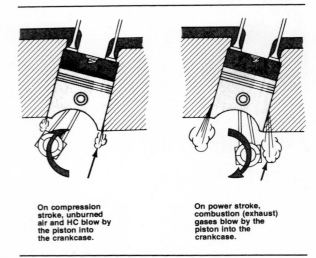

Figure 6-2. Crankcase blowby gases.

rings and cylinder walls. Contained in the blowby gases, figure 6-2, they pass by the piston rings and into the crankcase. From the crankcase, the particles circulate throughout the engine in the oil. Large amounts of abrasive particles in the oil can damage other moving parts of the engine. Figure 6-3 shows a camshaft and some valve lifters from an engine that was run for a long period of time without an air filter.

Although the basic airborne contaminants — dust, dirt, and carbon particles — are found wherever a car is driven, they vary in quantity according to the environment. For example, engine air intake of abrasive carbon particles will be far greater in constant bumper-to-bumper traffic. Dust and dirt particles will be greater in agricultural or construction areas.

THE AIR CLEANER AND FILTRATION

Cleaning of the intake air is done by a filter in the air cleaner, figure 6-1. The 2-piece, circular air cleaner housing of stamped steel is usually located on top of the carburetor and generally has a snorkel or air intake tube. The snorkel draws air from the engine compartment or from outside the car into the air cleaner; from there it passes to the carburetor. The air cleaner housing has a removable top section, figure 6-4.

Most air filter elements are the throw-away type, but some can be cleaned and reused. Certain 4-cylinder engines, such as those used

Carburetor Icing: A condition that is the result of the rapid vaporization of fuel entering a carburetor; the temperature drops enough to freeze the water particles in the airflow.

Air Cleaners and Filters

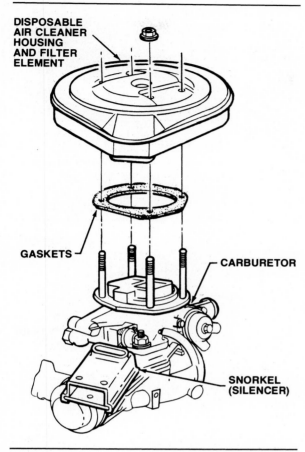

Figure 6-5. Disposable air cleaner with long-life filter. (Chevrolet)

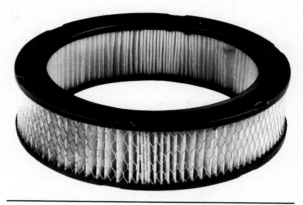

Figure 6-6. Paper air filter element. (AC Delco)

in the Chevette, Vega, and other GM small cars, have a throw-away air cleaner containing a long-life filter, figure 6-5. After 50,000 miles, the 1-piece, welded air cleaner housing is disconnected from the air intake snorkel and carburetor airhorn, and a new one is installed.

Replacement Requirements

Carmakers recommend cleaning or replacing air filter elements at periodic intervals, usually listed in terms of mileage or number of months of service. The mileage and time intervals are based on average, or normal, driving. Air filter element replacement may be necessary more often when the vehicle is driven under dusty, dirty, or other severe conditions.

It is best to replace a filter element before it becomes too dirty to be effective. A dirty air filter element will let contaminants through which causes engine wear. A dirty air filter element also can change the carburetor air-fuel ratio and thus affect engine performance. The higher the engine speed, the greater the airflow required. Restricted or clogged filters will greatly affect high speed operation of the engine. If the element becomes so clogged that it does not let through enough air, it can act as a choke to increase fuel consumption. In severe cases, a clogged air filter can keep the engine from running.

AIR FILTER ELEMENTS

Three general types of air filter elements are used on passenger cars and light trucks.
1. Paper filters
2. Polyurethane filters
3. Oil bath filters.

Paper Filter Element

The paper air filter element, figure 6-6, is the most common type used on late-model cars and trucks. It is made of a chemically treated paper stock that contains tiny passages in the fibers. These passages form an indirect path for the airflow to follow. This causes the airflow to pass through several fiber surfaces, each of which traps microscopic particles of dust, dirt, and carbon.

The paper is pleated and formed into a circle, with the top and bottom edges sealed with heat-resistant plastic. A fine wire mesh screen may be used on the inside of the filter ring to reduce the possibility of the element catching fire from an engine backfire. A similar, but coarser, wire mesh screen may be used on the outside of the filter ring for extra strength.

While these filter elements are generally made of dry paper, an oil-wetted paper stock may also be used. The light oil coating helps prevent contaminants from working their way through the paper. It also increases the dirt-holding capacity over the same area of dry paper stock. An outer wrapper of polyurethane sometimes is used to make the filter work bet-

Chapter Six

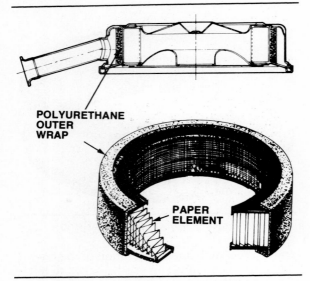

Figure 6-7. Paper air cleaner element with polyurethane wrapper. (AC Delco)

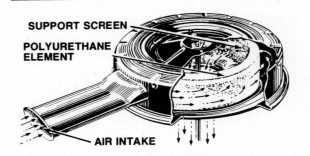

Figure 6-8. Air cleaner with polyurethane filter element over a metal support screen.

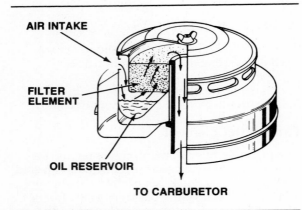

Figure 6-9. Oil bath air filter.

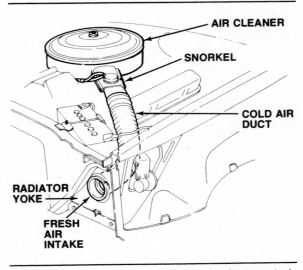

Figure 6-10. Air cleaner with fresh air intake mounted on the radiator yoke.

ter, figure 6-7. Paper element filters are disposable and should be replaced at the recommended intervals.

Polyurethane Filter Element

Sometimes called a foam filter, the polyurethane element is similar to the paper element type. It consists of a polyurethane wrapper stretched over a metal support screen, figure 6-8. Polyurethane contains thousands of pores and interconnecting strands that create a maze-like dirt trap, while allowing air to flow through. Properly maintained, the polyurethane element has a capacity and efficiency equal to that of the paper type. It may be used dry, or it can be lightly oiled. It can be cleaned and reused or replaced if necessary. Polyurethane filter elements are not common as original equipment on late-model vehicles, but some are sold as aftermarket equipment.

Oil Bath Filter

This kind of filter is used mainly on older cars, some trucks, and some off-road vehicles. It is most efficient at high airflow rates. The oil bath cleaner, figure 6-9, contains a wire mesh filter which rests in an oil reservoir in the air cleaner housing. Air entering the housing is deflected down, where it strikes the oil in the reservoir and deposits heavier particles of dirt. Picking up an oil mist from the reservoir, the air flows back up and across the surface of the filter, where it leaves the mist with finer particles entrapped. The oil then drains back to the reservoir from the filter, carrying the entrapped dirt with it in a self-cleaning action.

AIR INTAKE DUCTS AND FRESH AIR INTAKES

The main source of air intake to the carburetor is the air cleaner snorkel. Some air cleaners use a second snorkel to provide additional air intake

Air Cleaners and Filters

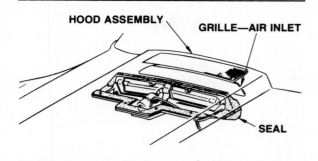

Figure 6-11. Ducted air door located on the rear area of the Corvette hood. (Chevrolet)

at full throttle. The snorkel passes air to the filter and then to the carburetor from the engine compartment. The snorkel also increases the velocity of the air entering the air cleaner. Temperatures in the engine compartment will often exceed 200° F on a hot day, and hot air can thin out the air-fuel mixture enough to cause detonation and possible engine damage.

Allowing the engine to breathe cooler air from outside the engine compartment will prevent such problems. Cooler air is provided by a cold air duct, or induction (zip) tube. The tube runs from the snorkel to a fresh air intake at the front of the car, figure 6-10. The fresh air intake is normally open at all times, but may have a screen to prevent insects and other foreign matter from being drawn into the air cleaner.

Some fresh air intakes are in the cowl or in the rear area of the hood. Ducted hood air doors used on Corvettes, figure 6-11, open electrically at full throttle. Pedal linkage closes a switch when the accelerator is pushed to the floorboard. This in turn operates a solenoid attached to the air door linkage. The air door provides more intake air at wide-open throttle, just as the second snorkel does on some air cleaners.

THERMOSTATICALLY CONTROLLED AIR CLEANERS

Late-model cars all have **thermostatic** controls on their air cleaners to regulate intake air temperature. This is important to maintain the precise air-fuel ratios required for exhaust emission control.

The thermostatically controlled air cleaner has the usual sheet metal housing mounted on the carburetor air horn. Another sheet metal duct, called a heat stove or shroud, is fastened around the exhaust manifold. The heat stove is connected to the air cleaner intake snorkel by a flexible hose or metal tube, figure 6-12. Heat radiating from the exhaust manifold is kept in

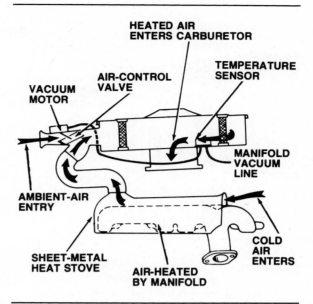

Figure 6-12. A thermostatically controlled air cleaner with a heat stove.

by the heat stove, and sent to the air cleaner snorkel to provide heated air to the carburetor.

An air control valve, or damper, in the snorkel permits the intake of:

Thermostatic: Referring to a device that automatically responds to temperature changes, in order to activate switches.

■ Air Cleaner Filter Maintenance

What causes engine wear? Contrary to popular belief, it is not how many miles the engine has been driven, or how old the engine is. Providing that the engine is kept properly lubricated, wear is mainly caused by the dust and dirt which enters the engine. A teaspoonful of gritty dirt will ruin the piston rings; a cupful will virtually destroy the entire engine. A properly maintained air cleaner filter is the primary line of defense against the gallons of dirty air sucked into the engine with each tiny sip of fuel.

To prevent dirt from entering the engine, some mechanics say paper filter elements should be discarded when dirty. Others say the element can be reused if properly cleaned with low pressure compressed air (under 100 psi nozzle pressure).

Avoid rough handling or damage to filters. Never try to clean a paper filter by striking or rapping the filter assembly. Washing the filter element is also out, since even drying a wet filter will not restore the air passages to normal.

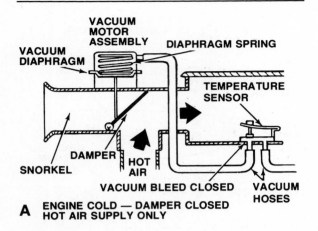
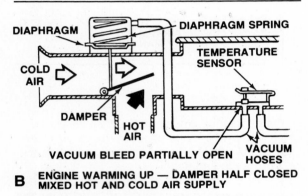
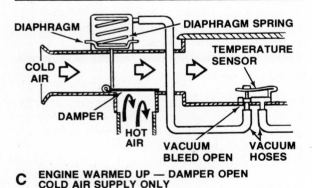

Figure 6-13. A thermostatically controlled air cleaner with vacuum motor control.

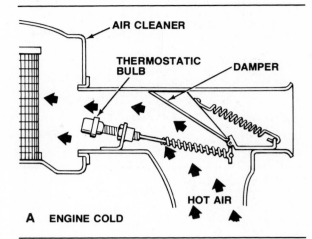
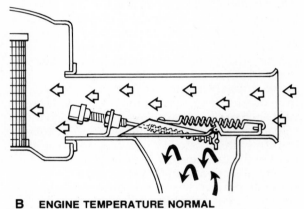

Figure 6-14. A thermostatically controlled air cleaner with thermostatic-bulb control.

1. Heated air from the heat stove
2. Cooler air from the snorkel or cold-air duct
3. A combination of both.

The damper maintains intake air at a specified temperature, usually between 90° and 100° F. The damper may be operated by a vacuum motor or diaphragm or by a thermostatic bulb.

In an air cleaner with a vacuum motor, a **bimetal temperature sensor** and a vacuum bleed in the air cleaner housing regulate vacuum supply to the vacuum motor. Vacuum is supplied from the intake manifold. When intake air temperature is below about 100° F, the temperature sensor holds the vacuum bleed closed and full manifold vacuum is applied to the vacuum motor. The motor holds the air control valve in the full-hot-air position, figure 6-13, position A.

As intake air warms up, the sensor begins to open the vacuum bleed. This decreases the vacuum going to the motor. A spring in the motor then starts to move the air-control valve from the hot-air to the cold-air position, figure 6-13, position B.

As air temperature continues to rise, the vacuum bleed continues to open. Vacuum to the motor is further reduced. At high air temperatures, vacuum to the air cleaner motor is completely shut off, and the air control valve is in its full-cold-air position, figure 6-13, position C. The air control valve will also open to the full-cold-air position during heavy acceleration, regardless of air temperature. This provides maxi-

Air Cleaners and Filters

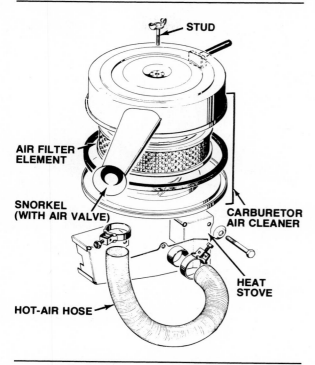

Figure 6-15. AMC's air cleaner with thermostatic-bulb control. (AMC)

mum airflow through the air cleaner to the carburetor when it is needed the most.

The thermostatic bulb, used to control some air cleaners, is inside the air cleaner snorkel and connected by linkage to a spring-loaded air control valve, or damper. The air control valve is normally held in its closed position by the spring, allowing heated intake air to enter the snorkel, figure 6-14, position A. As the temperature rises, the thermostatic bulb begins to expand. This expansion is more than air valve spring tension, and the valve gradually opens to its cold-air position, figure 6-14, position B.

American Motors Thermostatically Controlled Air Cleaner (TAC) System

Thermostatic bulb control is used on AMC 6-cylinder and early V-8 engines. The 6-cylinder system, figure 6-15, includes a 2-piece heat stove, a hot-air hose, and an air valve assembly located in the air cleaner snorkel.

The system used on early V-8 engines includes a heat stove on the right-hand exhaust manifold, a hot air hose, and the air valve assembly in the air cleaner snorkel.

Vacuum-motor control is used on late-model AMC V-8 engines. This system includes a heat stove on the right-hand exhaust manifold, a hot-air hose, a vacuum motor and air valve assembly, and a temperature sensor inside the air cleaner, figure 6-16.

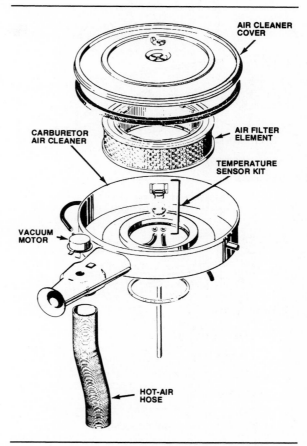

Figure 6-16. AMC's vacuum-motor control used on late-model V-8 engines. (AMC)

System operation

The AMC air cleaners with thermostatic bulbs operate as shown in figure 6-17. When air entering the snorkel is less than 105° to 110° F, the air control valve is held in the closed (heat on) position, figure 6-17, position A. Heated air from the heat stove passes through the hot-air hose to the snorkel.

At temperatures from 110° to 125° F, the air control valve is partially open. This allows a mixture of heated manifold air and cooler engine compartment air to enter the air cleaner duct.

Bimetal Temperature Sensor: A device made of two strips of metal welded together. When heated, one side will expand more than the other, causing it to bend.

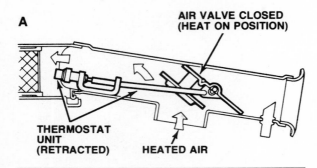

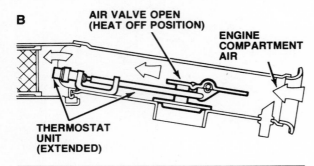

Figure 6-17. Operation cycle of AMC's thermostatically controlled air cleaner. (AMC)

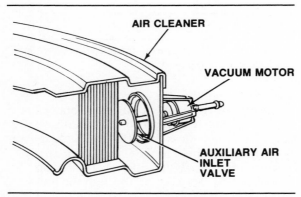

Figure 6-19. Ford Motor Company's air cleaner with an auxiliary air inlet valve controlled by a vacuum motor.

When intake air temperature reaches 125° to 130° F, the air control valve is held in the fully open (heat off) position, figure 6-17, position B. Air now enters the air cleaner snorkel from the engine compartment or from the fresh air duct if one is used.

Operation of an AMC V-8 air cleaner with a vacuum motor is shown in figure 6-13. The air control valve is held in the closed (heat on) position, until intake air temperature reaches approximately 115° F, at which time the valve begins to open.

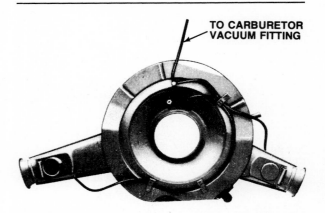

Figure 6-18. The Chrysler Corporation dual snorkel air cleaner. (Chrysler)

Chrysler Heated Air Inlet System

Vacuum motor control is used on all Chrysler engines with thermostatically controlled air cleaners. This system includes a sheet metal heat stove on the exhaust manifold, a flexible hot-air hose, a vacuum motor and air valve assembly on the snorkel, and a temperature sensor inside the air cleaner, figure 6-12.

System operation

Chrysler thermostatically controlled air cleaners work as we outlined in the first part of this chapter, discussing vacuum-motor-controlled air cleaners. The temperature range in which the temperature sensor opens the vacuum bleed is from 10° to 100° F on 1974 and earlier cars, and from 10° to 90° on 1975 and later models.

The dual-snorkel air cleaner, figure 6-18, used on some Chrysler engines, works in the same way as the single-snorkel air cleaner. However, only one snorkel receives heated air from the heat stove. The vacuum motor in this snorkel is controlled by a temperature sensor as we already described. The vacuum motor in the second snorkel receives vacuum directly from the intake manifold without passing through a temperature sensor. The damper in the second snorkel is held closed under all operating conditions except heavy acceleration. Under acceleration, manifold vacuum drops, and the springs in the vacuum motors of both snorkels open the dampers to provide the maximum flow of cooler air.

Ford Inlet Air Temperature Regulation System

Thermostatic-bulb control is used on all Ford Vehicles through 1972 (except those with 351C, 390GT, and 400-cid engines) and on all Pintos

Air Cleaners and Filters

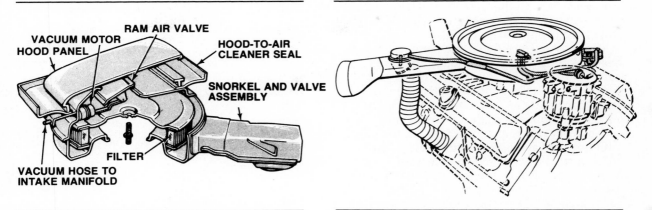

Figure 6-20. Ford Motor Company's ram air system. (Ford)

Figure 6-22. The vacuum motor control on a GM air cleaner. (Chevrolet)

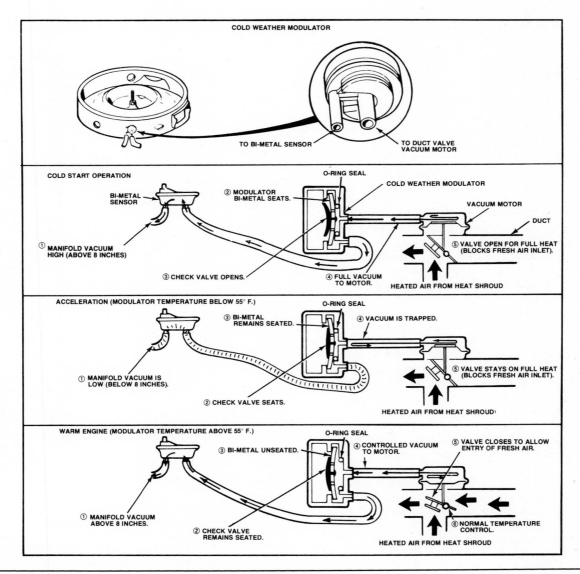

Figure 6-21. A thermostatically controlled vacuum modulator between the temperature sensor and the vacuum motor. (Ford)

through 1976 with the 1,600-cc and 2,000-cc engines. Vacuum motor control is used on all other Ford products.

The basic parts of these inlet air temperature systems are similar to those of other thermostatically controlled air cleaners. However, some Ford air cleaners have an auxiliary air inlet valve, controlled by a vacuum motor, figure 6-19. It is on the rear of the air cleaner and opens to give more airflow under full throttle.

Other Ford air cleaners with thermostatic-bulb control have an auxiliary vacuum motor under the air cleaner snorkel. It is linked to the damper by a piston rod. Under full throttle, the vacuum motor takes over from the thermostatic bulb and opens the damper to air from both the heat stove and the cold air intake.

A ram-air system, figure 6-20, is used on some 1973 and earlier Ford products. Under full throttle, a vacuum motor opens an air valve in the functional hood scoop to let in extra outside air.

System operation
On 1966-67 Ford cars, the thermostatic bulb holds the air cleaner damper in the closed (heat on) position when air temperature is below 75° F. On 1968 and later models, the bulb holds the damper closed until intake air temperature reaches 95° to 100° F. At temperatures between 85° and 105° F on 1966-67 models, or 100° and 135° F on later models, the bulb opens the damper to allow a mixture of hot air and colder air through the snorkel. Above these temperatures, the bulb opens the damper to the full cold-air position.

The auxiliary air inlet valve, used on some Ford air cleaners, is held closed by a vacuum motor that receives vacuum directly from the intake manifold. When vacuum drops under full throttle, a spring in the motor opens the air valve to allow extra airflow into the air cleaner.

The ram-air system on some Ford products is operated by a vacuum motor similar to the one used with the auxiliary air valve. Under full-throttle or other heavy load conditions, intake manifold vacuum drops, and a spring in the vacuum motor opens the ram-air valve.

On Ford vacuum-motor-operated air cleaners used through 1976, the damper is held in the closed (heat on) position until intake air temperature is about 120° F. At that temperature the valve starts to open. Since 1977, the temperature sensors in the air cleaner differ in heat ranges according to car model. The start-to-open temperature is indicated by a colored stripe on the sensor, as follows:
- Brown — 90° F
- Black or pink — 100° F
- Blue or yellow — 115° F.

CAR MODEL	OPERATING MODE		
	Heat On	Damper Starts To Open (Regulating)	Heat Off
AMC thermo bulb	105°-110°①	110°-125°	125°-130°①
AMC vacuum motor	110°	115°	130°
Chrysler, all	10°	10°-90°	90°
Ford thermo bulb			
1966-67	75°	85°-105°	105°
1968-74	95°-100°	100°-135°	135°
1975-76	95°-100°	105°-130°	150°
Ford vacuum motor			
to 1974	120°	120°-135°	135°
1975-on	②	②	②
Buick:			
All	85°	85°-135°	135°
Chevrolet:			
All I-6 and V-8	85°	85°-135°	135°
L-11 and L-13 4-cylinder	50°	50°-110°	110°
Cadillac:			
1968-71	85°	85°-135°	135°
1972-on	85°	85°-105°	105°
Oldsmobile:			
1968-72	85°	85°-128°	128°
1973-75 Blue Sensor	79°	79°-123°	123°
1973-76 Green Sensor	107°	107°-151°	151°
1977 and later	75°	75°-151°	151°
All L-11 and L-13 4-cylinder	50°	50°-110°	110°
Pontiac:			
1968-74	85°	85°-135°	135°
1975-76	80°	80°-125°	125°
1977 and later	79°	79°-153°	153°
All L-11 and L-13 4-cylinder	50°	50°-110°	110°

①1977 California Gremlin, 75° and 90°F.
②Start-to-open temperature varies according to sensor color.
1975-76 Blue, Yellow, Green — 110°F.
1975-76 Orange — 80°F.
1977 Brown stripe — 90°F.
   Black or pink stripe — 100°F.
   Blue or yellow stripe — 115°F

Figure 6-23. Thermostatic air cleaner temperature ranges in degrees Fahrenheit.

Many 1975 and later Ford engines have a thermostatically controlled vacuum modulator between the temperature sensor and the vacuum motor. This keeps the spring in the vacuum motor from opening the damper under full throttle with a cold engine. The vacuum modulator, figure 6-21, uses a bimetal thermostatic disc and a check valve to trap vacuum in the vacuum motor. Above 55° F, the bimetal thermostat in the vacuum modulator opens the vacuum passage through the modulator so that the air cleaner vacuum motor works normally.

GM Carburetor Heated Air (CHA) and Thermac Air Cleaner (TAC) Systems

General Motors introduced carburetor heated air systems on most engines in 1968. A cold-air intake was added to V-8 engines in 1973 and to 6-cylinder engines in 1975. The cold-air intake

Air Cleaners and Filters

consists of a flexible hose and metal duct from the car grille to the air cleaner snorkel. Most GM air cleaners use vacuum-motor control, figure 6-22. General Motors calls this the Thermac Air Cleaner (TAC) system. The 4-cylinder engines used in 1976 and earlier GM small cars have a thermostatic coil to control the air cleaner damper. This works similarly to a thermostatic bulb.

System operation

On GM air cleaners with thermostatic-coil control, the damper is in the closed (heat on) position at temperatures below 50° F. Between 50° and 110° F, the damper is partly open to allow a mixture of warm and cold air. Above 110° F the damper is fully open (heat off).

On GM air cleaners with a vacuum motor, the damper is closed (heat on) until intake air reaches the "heat on" temperature listed in figure 6-23 for a particular car model. When the air temperature is in the "regulating" range, the damper is partly open to allow a mixture of warm and cold air. When the intake air reaches the "heat off" position, the damper is fully open.

At full throttle, manifold vacuum drops, and the vacuum motor spring opens the damper no matter what the temperature is. Some 1975 and later GM engines have a temperature control valve (TCV) to improve engine warmup and performance with a cold engine. The TCV is located in the air cleaner and works similarly to Ford's vacuum modulator. At temperatures below 80° F, the valve traps vacuum in the vacuum motor to hold the air cleaner damper closed, even at full throttle. Above 95° F, the TCV is fully open to permit normal air cleaner operation.

SUMMARY

Like fuel, the air used by an engine has tiny particles of dirt and other materials that damage an engine if they are allowed to be breathed in. Air cleaners and their filters screen out this material, much like fuel filters clean the fuel before it gets to the carburetor. Air cleaners are also part of the emission control system, since they help reduce emissions and increase performance and fuel economy. Since 1968, most domestic passenger cars and light trucks have used thermostatically controlled air cleaners, which provide warm air to the carburetor at low temperatures. Each carmaker has a slightly different design, but all these devices work about the same.

Review Questions

Choose the single most correct answer.
Compare your answers to the correct answers on page 248.

1. Airborne contaminants can:
   a. Change air-fuel ratios
   b. Damage piston rings and cylinder walls
   c. Enter the engine oil
   d. All of the above

2. Which of the following is *not* true about a dirty air filter element:
   a. It will pass contaminants
   b. It does not affect fuel consumption
   c. It will change air-fuel ratio
   d. It may keep the engine from running

3. Of the following, which type of filter is *not* late-model original equipment:
   a. Polyurethane
   b. Paper
   c. Oil bath
   d. None of the above

4. To prevent the paper air filter element from catching fire:
   a. It is chemically treated
   b. It is soaked in oil
   c. A fine wire mesh screen is used on the inside of the filter ring
   d. A coarse wire mesh screen is used on the outside

5. Oil bath filters:
   a. Are efficient at all speeds
   b. Cause large particles to deposit in an oil reservoir
   c. Have an auxiliary paper element
   d. Pass oil through the carburetor

6. The primary source of air intake to the carburetor is the:
   a. Air cleaner snorkel
   b. Venturi
   c. Fuel bowl
   d. Heat stove

7. A heat stove may be located on the:
   a. Carburetor
   b. Snorkel
   c. Exhaust manifold
   d. Intake manifold

8. When intake air temperature is low in a thermostatic air cleaner:
   a. The vacuum bleed is closed
   b. The vacuum bleed is open
   c. The air control valve is in the cold-air position
   d. Little vacuum is applied to the vacuum motor

9. In AMC cars, air cleaner vacuum-motor control is used on:
   a. Early 6-cylinder engines
   b. Early V-8 engines
   c. Late-model V-8 engines
   d. All of the above

10. On AMC air cleaners with thermostatic bulbs, the air control valves are partially open at temperatures of:
    a. 125° to 200° F
    b. 110° to 125° F
    c. 55° to 110° F
    d. 32° to 55° F

11. In the Chrysler dual snorkel air cleaner, the damper in the second snorkel is:
    a. Always closed
    b. Always open
    c. Partially open
    d. Open only under full throttle

12. CHA systems were introduced in GM cars in which model year:
    a. 1968
    b. 1969
    c. 1970
    d. 1971

Chapter 7
Basic Carburetion

So far in our study of a car's fuel system, we have discussed fuel tanks, fuel lines, and fuel filters. That is enough to be able to dump raw gasoline into an engine. But putting a spark to this gasoline will not produce the combustion we need to create the power to move the car.

We need something more: we need to be able to change the fuel to a vapor by mixing it with air, and then feeding it to the cylinders. This is the job of the carburetor, figure 7-1.

There is a tremendous variation in carburetor designs, from the simple devices used on small passenger cars to the wildly complex and expensive versions used on racing engines. Regardless of the design, though, all carburetors use the same basic principle: differences in air pressure.

In this chapter, you will learn how these differences in air pressure apply to a caruretor, and how carburetors operate under all types of driving conditions. We will cover the similarities in all carburetor designs, and discuss how proper carburetor adjustments can improve driveability and lower the polluting emissions in the car's exhaust.

Once you know how the carburetor works, and what the similarities are, then you will be able to make carburetor adjustments properly and diagnose carburetor problems more accurately.

PRESSURE DIFFERENTIAL

Since air is a substance with weight, the air outside an engine has a specific weight, and so does the air inside the engine. The weight of air exerts pressure on whatever it touches; the greater this weight, the greater the pressure. When the weight of air outside the engine is greater than the weight of air inside the engine, we say that there is a pressure difference, or differential, between the two.

Atmospheric Pressure

The weight of air is not always the same, but changes with temperature and height above sea level. For this reason, we must have a reference point when we talk about air in terms of the pressure it exerts. Air pressure is measured in pounds per square inch (psi) at sea level at an average temperature of 68° F. In metric units, atmospheric pressure at sea level is measured in millimeters of mercury (mm Hg) at 20° C.

At sea level and at an average temperature, one cubic foot of air weighs about 1¼ ounce. This seems light enough, but remember that the blanket of air surrounding the earth extends many miles into the earth's atmosphere, figure 7-2. These many cubic feet of air piled one on

Basic Carburetion

Figure 7-1. The modern carburetor is a complex device, but it works on simple principles of airflow and pressure differential.

top of another all have weight. This increases the weight of air pressing down on an object at sea level to about 14.7 psi or 760 mm Hg.

Effects of temperature
Air expands and becomes lighter as its temperature rises. This reduces pressure it exerts. As its temperature falls, air contracts. This makes it heavier and increases its pressure. Variations in air temperature account for changing weather conditions. Direct heat from the sun and reflected heat from the earth's surface warm the air. As its temperature increases, air becomes lighter and rises. Cooler air sinks and takes its place, resulting in a constant motion. This motion creates wind and weather patterns.

Effect of height
As you climb above sea level, the amount of air pressing down on you becomes smaller. Since this air weighs less, it exerts less pressure. Air pressure gradually decreases with increased distance above sea level. At 30,000 feet above sea level, air pressure is reduced to about 5 psi. A few hundred miles above the earth, the atmosphere ends and is replaced by a vacuum, or a complete lack of pressure.

Manifold Pressure — Vacuum
With each intake stroke of an engine piston in its cylinder, a partial vacuum is produced. As the piston moves down, it creates a larger space in which the air molecules can move. Since the molecules spread out to occupy this increased

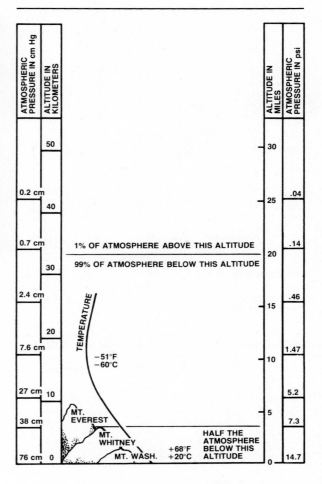

Figure 7-2. The blanket of air surrounding the earth extends many miles into the atmosphere. Atmospheric pressure decreases at higher altitudes.

space, the distance between them increases. The greater the space between the air molecules, the greater the vacuum created.

As the piston moves further down, it increases the vacuum and lowers the air pressure in the cylinder and intake manifold above it. This causes a *pressure differential* between the air inside and the air outside the engine. To offset this differential, outside air rushes into the engine. As it passes through the carburetor, it is mixed with gasoline to form an air-fuel mixture. This combustible vapor is then drawn by vacuum through the intake manifold and open intake valve into the cylinder, figure 7-3. Here it is compressed, burned, and exhausted.

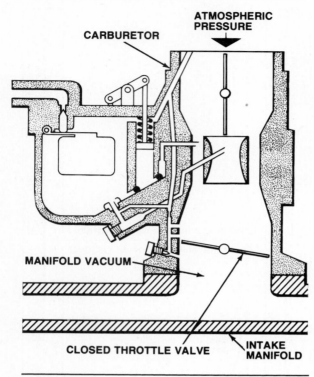

Figure 7-3. Manifold vacuum is the low pressure created in the intake manifold by the downward movement of the engine's pistons.

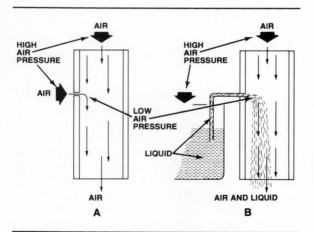

Figure 7-4. Airflow through a tube creates low pressure along the sides of the tube that can draw in more air (A). The low pressure can also draw liquid into the tube (B).

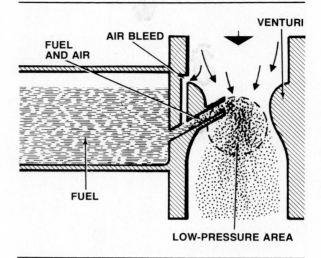

Figure 7-5. Air flowing through the venturi increases in speed. This lowers the pressure within the venturi to draw in more fuel.

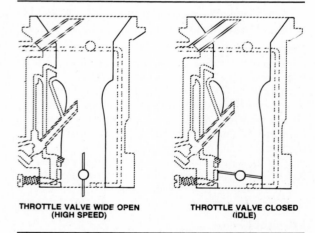

Figure 7-6. The carburetor throttle valve controls engine speed and power by regulating the amount of air and fuel entering the engine.

AIRFLOW AND THE VENTURI PRINCIPLE

Opening the carburetor throttle valve causes air to move from the higher pressure area outside the engine, through the carburetor, to the lower pressure area of the manifold. How much and how fast the air will travel is determined by the opening of the throttle valve.

The pressure of air passing rapidly through a carburetor barrel is lower along the sides of the barrel than it is in the center of the airflow. By putting a small hole in the side of the barrel, more air can be drawn into the stream of air rushing through the barrel, figure 7-4, position A. If a hose is used to connect the hole to a

Basic Carburetion

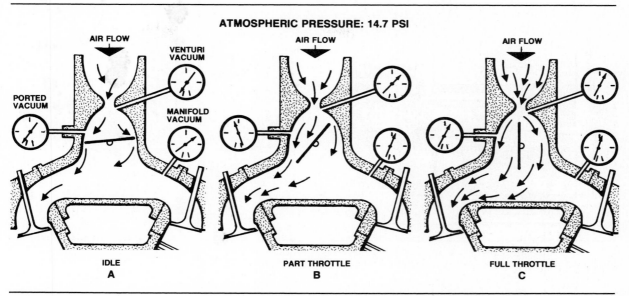

Figure 7-7. At idle, manifold vacuum is high, but vacuum in the carburetor barrel is low because the throttle is closed. At part throttle, venturi vacuum is high; ported vacuum is slightly lower, almost equal to manifold vacuum. At full throttle, venturi vacuum is still slightly higher, but all three are almost equal.

liquid-filled container or bowl, then the liquid will be forced through the hose and into the stream of air rushing by, figure 7-4, position B.

This is caused by the higher air pressure on the liquid, which forces it to the lower air pressure area inside the barrel. How much liquid passes through the hose will depend on the airflow velocity, or how fast the air is flowing through the inside of the barrel. The higher the airflow velocity, the lower the pressure will be at the inlet hole, and the more liquid will flow.

If we wish to make the carburetor work better, we must increase the air velocity through the barrel. This can be done by placing a restriction called a **venturi** inside the barrel, figure 7-5. When air flows through the venturi restriction, it speeds up. This speed increase lowers the pressure inside the carburetor barrel and permits more liquid fuel to be drawn into the airflow.

In addition to mixing liquid fuel with air, the carburetor must also vaporize the liquid as much as possible. To help break up the liquid fuel for better vaporization, a small opening called an air bleed is put in the fuel inlet passage, figure 7-5. This mixes air with the liquid fuel as it is drawn into the main airflow.

The carburetor must also change the air-fuel mixture automatically. It must deliver a rich mixture for starting, idle, and acceleration; and a lean mixture for part-throttle operation. Engine speed and power are regulated by the position of the carburetor throttle valve, which controls the flow of the air-fuel mixture, figure 7-6.

Carburetor Vacuum

There are four measurements of air pressure, or vacuum, that are important here:
1. Atmospheric pressure — the pressure of the air outside the carburetor
2. Manifold vacuum — the low pressure beneath the carburetor throttle valve
3. Venturi vacuum — the low pressure area created by airflow through the venturi restriction in the carburetor barrel
4. Ported vacuum — the low pressure area in the carburetor just above the throttle valve.

Pressure and vacuum measurements at these four areas are shown in figure 7-7.

Atmospheric pressure is always present outside the carburetor and engine. It remains constant at any given altitude and temperature. **Manifold vacuum** is produced by the engine and is always present when the engine is running. **Venturi vacuum** is created by airflow through the venturi and increases with the

---

Venturi: A restriction in an airflow, such as in a carburetor, that speeds the airflow and creates a vacuum.

Manifold Vacuum: Low pressure, in an engine's intake manifold, below the carburetor throttle.

Venturi Vacuum: Low pressure in the venturi of a carburetor, caused by fast airflow through the venturi.

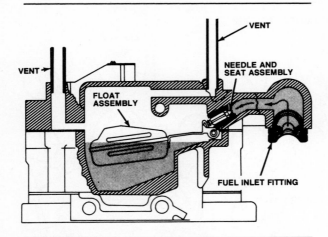

Figure 7-8. Fuel level in the fuel bowl is controlled by the float and needle valve acting against fuel pump pressure.

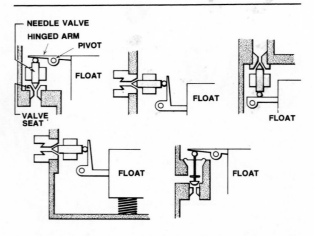

Figure 7-9. Float and needle valve designs vary with different carburetors.

speed of the airflow. Venturi vacuum is present whenever the throttle valve is open to allow air to flow through the carburetor. Because the throttle valve regulates the speed of the airflow, venturi vacuum increases as the throttle is opened. Ported vacuum is present whenever the throttle is opened to expose the lower portion of the carburetor barrel to manifold vacuum.

Ported vacuum exists just above the throttle valve. Vacuum taken from this point often is used to operate distributor vacuum advance units and other vacuum-operated devices. Small ports, or holes, in the sides of the carburetor are connected to hoses or tubes which are connected to the vacuum devices.

BASIC CARBURETOR SYSTEMS

To mix the fuel and regulate the speed, the carburetor has a series of fixed and variable passages, jets, ports, and pumps which make up the fuel-metering systems or circuits. There are seven basic systems common to all carburetors:
1. Float system
2. Idle system
3. Low-speed system
4. High-speed, or main metering, system
5. Power system
6. Accelerator pump system
7. Choke system.

The Float System

Gasoline from the fuel tank is delivered by the fuel pump to the carburetor fuel bowl, where it is stored for use. Once in the fuel bowl, the gasoline must be kept at a precise, nearly constant level. This level is critical, since it determines the fuel level in all the other passages and circuits within the carburetor. A fuel level that is too high in the bowl will produce an air-fuel mixture that is too rich. One that is too low will produce an overly lean mixture. For this reason, fuel level is one of the most critical adjustments needed on a carburetor.

The main fuel discharge nozzle for the high-speed system is connected directly to the bottom of the fuel bowl. Because liquids seek their own level in any container, the fuel level in the bowl and in the nozzle is the same. If the level is too high, too much fuel will be drawn into the high-speed system. If the fuel level is too low, too little fuel will be drawn in.

Fuel level is controlled by the float and the inlet needle valve, figure 7-8. As gasoline is drawn from the bowl, the float lowers in the remaining fuel. Fuel pump pressure then opens the needle valve and allows more fuel to enter the bowl. As the fuel level rises, so does the float, until it forces the inlet needle back against its seat. This closes the inlet valve and shuts off both fuel pump flow and pressure to the carburetor bowl. During many operating conditions, fuel flow into and out of the fuel bowl is about equal. The needle stays in a partly open position to maintain the required flow rate.

The float and needle valve regulate fuel flow, as well as fuel level. Since the needle valve is like a door between the carburetor fuel bowl and the fuel pump, it maintains an air space

Basic Carburetion

above the fuel in the bowl. This reduces pressure on the fuel to atmospheric pressure. Atmospheric pressure is maintained in the fuel bowl by a vent or balance tube venting the bowl to the carburetor airhorn. Atmospheric pressure pushing down on the fuel in the bowl provides the pressure differential needed for precise fuel metering into the venturi vacuum area of the carburetor barrel.

Float and needle valve design and location in the fuel bowl vary with different carburetor designs, figure 7-9. Some floats have small springs to keep them from bobbing up and down when the car travels over rough roads. Many fuel bowls have baffles, which keep the fuel from sloshing on rough roads and sharp turns. The needle valves and their seats in older carburetors were usually made of stainless steel. The steel often attracted metallic particles in the fuel. The particles would collect between the needle and seat and allow the valve to leak. The needles and seats in most modern carburetors are made of brass, and the needles often have plastic tips that conform to any rough spots on the seat and still provide a good seal when the valve is closed. If the float and the needle valve do not keep the correct fuel level in the bowl and too much fuel enters, the carburetor will flood.

When the engine is shut off, engine heat causes the fuel in the bowl to evaporate. This was no problem in pre-emission control days, but with the installation of vapor canister systems, the amount of evaporation from a large fuel bowl can easily overload the canister. Thus, late-model carburetors use a somewhat smaller float bowl. Some carburetors use a molded plastic float bowl to reduce heat evaporation because plastic does not conduct heat as much as metal does. Others use an insulator, figure 7-10, between the intake manifold and the carburetor to reduce heat.

The Idle System

When an engine is idling, the throttle is just cracked open and airflow through the carburetor barrel and venturi is reduced. Since there is little or no venturi effect, no fuel flows from the main discharge nozzle. The idle system, figure 7-11, supplies enough air and fuel to keep the engine running under these conditions.

Intake manifold vacuum is high at idle, so idle ports are located just below the closed throttle. The pressure differential between the fuel bowl and the vacuum at the idle ports forces fuel through the ports. Gasoline flows from the bowl, through the main jet, and to the idle tube. Because the fuel must be well mixed with air for proper distribution, air bleeds in the

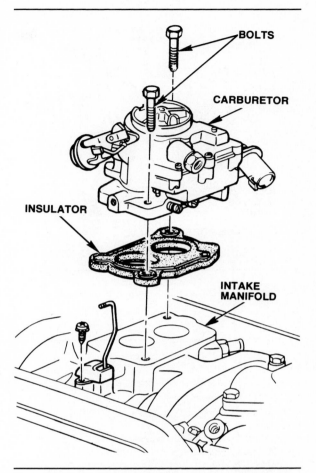

Figure 7-10. This insulator between the carburetor and the intake manifold reduces heat that causes fuel evaporation in the fuel bowl.

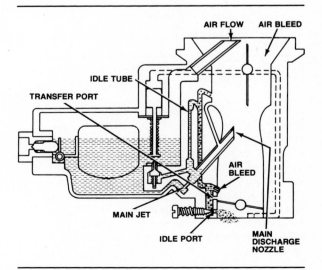

Figure 7-11. Air and fuel for the idle system are mixed inside the carburetor passages and delivered to the idle port below the throttle.

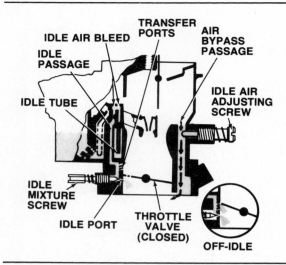

Figure 7-12. Air for the idle circuit in this carburetor passes through a bypass passage and is controlled by an idle air adjusting screw.

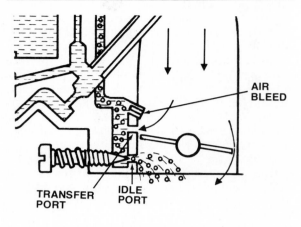

Figure 7-13. At idle, air flows in through the transfer port to mix with the idle air-fuel mixture. As the throttle opens, flow reverses through the transfer port. Fuel and air now flow out for low-speed operation.

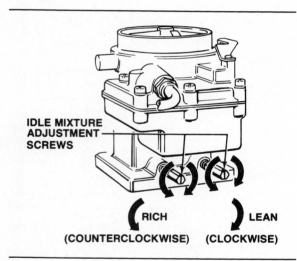

Figure 7-14. Idle mixture adjustment screws control the amount of gasoline in the air-fuel mixture.

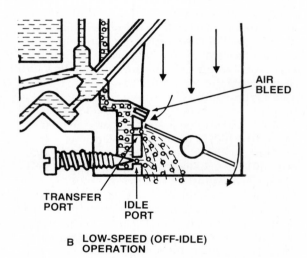

idle tube let in air for the idle mixture. The air bleeds also prevent fuel **siphoning** at high speeds or when the engine is stopped.

Extra air for the idle air-fuel mixture is provided in several different ways, depending on the carburetor design. In many carburetors, the throttle valve does not close completely, but remains slightly open to let in a small amount of air, figure 7-13. A few designs draw air for the idle circuit through a separate air passage in the carburetor body called the idle air bypass, figure 7-12.

Additional small openings called transfer ports, figure 7-13, are located just above the closed throttle valve in the carburetor barrel. At idle, the transfer ports suck air from the barrel into the fuel flow in the idle system. A small amount of air and fuel is released just below the throttle valve. When the engine is under slight acceleration, the throttle valve opens a little and exposes the transfer port to manifold vacuum. This draws the fuel out into the barrel to mix with the air. We will discuss this more under The Low-Speed System.

Adjustable needle valves, called idle mixture screws, figure 7-14, control the amount of gasoline used in the idle air-fuel mixture. One adjustment screw generally is used for each pri-

Basic Carburetion

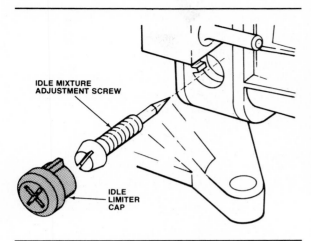

Figure 7-15. Idle limiter caps restrict the amount of adjustment allowed for the idle mixture.

Figure 7-16. This large idle air bypass screw is used to adjust idle speed by varying the airflow through the bypass passage.

mary barrel. The screw tips stick out into the idle system passages and are turned inward (clockwise) to create a lean mixture, or outward (counterclockwise) to richen the mixture.

Some carburetor mixture screws have plastic limiter caps, figure 7-15. These caps restrict the amount of adjustment to prevent excessively rich idle mixtures. They are required on some engines in order to meet emission standards and must be put back on whenever a carburetor is overhauled.

Engine idle speed is adjusted by changing the amount of air going to the idle system. This idle speed adjustment is usually done by a screw that changes the position of the throttle valve in the carburetor. Carburetors which use idle air bypass passages are adjusted by a large screw that varies the opening in the air passage to change the airflow, figure 7-16.

The Low-Speed System

Once the throttle valve begins to open for low-speed operation, the engine needs more fuel than the idle port alone can provide. The airflow passing through the venturi is still not strong enough to develop fuel flow through the main discharge nozzle. To provide more fuel, the transfer port comes into operation as the low-speed system, figure 7-13.

The transfer port is located above the throttle at idle, and the air pressure there is about equal to atmospheric pressure. Air from the barrel flows *into* the transfer port to mix with the fuel going to the idle port. As the throttle opens, the transfer port is exposed to intake vacuum, and the flow reverses. Extra fuel flows *out* of the transfer port to meet the engine's needs during the switch from idle to low-speed operation.

Fuel continues to flow from the idle port, but at a reduced rate. This permits an almost constant air-fuel mixture during this transition period.

The High-Speed (Main Metering) System

When the throttle valve opens wider, airflow increases through the carburetor. At the same time, the partial vacuum area of the intake manifold moves up in the carburetor barrel. This airflow and pressure change strengthens the

---

Siphoning: The flowing of a liquid as a result of a pressure differential, without the aid of a mechanical pump.

■ **Funny-Looking Gaskets Aren't Funny**

It's easy to overlook gaskets. They are small, not too expensive and they don't seem very important.

Don't you believe it. Gaskets that don't fit right will not work, and a bad gasket can rob your engine of a lot of power. Here is a tip from the professionals: be careful of how you store gaskets. Store them flat, never standing on edge. Don't hang them on a nail; that would most certainly pull them into funny shapes. Don't store gaskets near heat, since that also could cause them to warp into some interesting shapes. All non-metal gaskets have some water content, so they can't be allowed to dry up. If you do have a dry one, soak it in warm water for just a few minutes. Too much water will, again, cause it to warp.

Best thing to do is store gaskets flat in a protective wrapper. Leave them in a drawer or cabinet where they will not be disturbed, and won't have other parts set on top of them.

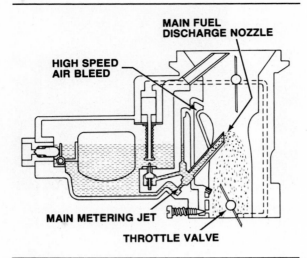

Figure 7-17. Fuel for the high-speed, or main metering system flows through the main jet and out the fuel discharge nozzle in the venturi.

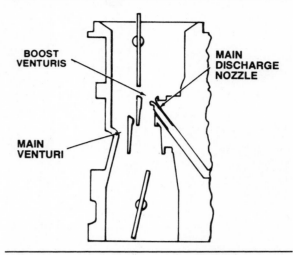

Figure 7-18. Most carburetors have multiple (boost) venturis for better air and fuel mixing.

venturi effect, causing gasoline to flow from the main discharge nozzle in the venturi. This is the high-speed or main metering system, figure 7-17.

For better mixing of the fuel and air, most carburetors have multiple, or boost, venturis placed one inside another, figure 7-18. The main discharge nozzle is located in the smallest venturi to increase the partial vacuum effect on the nozzle. Fuel flows from the bowl, through the main jet and main passage, and into the discharge nozzle. A high-speed air bleed, figure 7-17, mixes air into the fuel before it is discharged from the nozzle.

The primary or upper venturi produces vacuum, which causes the main discharge nozzle to spray fuel. The secondary venturi creates an airstream which holds the fuel away from the barrel walls where it would slow down and condense. The result is air turbulence, which causes better mixing and finer atomization of the fuel.

As the throttle continues to open wider, fuel flow from the low-speed system tapers off, while flow from the high-speed system increases. The engine's fuel needs are now supplied entirely by the main discharge nozzle during high-speed, light-load cruising.

The Power System

The high-speed system delivers the leanest air-fuel mixture of all the carburetor systems. When engine load increases during high-speed operation, this mixture is too lean to deliver the necessary power required by the engine. The extra fuel needed is provided instead by another system called the power system, or power valve. It supplements main metering fuel delivery. The power system or valve can be operated by vacuum or mechanical linkage. The exact type differs according to carburetor design, but all provide a richer air-fuel mixture.

One type of power valve, figure 7-19, is located in the bottom of the fuel bowl with an opening to the main discharge tube. A spring holds a small poppet valve closed, while a vacuum piston holds a plunger above the valve. Since manifold vacuum decreases as the engine load increases, a large spring moves the plunger downward. This opens the valve and lets more fuel pass to the main discharge nozzle.

Another type of vacuum-operated power valve uses a diaphragm, figure 7-20. Manifold vacuum against the diaphragm holds the valve closed. As vacuum decreases under an increased load, a spring opens the valve. This sends more fuel through the power system and main discharge nozzle.

Metering rods also can be used as a power system, figure 7-21. These may be controlled by vacuum pistons and springs, or by mechanical linkage connected to the throttle. The ends of the rods are tapered, or stepped, and installed in the main jet opening. The rods restrict the area of the main jets and reduce the amount of fuel that flows through them during light load

Basic Carburetion

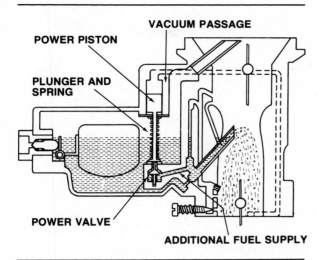

Figure 7-19. This power valve is operated by a vacuum-controlled piston and plunger. When vacuum decreases, the spring moves the plunger to open the power valve.

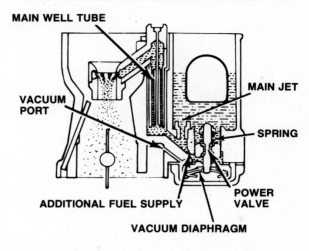

Figure 7-20. The vacuum diaphragm holds the power valve closed. When vacuum decreases, a spring opens the valve to allow more fuel into the main passage.

operation of the main metering system. Extra fuel for full-throttle power is provided by moving the rods out of the jets to increase the flow through the jets.

Vacuum-controlled metering rods, also called step-up rods, are held in the jets by manifold vacuum applied to pistons attached to the rods. When vacuum drops under heavy load, springs, working against the pistons, move the rods out of the jets.

Mechanically operated metering rods are controlled directly by mechanical linkage connected to the throttle linkage. Most metering rods are tapered, or stepped, to increase the extra fuel flow gradually.

The Accelerator Pump System

This system provides additional fuel for some engine operating conditions. If the throttle is opened suddenly from a closed, or nearly closed, position, airflow increases more rapidly than fuel flow from the main discharge nozzle. This "dumping" of air into the intake manifold reduces manifold vacuum suddenly and causes a lean air-fuel mixture. This excessively lean mixture results in a brief hesitation, or stumble, sometimes called a flat spot. To keep the mixture rich enough, extra fuel must be provided by the accelerator pump.

The accelerator pump, figure 7-22, is a plunger or diaphragm in a separate chamber in the carburetor body. It is operated by a linkage connected to the carburetor throttle linkage, figure 7-23. When the throttle closes, the pump

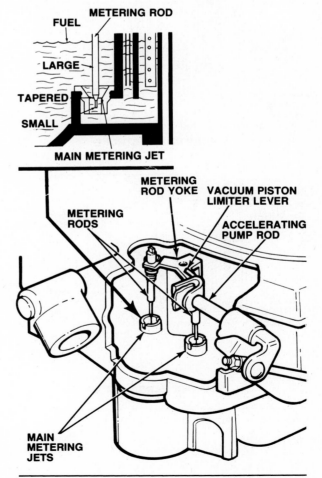

Figure 7-21. Some power systems consist of metering rods placed in the main jets. Mechanical or vacuum linkage moves the rods upward to allow more fuel to flow through the jets when required.

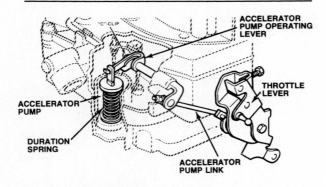

Figure 7-22. Typical plunger-type accelerator pump. (Ford)

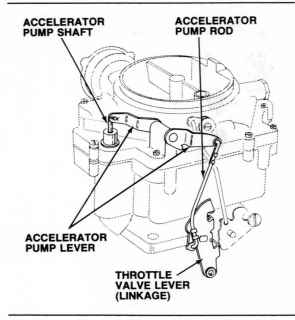

Figure 7-23. The accelerator pump linkage is connected to the throttle linkage.

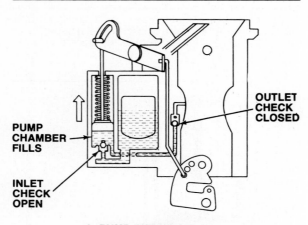

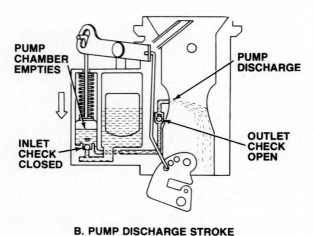

Figure 7-24. Accelerator pump operation.

draws fuel into the chamber. An inlet check valve opens to allow fuel into the chamber, figure 7-24, position A, and an outlet check valve closes so that air will not be drawn through the pump nozzle. When the throttle is opened quickly, the pump moves down or inward to deliver fuel to the nozzle in the barrel, figure 7-24, position B. The pump outlet check opens, and the inlet check closes. The inlet check ball is usually (but not always) larger than the outlet check ball.

The pump outlet check may be a steel ball or plunger. The inlet check may be a steel ball, a rubber diaphragm, or part of the pump plunger. Not all pumps have inlet checks. Some rely on an inlet slot in the pump well, or chamber, that is closed by the plunger on the downward stroke.

Most pump plungers or diaphragms are operated by a duration spring, figure 7-25. The throttle linkage holds the pump in the returned position. When the throttle opens, the linkage releases the pump, and the spring moves the plunger for a steady and uniform fuel delivery. The accelerator pump operates during the first half of the throttle travel from the closed to the wide-open position.

During high-speed operation, the vacuum at the pump nozzle in the carburetor barrel may be strong enough to unseat the outlet check and siphon fuel from the pump. This is called pump pullover, or siphoning. In a few carburetors, this extra fuel is included in the high-speed system adjustment. In most carburetors, air

Basic Carburetion

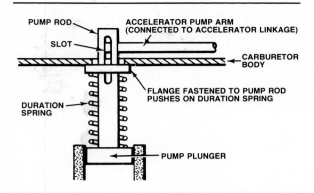

Figure 7-25. The duration spring provides uniform pump delivery regardless of the speed at which the throttle linkage moves.

bleeds are placed in the pump discharge passages to prevent the siphoning. In still other carburetors, an extra weight is added to the outlet check to resist the siphoning. The pump plungers in some carburetors have anti-siphon check valves.

The Choke System

The choke provides a very rich mixture for starting a cold engine. This extra-rich mixture is needed because:
- Engine cranking speed is slow.
- Airflow speed is slow.
- Cold manifold walls cause gasoline to condense from the air-fuel mixture, and less vaporized fuel reaches the combustion chambers.

To make the mixture richer, a choke plate, or butterfly valve, is placed above the venturi in the carburetor barrel. This choke plate can be tilted at various angles to restrict airflow, figure 7-26. Cranking the engine with the choke plate closed creates a partial vacuum throughout the carburetor barrel below the plate. This airflow reduction and partial vacuum area work together to allow more fuel to be drawn into the mixture.

The choke plate can be controlled manually by a cable running to the driver's compartment or automatically by a thermostatic spring. Chokes on most domestic carburetors for the past 25 or 30 years have been operated by a bimetal thermostatic spring. The choke plate shaft is connected to the spring by linkage. The bimetal spring is normally located in one of two places:
1. In a round housing on the carburetor airhorn, figure 7-27. This is called an integral, or piston-type, choke.

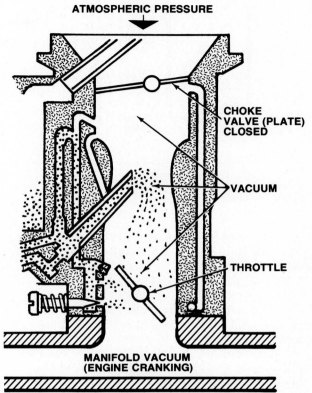

Figure 7-26. Vacuum is present throughout the carburetor barrel below the closed choke. This draws fuel from the idle, low-speed, and high-speed circuits for starting the engine.

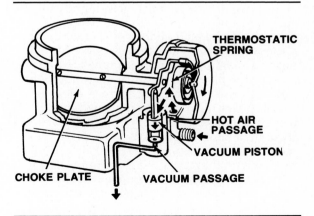

Figure 7-27. In an integral (piston-type) choke, the thermostatic bimetal spring is in a housing on the carburetor airhorn.

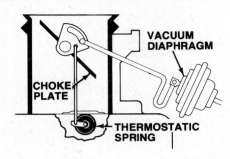

Figure 7-28. In a remote (well-type) choke, the thermostatic bimetal spring is in a heated well on the intake manifold.

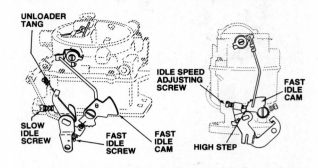

Figure 7-29. The fast-idle cam opens the throttle wider for faster engine speed when the choke is operating. It may work on the slow-idle speed adjusting screw or on a separate fast-idle screw.

2. Off the carburetor in a well on the intake manifold, figure 7-28. This is called a remote, a well-type, or a vacuum-break, choke.

Regardless of type and location, the thermostatic spring forces the choke closed when the engine is cold. Running the engine heats up the spring, which then opens the choke. With an integral choke, hot air from a source near the exhaust manifold, or hot coolant from the cooling system, is routed to the choke housing to heat the spring. Remote, well-type chokes are normally heated by exhaust routed through a crossover passage in the intake manifold. Both integral and remote chokes also may have electric heating elements to heat the spring faster and speed the choke opening.

When a cold engine is cranked, the choke must be completely closed. As soon as the engine starts, the choke must open slightly for sufficient airflow. This is done in two ways. First, the choke plate shaft is offset in the carburetor so that airflow will tend to open the plate. Second, manifold vacuum is applied to a vacuum piston or a vacuum-break diaphragm that pulls the choke open a few degrees. Most integral chokes have a vacuum piston built into the choke housing, figure 7-27, which gives them their name of piston-type chokes. A remote choke has the vacuum-break diaphragm mounted on the side of the carburetor, figure 7-28. Some integral chokes have a vacuum-break diaphragm rather than a piston to open the choke.

A cold engine must idle faster than a warm engine, or the lack of air and fuel flow will cause it to stall. A fast-idle cam and screw, figure 7-29, provide enough air and fuel to prevent engine stalling. The cam is linked to the choke plate, and the screw is located on the throttle valve shaft. Pushing down on the accelerator to start a cold engine allows the choke to close. This moves the fast-idle cam to allow the screw to rest against a high step of the cam. The cam may contact the normal slow-idle adjusting screw or a separate fast-idle screw. In either case, the throttle is held open slightly more than for a normal slow idle, and idle speed increases between 400 and 800 rpm.

A mechanical link or choke unloader, figure 7-29, opens the choke about halfway when the throttle is fully open. If the engine is accidentally flooded during starting, this provides the extra airflow necessary to help clear out the fuel. Because the choke system provides a very rich mixture, external devices are used to modify its operation in various ways to improve driveability and reduce emissions. These will be discussed later.

CARBURETOR TYPES

The operation of the basic carburetor systems has been explained in terms of a carburetor which uses a single barrel and throttle valve. But carburetors also are made with two or more barrels. Various carburetor types are used to match fuel flow to engine requirements. Domestic engines all use downdraft carburetors, while some imports use sidedraft carburetor designs. Carburetors are usually classified by the number of barrels or venturis used. The differences are detailed below.

One-Barrel

The 1-barrel carburetor has a single outlet through which all systems feed to the intake manifold. This type of carburetor may also be known as a single-venturi design. These carburetors are generally used on 4- and 6-cylinder engines.

Basic Carburetion

Figure 7-30. A two-barrel carburetor has one airhorn but two venturis and throttle plates.

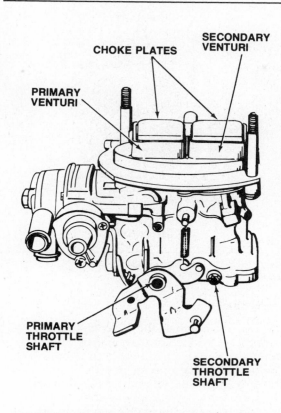

Figure 7-31. The 2-stage, 2-barrel carburetor was developed in response to emission control requirements.

Single-Stage Two-Barrel

This carburetor contains two barrels and two throttles which operate together, figure 7-30. Since the various fuel discharge passages in each barrel operate at the same time, it can be considered as two 1-barrel carburetors sharing the same body. The two throttle plates are mounted on the same shaft, and operate together. The two barrels share a common float choke, power system, and accelerator pump. Single-stage 2-barrel carburetors are used on many 6- and 8-cylinder engines.

Two-Stage Two-Barrel

This carburetor, figure 7-31, is a relatively recent development, brought about by emission control requirements. It differs from the single-stage, 2-barrel design in that its two throttles operate independently. The primary barrel is generally smaller than the secondary, and handles engine needs at low-to-moderate speeds and loads. The larger secondary opens as necessary to handle heavier load requirements.

The primary stage usually includes the idle, the accelerator pump, the low-speed, the main metering, and the power systems. The secondary stage usually has a transfer, a main metering, and a power system. Both stages draw fuel from the same fuel bowl. Some designs use a common choke for both barrels. In others, only the primary stage is choked.

A 2-stage 2-barrel carburetor is used primarily on 4- and 6-cylinder engines and a few V-8's.

■ **Check For A Cold Carburetor**

Cold outside air can cause carburetors to freeze up, especially when low temperature combines with high humidity. This carburetor icing can, in turn, cause the choke valve to stick or bind in the carburetor. If you are servicing a car whose owner has complained about poor engine performance during cold weather, always be sure to check for carburetor icing. If it's your car, you can throw a blanket over the engine and carburetor when it is left to sit for a long time in cold weather.

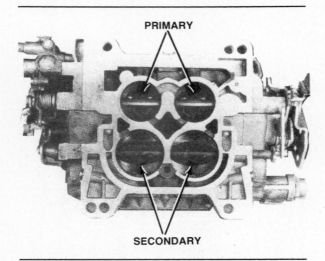

Figure 7-32. A four-barrel carburetor is made of two 2-barrel carburetors.

Four-Barrel

Used on V-8 engines, the 4-barrel, or quad, carburetor uses two primary barrels and two secondary barrels in a single body, figure 7-32. The two primaries operate like a 1-stage 2-barrel at low-to-moderate engine speeds and loads. The secondary barrels open at about half to three-quarter throttle to provide the increased fuel and airflow required for high-speed operation. The primary barrels contain the choke, the idle, the low-speed, and the high-speed systems, as well as an accelerator pump and a power system. The secondary barrels have their own high-speed and power systems and may use their own accelerator system.

Two methods provide airflow through the secondary barrels: venturi action or air velocity valves. Air velocity valves look like large choke plates located in the secondary barrels. They are opened by the low pressure created in the secondary barrels when the throttles are opened. Venturis and air valves also may be combined in one carburetor to modulate the airflow through the barrels.

The primary barrels supply all eight cylinders during low-to-moderate speeds and loads. The secondary barrels provide additional fuel and airflow for high speeds and heavy loads.

VARIABLE-VENTURI (CONSTANT DEPRESSION) CARBURETORS

As we have seen, carburetors meter fuel by using a venturi to create a partial vacuum in the barrel. Airflow through the venturi increases velocity which decreases pressure. This pressure drop causes fuel to flow through the discharge nozzle into the barrel. Since both the carburetor barrel and venturi are fixed in size, the volume and velocity of air passing through will be correct for some operating conditions but not correct for others. To produce better performance under *all* operating conditions, auxiliary circuits such as the choke, power, and idle systems must be added to the main metering systems.

A carburetor with a venturi whose size changes according to the demands of engine speed and load does not need these extra systems. At the same time, air-fuel mixtures can more closely be controlled for better fuel economy and emission control. Changing the size of the venturi in relation to engine speed and load results in an even pressure drop across the venturi under all operating conditions. This gives a variable venturi carburetor its other name, a "constant depression" carburetor. Variable-venturi carburetors, such as the SU, Hitachi, and Stromberg CD models, have been used on some imported cars for many years. In 1977, Ford Motor Company introduced the first variable-venturi carburetor used on domestic cars in 45 years. (Ford's first V-8 engines in 1932 had variable-venturi, 1-barrel carburetors.)

Called the Motorcraft 2700 VV, figure 7-33, Ford's variable-venturi, 2-barrel carburetor was first used on V-6 and 302-cid V-8 engines. It has a fuel inlet system with a replaceable filter in the inlet housing. Throttle plates and an accelerator pump are also used. However, the variable venturis and the different fuel metering systems make this carburetor unique.

The variable venturis are formed by two rectangular valve plates (actually one casting) that slide back and forth across the tops of the two barrels. Movement is controlled by a spring-loaded vacuum diaphragm, regulated by a vacuum signal taken below the venturis, but above the throttle plates, in the carburetor barrels. As the throttle opens, the vacuum increases, opening the venturis and allowing more air to enter.

The front edge of each venturi valve has a tapered metering rod. Each rod moves in and out of a fixed main jet on the other side of the barrel, figure 7-34. This arrangement meters fuel in proportion to the airflow through the venturis. Because of the variable venturis, there are fewer fuel metering systems than in a fixed-venturi carburetor. However, the main meter-

Basic Carburetion

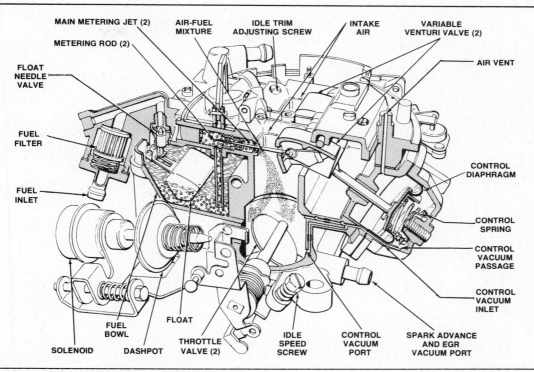

Figure 7-33. The Motorcraft 2700 VV variable-venturi carburetor uses vacuum to control the movement of the venturi valves.

■ Ford's First Variable Venturi Carburetor

Ford's first variable venturi carburetor was a one-barrel model made by Detroit Lubricator. It was used on the original flathead, Model 18 V-8's in 1932 and early '33. The variable venturi was formed by two air vanes in the barrel that responded directly to airflow, rather than to vacuum linkage control as in the current Motorcraft 2700 VV carburetor.

The air vanes were linked to a movable main jet (metering valve) that slid up and down on a fixed metering rod (pin) in the center of the venturi. The metering rod was adjustable to obtain the proper idle mixture and main system fuel flow.

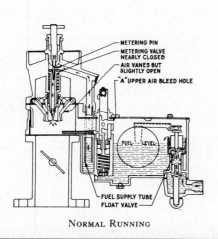
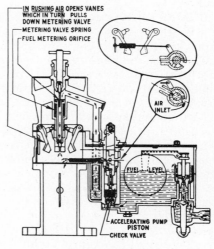

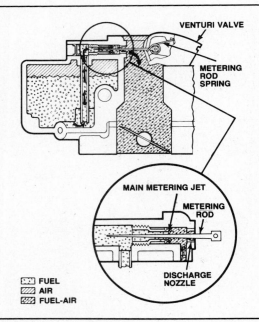

Figure 7-34. Motorcraft 2700 VV main metering system. (Ford)

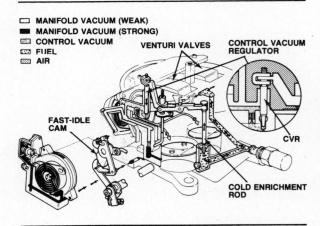

Figure 7-35. Motorcraft 2700 VV cold enrichment system operation with a cold engine. (Ford)

ing system of the 2700 VV carburetor cannot handle all operating conditions without help from secondary systems. For example, at full throttle under heavy load, vacuum may not be strong enough to override the diaphragm spring. In this case, a limiter lever on the throttle shaft pushes the venturi valves fully open. Other auxiliary systems in this carburetor are the accelerator pump, the idle trim system, the cranking enrichment system, and the cold enrichment system.

Cold Enrichment System

The 2700 VV carburetor does not have a conventional choke. Instead, a bimetal thermostatic choke control, with Ford's electric-assist heater (explained later in this chapter), controls the cold enrichment system, figure 7-35. This system contains a fast-idle cam, a cold enrichment auxiliary fuel passage and metering rod, and a control vacuum regulator rod. A unique feature of the fast-idle cam is a vacuum-operated "high cam speed positioner" (HCSP). When a cold engine is started, a lever slides between the fast-idle cam and the fast-idle lever to provide more throttle opening. When the engine starts, vacuum is applied to the HCSP diaphragm and the lever is retracted.

When the engine is started below 95° F, the choke spring pushes the control vacuum regulator rod to block the ported control vacuum and send manifold vacuum to the venturi valve dia-

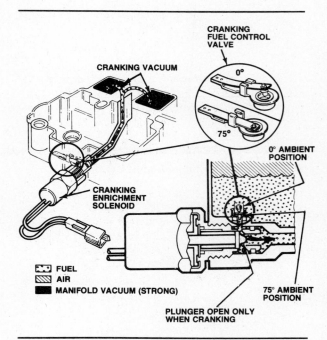

Figure 7-36. Motorcraft 2700 VV cranking enrichment system. (Ford)

phragm. This opens the venturis wider than normal. The fast-idle cam is touched by the HCSP for a wider throttle opening.

Cranking Enrichment System

The cranking enrichment system, figure 7-36, also provides extra fuel only for starting a cold engine. It uses an electric solenoid, energized by the ignition switch, to open an auxiliary fuel passage. When the engine starts, the solenoid is deenergized and closes the cranking fuel passage. The main metering system and the cold enrichment system then maintain the fuel flow.

Basic Carburetion

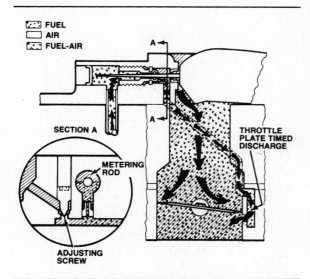

Figure 7-37. Motorcraft 2700 VV idle trim system. (Ford)

Idle Trim System

The idle fuel flow is controlled by the main jets, but an additional small amount of fuel is drawn by manifold vacuum through internal passages to discharge ports below the throttles. This is called the idle trim system, figure 7-37.

CARBURETOR LINKAGE

Car speed is controlled by the accelerator pedal, which moves rods, cables, levers, and springs to operate the carburetor throttle valve. The throttle linkage on some cars is a combination of solid rods and levers, figure 7-38. Other cars use a cable to link the accelerator to the carburetor throttle, figure 7-39.

Other carburetor linkage pieces operate the accelerator pump, the automatic choke, the fast-idle cam, secondary throttle valves, and the automatic transmission downshift points on some cars.

Accelerator Pump Linkage

The accelerator pump piston or plunger is connected to the throttle by a small rod, or rods and levers. On some carburetors, the linkage holds the pump in the retracted position, against a compressed duration spring. When the throttle opens all the way, the linkage releases the pump, and the duration spring moves the pump piston through its stroke. On some carburetors, the accelerator pump linkage also opens a vent for the fuel bowl when the throttle is closed.

Figure 7-38. This throttle linkage is a combination of solid rods and levers.

Figure 7-39. A cable from the accelerator to the carburetor is used for this throttle linkage.

On other carburetors, the accelerator pump linkage operates a diaphragm to deliver fuel to the pump system. The accelerator pumps on other carburetors are operated directly by mechanical linkage. Pump operation may be balanced by a duration spring and a return spring.

Accelerator pump and linkage designs vary from carburetor to carburetor, but all work on these same principles. On most carburetors, the accelerator pump ends its stroke at the half-throttle position. From this point to the full-throttle position, the high-speed and the power systems can supply enough fuel. The pump linkage on some carburetors can be installed in two or three different positions on the carburetor throttle linkage to provide different pump strokes.

Automatic Choke Linkage

Automatic chokes vary in design, but all require connecting linkage between the thermostatic spring that closes the valve and the vacuum piston or diaphragm that opens it. A remote, well-type choke with a vacuum-break diaphragm is shown in figure 7-40. The rod linkage from the spring to the choke valve closes the choke. The vacuum-break diaphragm opens the choke through its linkage as soon as the engine starts. With most integral, piston-type chokes, the thermostatic spring acts directly on a lever on the end of the choke shaft, inside the choke housing, figure 7-27. The vacuum piston also acts directly on this lever to open the choke

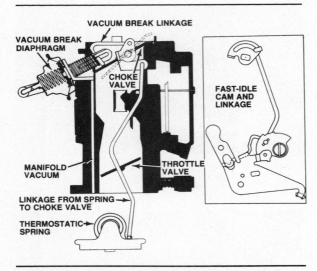

Figure 7-40. Several pieces of linkage are used on this choke system. (Chevrolet)

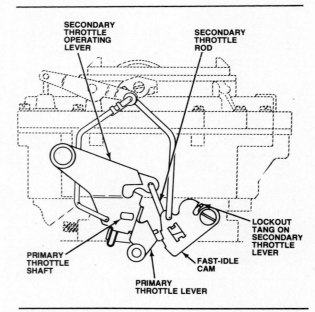

Figure 7-41. Mechanical secondary throttle linkage on a four-barrel carburetor.

Figure 7-42. Air velocity valves in a 4-barrel carburetor.

when the engine starts. Some carburetors are built with an integral-type of choke housing mounted away from the choke valve. This arrangement requires an external linkage rod from the thermostatic spring to the choke valve, as well as a separate vacuum-break diaphragm.

Fast-Idle Cam Linkage

The fast-idle cam is linked to the choke valve, figure 7-40, to provide a faster than normal idle and prevent stalling when the engine is cold. A fast-idle operating lever is attached to the choke shaft and connected to the fast-idle cam by a link. When the choke closes, the linkage turns the fast-idle cam so that a high step of the cam touches the idle speed adjusting screw or a separate fast-idle screw. As the choke opens, the fast-idle cam continues to follow the choke movement and reduces idle speed step by step. When the choke is fully open, gravity keeps the fast-idle cam away from the idle speed screw. However, when the engine is shut off, the idle speed screw blocks the cam and keeps it from turning back to the fast-idle position as long as the throttle is closed. This also keeps the choke thermostatic spring from closing the choke valve as the engine cools. To begin the choking operation with a cold engine, the accelerator must be pressed down to release the fast-idle cam and linkage and allow the choke to close.

Secondary Throttle Linkage

The secondary throttles of 4-barrel and 2-stage 2-barrel carburetors may be operated by mechanical linkage or by vacuum. Secondary throttle valves operated by mechanical linkage have an operating rod to connect the primary throttle shaft to the secondary throttle shaft, figure 7-41. Secondary throttles begin to open when the primary throttles are about one-half open. The secondaries continue to open along with the primaries, but at a faster rate. The primary and secondary throttles then reach the fully open position at the same time.

On some carburetors, the primary and secondary throttle shafts are mechanically linked, but the secondary throttles are not visible through the carburetor airhorn. This type of carburetor has secondary air velocity valves, called auxiliary throttle valves, figure 7-42. These valves have offset shafts and coun-

Basic Carburetion

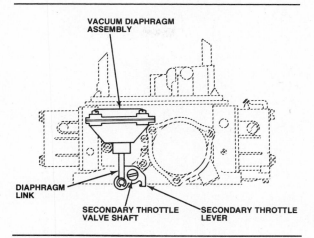

Figure 7-43. Vacuum-operated secondary throttles are controlled by a diaphragm like this.

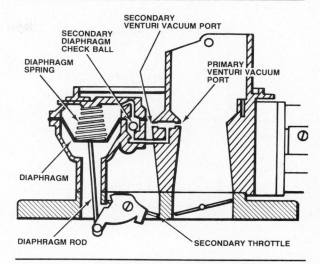

Figure 7-44. The secondary diaphragm responds to vacuum from ports within the carburetor barrels.

terweights so that they remain closed until air velocity through the carburetor barrels is strong enough to open them. These auxiliary throttle valves operate only when the mechanical secondary throttles are open, but they do not rely on the mechanical movement of the secondary throttles.

Vacuum-operated secondary throttle valves are controlled by a vacuum diaphragm that is mounted on the side of the carburetor, figure 7-43. At low cruising speeds, the secondary throttle valves are closed, and the engine's air-fuel requirements are met by the primary half of the carburetor. At higher speeds, when more fuel and air are needed, the secondary throttles are opened by linkage connected to the vacuum diaphragm.

The vacuum diaphragm responds to increasing vacuum within the primary venturi as engine airflow increases, figure 7-44. Linkage from the diaphragm opens the secondary throttles. Diaphragm action is changed by another vacuum port, or air bleed, in the secondary barrels. The amount and rate at which the secondary throttles open are determined by the vacuum signal at the diaphragm. When the secondaries are closed, the air bleed in the secondary barrels weakens the vacuum signal at the diaphragm a set amount. As the secondaries open, the air bleed becomes a vacuum port as vacuum develops within the secondary barrels. This vacuum signal is then added to the vacuum from the port within the primary venturis to open the secondary throttles completely. Sudden secondary throttle opening is prevented by a ball check valve in the vacuum chamber passage which allows a gradual vacuum buildup. As engine speed decreases, the weaker vacuum signal allows the diaphragm spring to close the secondary throttles.

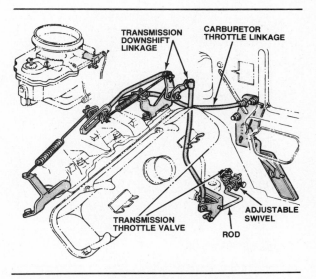

Figure 7-45. Chrysler products with automatic transmissions have mechanical linkage from the carburetor to the transmission throttle valve, which control shift points. (Chrysler)

All secondary throttle linkage, whether mechanical or vacuum operated, includes a secondary throttle or air valve lockout device to keep the secondary throttles from opening when the choke is closed.

Transmission Linkage

Chrysler Corporation cars with automatic transmissions use an adjustable throttle rod connection, figure 7-45, between the transmission and the carburetor to control shift points and shift quality. Other automatic transmissions accomplish this with vacuum control.

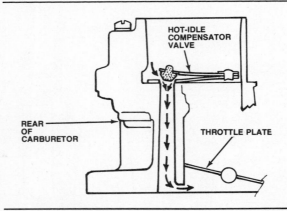

Figure 7-46. The hot-idle compensator is a thermostatic valve that opens at high temperature to admit more air to the idle circuit.

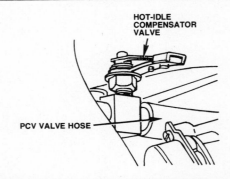

Figure 7-47. This hot-idle compensator is located in the PCV valve hose, away from the carburetor.

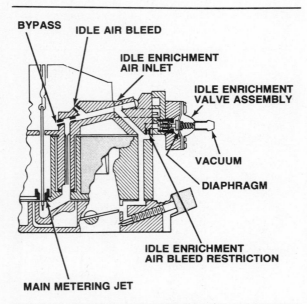

Figure 7-48. Chrysler idle enrichment valve.

CARBURETOR CIRCUIT VARIATIONS AND ASSIST DEVICES

Variations in the basic carburetor systems we just talked about are used by all carmakers. One or more add-on devices also may be used to improve economy, driveability, and emission control. Those most commonly used are discussed below.

Hot-Idle Compensator Valve

High carburetor inlet air temperature causes gasoline to vaporize rapidly, which can cause an overly rich idle mixture. To prevent this, many carburetors use a hot-idle compensator valve, figure 7-46. The compensator is a thermostatic valve consisting of a bimetal spring, a bracket, and a small poppet. The compensator valve usually is located either in the carburetor barrel or in a chamber on the rear of the carburetor bowl. A dust cover is placed over the chamber. A third location, used primarily in Autolite 2-barrel carburetors on older air-conditioned cars, is an external mounting in the PCV valve hose near the carburetor, figure 7-47.

The hot-idle compensator valve is normally closed by spring tension and engine vacuum. As temperature rises, the bimetal strip bends. This uncovers an auxiliary air passage, or air bleed, through which air enters the caburetor below the throttle plates. As this extra air mixes with excess fuel to lean out the idle mixture, it prevents stalling and rough idling. Once the carburetor temperature returns to normal, the compensator valve closes to shut off the extra air supply.

Idle Enrichment Valve

To reduce emissions, late-model carburetors are adjusted for leaner mixtures. For good cold-engine operation, the idle mixture must be enriched in some cases. All 1975 and later Chrysler cars with automatic transmissions use an idle enrichment system. This works opposite to a hot idle compensator valve.

A small vacuum diaphragm mounted near the carburetor top, figure 7-48, controls idle circuit air. When control vacuum is applied, the diaphragm reduces idle system air. This increases fuel and reduces the air in the air-fuel mixture. Diaphragm vacuum is controlled by a temperature switch in the radiator. As the engine warms, this switch stops the vacuum signal, returning the airfuel mixture to its normal lean level.

Basic Carburetion

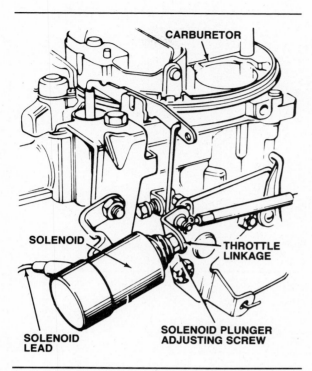

Figure 7-49. The throttle stop solenoid holds the throttle open for normal slow idle and allows the throttle to close farther when the engine is shut off.

Fast-Idle Pulloff (Choke Pulloff)

Long periods of choke and fast-idle can damage catalytic converters with rich air-fuel mixtures. Some converter-equipped GM cars use a fast-idle pulloff to avoid converter overheating. In one system, manifold vacuum acts on a vacuum-break diaphragm at the rear of the carburetor. This diaphragm will drop the fast-idle cam to a lower step 35 seconds after engine coolant temperature reaches 70° F. Vacuum to the diaphragm is controlled by a vacuum solenoid operated by a coolant temperature switch and delay timer.

Another system uses the front vacuum diaphragm to pull the throttle down one step on the fast-idle cam as soon as engine coolant temperature reaches 150° F. Vacuum to the diaphragm is controlled by a temperature vacuum switch.

A third method uses an electric solenoid instead of a vacuum diaphragm to open the choke. This pulls the fast-idle screw off the cam whenever the engine is started with the coolant below a specified temperature. A temperature switch on the engine and a firewall-mounted relay provide current for the pulloff solenoid.

Regardless of the system used, fast-idle pulloff has no effect on engine warmup during ordinary operation because normal throttle movement will disengage the fast-idle cam. The fast-idle pulloff system only operates when the car is warming up while parked.

Throttle Stop Solenoid

Engine **dieseling** or after-run results when combustion chamber temperatures remain hot enough to ignite an idle air-fuel mixture after the ignition is turned off. Dieseling is caused by several aspects of late-model engines: higher operating temperatures, faster idle speeds, retarded ignition timing at idle, and lean air-fuel mixtures. Closing the throttle more than it would close for the engine's normal slow idle speed will prevent dieseling. By shutting off airflow past the throttle valve, the idle circuit is closed.

A throttle stop solenoid, figure 7-49, provides the new stop position for the throttle during normal slow idle. Turning on the ignition energizes the solenoid, and its plunger moves out to contact the idle speed adjusting screw or a bracket on the throttle shaft. This holds the throttle open slightly for a normal slow idle until the ignition is shut off. The solenoid is then deenergized, its plunger retracts, and the throttle closes to block airflow.

Air Conditioning Throttle Solenoid

Late-model Ford, GM, and Chrysler engines with air conditioning may use a solenoid which looks exactly like a throttle solenoid, figure 7-50, and even may carry the same factory part number. But, it should not be confused with the throttle stop solenoid we just discussed, since it is energized *only* through the air conditioner switch. Its plunger moves forward to contact a bracket on the throttle shaft *only* when the air conditioning is turned on. This raises engine idle speed to prevent the engine from stalling due to the increased load. It also helps to prevent overheating, from the air conditioning condenser heat load, by speeding up the radiator fan.

Dieseling: A condition in which extreme heat in an engine's combustion chamber continues to ignite excess fuel after the ignition has been turned off.

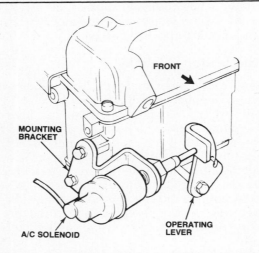

Figure 7-50. The air conditioning throttle solenoid opens the throttle slightly when the air conditioner is on. This maintains a uniform idle speed, even with the increased engine load.

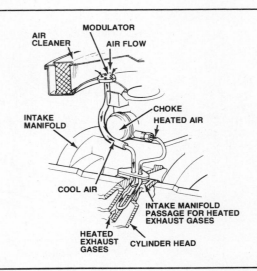

Figure 7-51. General Motors' choke hot air modulator.

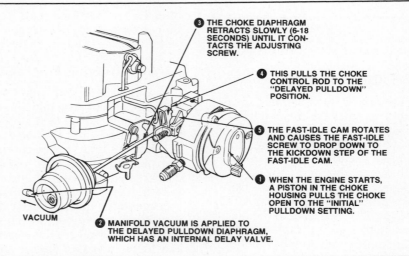

Figure 7-52. The delayed choke pulldown diaphragm provides rich initial choking, rapid choke release, and fast-idle modulation. (Ford)

Choke Hot Air Modulator

Some GM engines use a choke hot air modulator check valve (CHAM-CV) in the air cleaner, figure 7-51. At air cleaner temperatures below 68° F, this valve is closed. Air that is to be heated by the heater coil passes through a tiny hole in the modulator. This restricts hot airflow over the bimetal thermostatic coil and results in a slower choke warmup. When air cleaner temperature rises above 68° F, the modulator opens to permit more airflow for a faster choke warmup.

Staged Choke Pulldown

Found only on 1972 Ford engines, this system has a small double-chamber housing connected to the choke linkage at one end and to manifold vacuum at the other. Inside the housing, a divider with an orifice separates the two chambers. The chamber facing the carburetor contains a spring and silicone fluid; the other chamber holds a diaphragm and a bimetal valve. The valve controls the application of manifold vacuum to the diaphragm.

At temperatures below 60° F, the bimetal valve shuts off vacuum, and the spring in the other chamber keeps the operating rod extended. This allows the choke to close. When-

Basic Carburetion

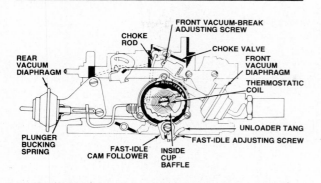

Figure 7-53. Temperature-controlled vacuum break on a GM V-8 engine. (Pontiac)

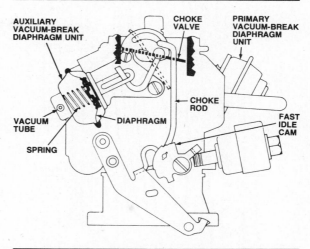

Figure 7-54. Temperature-controlled vacuum break on a GM 6-cylinder engine. (Chevrolet)

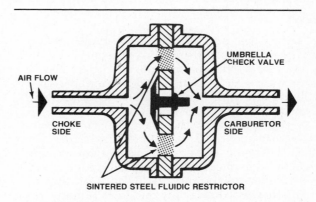

Figure 7-55. Choke delay valve.

ever the bimetal valve senses temperatures above 60° F, it permits manifold vacuum to operate on the diaphragm. This draws fluid from the front chamber, and the spring and operating rod are pulled back by the diaphragm to open the choke. Shutting the engine off causes the spring to push the operating rod out once more, sucking the fluid back into its chamber in the process. The orifice controls the time interval of choke opening and varies between 15 to 54 seconds, depending upon engine model.

Delayed Choke Pulldown

Some 1975 and later Motorcraft carburetors have a delayed choke pulldown operated by a vacuum diaphragm. This opens the choke to a wider setting 6 to 18 seconds after the engine starts. As the pulldown diaphragm operates, the fast-idle screw is pulled from the top to the second step of the fast-idle cam. This reduces cold engine idle speed. Figure 7-52 shows the exact sequence of operation.

Temperature-Controlled Vacuum Break

Since 1975, many Rochester carburetors use two vacuum-break diaphragms to provide better mixture control by choking the engine more when it is cold and less when warm. Slightly different systems are used on GM V-8 engines, figure 7-53, and 6-cylinder engines, figure 7-54.

On V-8 engines, the primary (front) diaphragm opens the choke to keep the engine from stalling when first started. The secondary (rear) diaphragm opens the choke wider when air cleaner temperature exceeds 62° F. The rear diaphragm operates on a manifold vacuum provided by a temperature vacuum valve on the air cleaner. The vacuum-break diaphragm has an inside restriction to delay diaphragm movement by several seconds.

On 6-cylinder engines, the primary diaphragm (choke coil side) opens the choke to keep the engine from stalling when first started. The auxiliary diaphragm (throttle lever side) opens the choke wider when engine coolant temperature is higher than 80° F. The auxiliary diaphragm operates on manifold vacuum provided by a two-nozzle temperature vacuum switch on the cylinder head. The dual vacuum-break diaphragm system was discontinued on several 1977 carburetors in favor of a single vacuum break.

Choke Delay Valves

Choke delay valves have various designs, but all delay the opening of the choke for a period of time to improve driveability and cold engine warmup.

The Ford choke delay valve, in a hose between intake manifold vacuum and the choke vacuum piston or diaphragm, figure 7-55, is the

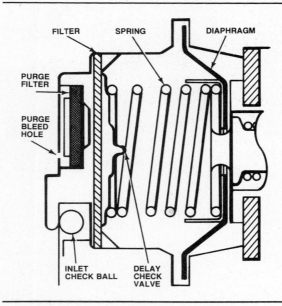

Figure 7-56. General Motors choke delay valve.

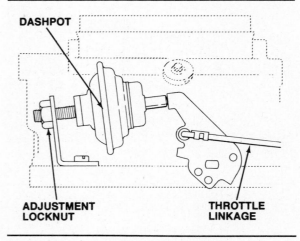

Figure 7-57. The dashpot slows throttle closing.

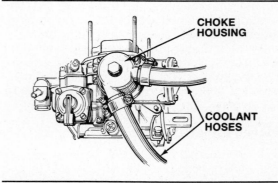

Figure 7-58. Hot-water-heated choke.

same valve used for spark delay in the ignition system. The black side must face the vacuum source, and the colored side the choke. The amount of delay is indicated by the valve color and is the same interval as when used as a spark delay valve.

GM uses an internal bleed check valve in the rear vacuum break diaphragm unit, figure 7-56. This delays choke opening beyond the amount allowed by the front vacuum break for a few seconds until the engine can run at a leaner mixture. Vacuum acting on the diaphragm draws filtered air through the bleed hole to purge any fuel vapor or contamination.

A slightly different system is used on the 151-cid, 4-cylinder GM engine. This engine has a choke vacuum-break and vacuum-delay valve. The valve delays manifold vacuum against the choke vacuum-break unit for approximately 40 seconds when starting the engine. If the engine should stall after being started, a relief feature in the valve permits immediate release of vacuum to let the choke close quickly.

Dashpots

These small chambers with a spring-loaded diaphragm and plunger have been used for over 20 years on some cars with automatic transmissions and on a few with manual transmissions. Hydraulic and magnetic dashpots also have been used on older cars.

A link from the throttle touches the dashpot plunger as the throttle closes, figure 7-57. As force is applied to the plunger, air slowly bleeds out of the diaphragm chamber through a small hole.

Dashpots were originally used to prevent an excessively rich mixture on deceleration, which can cause stalling. They now work as emission control devices on late-model cars, reducing HC emissions on deceleration.

Choke Heaters

Choking the engine can only be done for a brief time to keep emissions low. This can be done in several ways, all of which apply heat to the thermostatic spring to warm it up quickly and cause choke release.

Hot water choke

One way to apply heat to the thermostatic spring is by routing part of the engine coolant through the choke housing, figure 7-58. Once the engine reaches normal operating temperature, coolant heat helps release the choke. Older Ford models used an external coolant bypass

Basic Carburetion 101

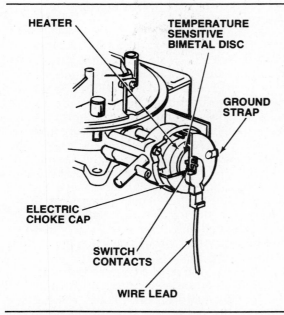

Figure 7-59. Ford and AMC electric heater for the automatic choke.

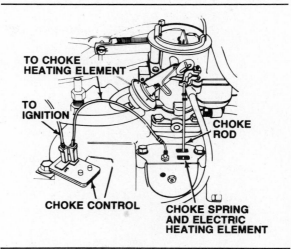

Figure 7-60. Chrysler electric-assist choke.

hose held against the choke housing cap by a spring clip. This supplemented the exhaust manifold air passed through the choke housing.

Hot water chokes reduce over-choking when an engine is restarted. Because water holds heat longer than air, the choke coil will stay warm longer when exposed to heat from hot coolant. This reduces the amount of choking on a restart.

Electric-assist chokes

Electrically assisted heater elements are used on many late-model cars. These speed up the choke opening when underhood temperature is above about 60° F, by heating the choke bimetal thermostatic spring.

AMC and Ford choke heater elements are located in the thermostatic spring housing on the side of the carburetor, figure 7-59. Electric current is constantly supplied from the alternator directly to the choke cover temperature sensing switch. At underhood temperatures above 60° to 65° F, the switch closes to pass current to the choke heater. The circuit is grounded through a strap connected to the carburetor.

The Chrysler electric-assist choke heater is located in the intake manifold choke well and is regulated by a control switch which receives power from the ignition switch, figure 7-60. The control switch energizes the heater at temperatures above 63° to 68° F, and deenergizes it when the switch warms to between 110° and 130° F.

A two-stage heater control, introduced on some cars in 1975, provides three levels of heat, depending on outside temperature. The low heat level is provided by a resistor on the control switch.

GM cars from 1975 on may use an electric choke cover heater. This two-stage heater receives current from the engine oil pressure switch. Below specified air temperatures (usually 50° to 70° F), a bimetal sensor in the cover turns off current to the larger heater stage. Both stages operate at higher temperatures. This choke heater receives current as long as the engine is running.

Fuel Deceleration (Decel) Valve

A decel valve, figure 7-61, is used with some Ford four-cylinder and V-6 engines to momentarily provide extra fuel and air during deceleration. This prevents cylinders from misfiring during deceleration and sending an unburned charge of hydrocarbons through the exhaust. The extra air and fuel provided by the valve ensure total combustion. The valve is located on or near the intake manifold. One end of the valve is connected to the manifold, either directly or by a hose. The other end connects to a deceleration section inside the carburetor. This contains a fuel pickup tube and an air bleed.

The decel valve operates only during deceleration, when manifold vacuum is high enough to operate the spring-loaded diaphragm and open the valve, figure 7-62. Extra fuel and air pass through the valve as long as vacuum remains above 20 inches of mercury, which is about five seconds. When vacuum drops, spring tension reseats the valve. A bleed hole in the valve allows constant atmospheric pressure to be applied to the diaphragm.

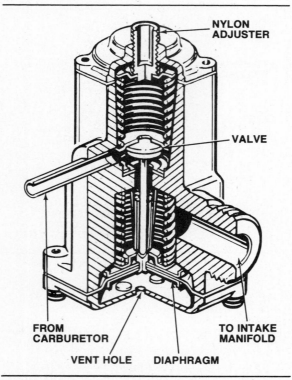

Figure 7-61. Fuel deceleration (decel) valve.

Some decel valves are adjustable. A plastic screw in the valve top can be adjusted to control the valve's opening time and duration. A round valve design, figure 7-63, was first used on 1974 2,300-cc engines. For that model year, it was not adjustable, but it was modified to permit adjustment on 1975 and later engines.

Decel valves on 1975 and later V-6 engines use a speed lockout to shut off valve operation at speeds below 11 mph. The lockout system consists of a vacuum solenoid, an electronic control module, and a speed sensor in the speedometer cable. Decel valves on 1977 Mustangs with V-6 engines are interconnected with the air conditioner, operating only when the air conditioning is on.

Deceleration Throttle Opener

This is a solenoid which holds the throttle open during deceleration to prevent too much fuel from being pulled through the idle system by high manifold vacuum. It is used on catalytic-converter-equipped cars to prevent converter overheating. GM cars with the combination emission control (CEC) system use a throttle-positioning solenoid that is energized on deceleration. Some Chrysler Corporation cars with converters use a solenoid to keep the throttle from closing completely at engine speeds above 2,000 rpm.

ALTITUDE-COMPENSATING CARBURETORS

Earlier in this chapter, you learned that atmospheric pressure is greatest at sea level. Suppose that you drive a car from a low elevation into an area where the elevation is 7,000 feet. As the altitude increases, atmospheric pressure decreases and less air enters the carburetor. This means that the air-fuel mixture passing into the engine becomes richer as altitude increases. The result is poor driveability and high CO emissions at the higher elevation. For the driver who is only passing through high elevation areas, the poor driveability is mostly a temporary inconvenience. But for driving at that elevation for an extended time, the car will run better if the engine is tuned for high elevation by leaning the air-fuel mixture.

However, if the retuned engine is driven back to lower elevations without once again adjusting the mixture, performance will suffer. As the car descends from the higher elevation to sea level, the air-fuel mixture receives increasing amounts of air, leaning the mixture too much. If the engine is to operate properly at the lower elevation, it will have to be retuned to restore the proper air-fuel mixture. Unfortunately, it is not always possible or even desirable to tune the engine for such driving conditions. As a result, performance suffers, driveability is impaired, and emissions are high under such circumstances.

To maintain an appropriate air-fuel mixture while the car is driven in an altitude other than that for which it is adjusted, GM and Chrysler introduced the altitude-compensating carburetor on some 1975 models. In the 1977 model year, the U.S. Federal EPA designated 167 counties in 10 western states (not including California) as high-altitude emission control areas. These counties are entirely above 4,000 feet in eleacation. These high-altitude emission control requirements were suspended in 1978 but will be reintroduced in a modified form in 1980.

Because the major problem at high altitude is CO emission due to richer mixtures, the heart of the special emission controls is the carburetor. High altitude carburetors have auxiliary systems or devices to provide more air or less fuel when operating at higher elevation than when operating at sea level. Many of the altitude-compensating systems are automatic, responding to changes in atmospheric pressure. Other devices require manual adjustment or operation.

Basic Carburetion

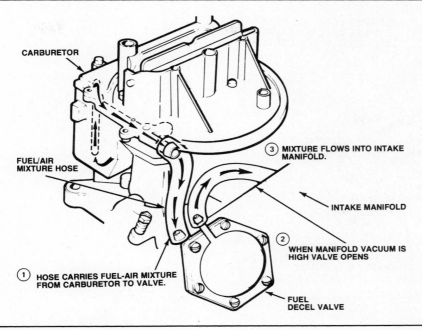

Figure 7-62. Decel valve installation. (Ford)

Automatic Compensation

The most widely used altitude-compensating device is the aneroid bellows. An aneroid bellows is an accordion-shaped bellows that responds to changes in atmospheric pressure by expanding and contracting. As pressure decreases at high altitude, the bellows expands.

Aneroid bellows are used on the following carburetors to open auxiliary air tubes into the main metering circuit at high altitude:
- Motorcraft 2150, 2-barrels, used on some AMC and Ford V-8's.
- Motorcraft 4350, 4-barrels, used on other Ford V-8's.
- Some Carter Thermo-Quads, used on Chrysler V-8's.

The air bypass passage on the Motorcraft carburetors has its own air intake and choke valve, figure 7-64. The aneroid bellows on these carburetors have adjusting screws and locknuts that appear to be for adjusting the tension on the bellows. However, these screws are for original factory adjustments only, and should not be changed while overhauling the carburetor. The bellows and air valve can be removed for carburetor cleaning without upsetting the adjustment.

The altitude compensator on the Carter Thermo-Quad, figure 7-65, is also automatic and requires no service. It was introduced on some 1975 cars, and its use has increased since then.

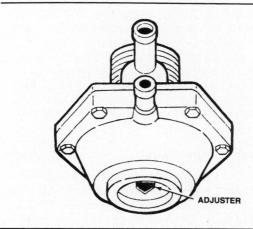

Figure 7-63. In 1974, the decel valve was redesigned to look like this. Operation remains unchanged. (Ford)

■ A Tube Tip

Sometimes, when you are working on a carburetor still in the engine, you'll bust your knuckles trying to reach the idle mixture screws. Most mechanics know that one way to get at those screws is to use a length of rubber tubing. Slip it tightly over the idle screws to make the adjustments. Similar tubing, but a bit larger on the inside diameter, can be used to loosen or tighten hard-to-get-at spark plugs.

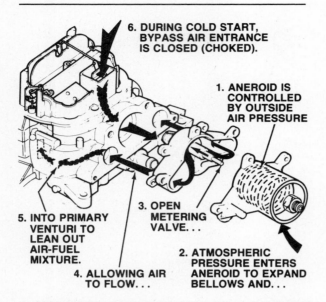

Figure 7-64. Motorcraft 2150 and 4350 automatic altitude compensator.

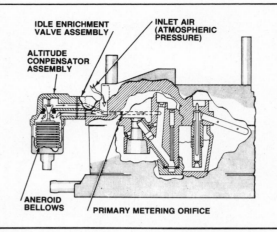

Figure 7-65. Carter Thermo-Quad automatic altitude compensator. (Chrysler)

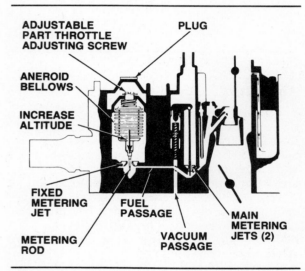

Figure 7-66. Rochester Quadrajet automatic altitude compensator.

The aneroid bellows is also part of the Rochester M4MEA Quadrajet used on 1976 and later high-altitude Cadillacs. On this carburetor, the aneroid bellows controls the position of the metering rods in the primary main jets, figure 7-66. At high altitude, the bellows expands and moves the rods into the jets to reduce fuel flow and keep the mixture from getting overly rich. This unit needs adjustment only if the unit is replaced.

Manual Compensation

American Motors' high-altitude 6-cylinder engines use a Carter YF 1-barrel carburetor with a manually adjustable auxiliary air bleed. The adjustment plug is located on the side of the airhorn, near the fuel inlet, figure 7-67.

For operation above 4,000 feet, use a screwdriver to turn the plug fully *counterclockwise*. This opens an extra air passage for the idle and main metering circuits. Below 4,000 feet, the plug must be turned fully *clockwise* to close the air passage. The ignition timing must be readjusted whenever this carburetor compensator is reset.

Carter BBD 2-barrel carburetors used on 318-cid Dodge and Plymouth high-altitude truck engines in 1977 also have manually adjustable air bleeds. The adjustment screw for this air bleed is inside the airhorn, above the venturi clusters, figure 7-68. For high-altitude operation turn the screw counterclockwise from the lightly seated position. For operation below 4,000 feet, lightly seat the screw to close the air bleed.

Holley 5200 2-stage, 2-barrel carburetors have a driver-controlled fuel valve to compensate for changes in altitude, figure 7-69. These carburetors are used on Ford's 2.3-liter engine in Pintos, Bobcats, Mustangs, and Capris sold at high altitude in 1977.

The carburetor valve is controlled by a lever mounted below the instrument panel. The lever has two positions. When it is set at SEA LEVEL, the fuel valve in the base of the carburetor is opened to meter extra gasoline from the fuel bowl to the primary and secondary metering wells. When the lever is set at ALTITUDE, the fuel valve is closed to maintin a leaner air-fuel mixture.

The driver's control lever is connected to the carburetor valve by a cable and swivel linkage. Under normal use, the system needs no adjustment.

Basic Carburetion

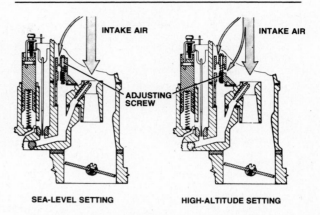

Figure 7-68. Carter BBD altitude compensator. (Chrysler)

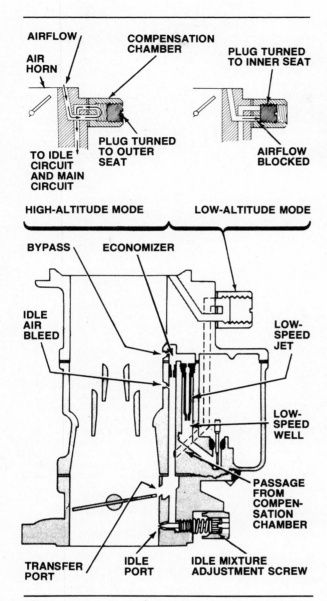

Figure 7-67. Carter YF altitude compensator. (AMC)

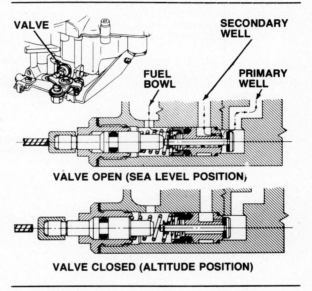

Figure 7-69. Motorcraft 5200 altitude compensator.

SUMMARY

The carburetor is the all-important device that converts air and gasoline into an air-fuel mixture that can be burned in the cylinders. Carburetors operate under all types of conditions and in all temperatures. Yet despite the dozens of designs available, they all operate on the same basic principle of pressure differential: the difference in air pressure between the relatively high pressure of the atmosphere outside the engine and the low-pressure of the intake manifold and the carburetor. The vacuum is created by the downward stroke of the piston, which sucks air in through the manifold and the carburetor.

In spite of different designs, most carburetors have the same seven basic systems of passages, ports, jets, and pumps. Venturis, or restrictions, help speed the airflow and draw more fuel into the airflow. For each barrel, or throat, on a carburetor, there is also a throttle valve. Carburetors can be used with various assist devices to improve driveability, reduce emissions, help in cold-weather starting, or avoid overheating. Other assist devices provide better fuel mixture during the full range of driving conditions, including high-altitude driving.

Review Questions

Choose the single most correct answer. Compare your answers to the correct answers on page 248.

1. Which of the following is *not* true of air pressure:
   a. If is measured in millimeters of mercury (mm Hg)
   b. It is measured in pounds per square inch (psi)
   c. It is always constant
   d. It results from the weight of air pressing on a surface

2. Air pressure:
   a. Increases with height
   b. Decreases with height
   c. Remains the same regardless of height
   d. Increases with warming temperature

3. A pressure differential is created between outside air and by engine air by:
   a. An increase in cylinder volume
   b. A decrease in cylinder volume
   c. Air rushing through the carburetor
   d. The intake manifold

4. The restriction in an airflow tube or barrel is called:
   a. A throttle
   b. An air bleed
   c. A vaporizer
   d. A venturi

5. The air pressure along the sides of a barrel in which there is an airflow:
   a. Is higher than at the center of the flow
   b. Is higher than atmospheric or outside pressure
   c. Is lower than at the center of the flow
   d. Increases with airflow velocity

6. Ported vacuum:
   a. Is the low-pressure area in the carburetor just above the throttle valve
   b. Is the low pressure beneath the throttle valve
   c. Is the air pressure in the venturi
   d. Is equal to atmospheric pressure

7. At idle:
   a. Venturi vacuum is high
   b. Manifold vacuum is high
   c. Ported vacuum is high
   d. All of the above

8. Which is *not* a fuel-metering system:
   a. Float system
   b. Idle system
   c. Power system
   d. EEC system

9. The carburetor float controls fuel level:
   a. By closing the needle valve when the level is low
   b. By opening the needle valve when the level is high
   c. By closing the needle valve when the level is high
   d. All of the above

10. At engine idle:
    a. Venturi vacuum is high
    b. Fuel enters the carburetor barrel above the throttle
    c. Air is provided by an air bleed in the idle tube
    d. The throttle is one-third open

11. Which is *not* used to provide extra air at idle:
    a. The throttle valve
    b. The choke
    c. Transfer ports
    d. Idle air bypass

12. At low off-idle speeds, extra fuel is provided by:
    a. The transfer port
    b. The main nozzle
    c. The idle air bleed
    d. The idle air adjust screw

13. In a high-speed system, better fuel and air mixtures are obtained with:
    a. Transfer ports
    b. A single venturi
    c. Multiple venturis
    d. None of the above

14. Power circuits are operated by:
    a. Vacuum diaphragms
    b. Vacuum pistons
    c. Mechanical metering rods
    d. All of the above

15. Which is *not* part of the accelerator pump circuit:
    a. The metering rod
    b. The inlet check
    c. The outlet check
    d. The duration spring

16. Which carburetor circuit makes the fuel mixture richer when starting an engine:
    a. Power circuit
    b. Choke circuit
    c. High-speed circuit
    d. Accelerator pump circuit

17. The illustration shows:
    a. A power valve diaphragm
    b. An integral choke
    c. A remote choke
    d. None of the above

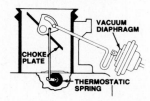

18. The choke unloader:
    a. Releases the fast idle cam after a specified time
    b. Directs warm air to the thermostatic spring
    c. Opens a vacuum bleed in the vacuum break diaphragm
    d. Opens the choke valve when the throttle is open fully

19. The fast-idle cam:
    a. Opens the throttle
    b. Closes the throttle
    c. Opens the choke valve
    d. Closes the choke valve

20. Independently operated throttles are found in:
    a. One-barrel carburetors
    b. Single-stage two-barrel carburetors
    c. Two-stage two-barrel carburetors
    d. All of the above

21. Which is *not* true of four-barrel carburetors:
    a. The primaries act like single-stage two-barrel at low speeds
    b. Airflow through the secondary barrels is through venturi action *or* air velocity valves
    c. The primaries feed four cylinders and the secondaries feed the other four
    d. The secondary barrels open between half and three-quarter throttle

22. Variable-venturi carburetors:
    a. Maintain constant throttle
    b. Maintain a uniform pressure drop across the venturi
    c. Have been used on domestic cars for 45 years
    d. Are formed by two parallel valve plates

23. Main fuel metering in a variable-venturi carburetor is done by:
    a. Metering rods
    b. A dashpot
    c. EGR port vacuum
    d. The idle trim system

24. Which is *not* part of the cold enrichment system of a 2700 VV carburetor:
    a. A fast-idle cam
    b. A solenoid
    c. A control vacuum regulator rod
    d. An auxiliary fuel passage and metering rod

25. The accelerator pump is controlled by:
    a. Vacuum break linkage
    b. Thermostatic spring linkage
    c. Throttle linkage
    d. All of the above

Chapter

8
Intake and Exhaust Manifolds

The intake manifold is a series of enclosed passages that routes the vaporized air-fuel mixture from the carburetor to the intake ports in the engine cylinder head, or heads. Another series of enclosed passages in the exhaust manifold routes the hot gases resulting from combustion to the exhaust system.

In this chapter you will learn:
• What the intake and exhaust manifolds do
• How their designs affect engine performance and emission control
• How add-on devices are used to change manifold operation for better driveability and lower emissions.

INTAKE MANIFOLD PRINCIPLES

The vaporized air-fuel mixture flowing through the carburetor must be evenly distributed to each cylinder. The intake manifold does this with a series of carefully designed passages to connect the carburetor with the engine's intake valve ports, figure 8-1. To do its job, the intake manifold must be designed for efficient vaporization and air-fuel delivery.

As you learned in Chapter 3, liquid gasoline is composed of various hydrocarbon compounds, which vaporize at different temperatures. If all of the hydrocarbons in gasoline vaporized at the same rate, the job of the intake manifold would be simple. But they do not, and so the intake manifold must be heated to keep the air-fuel mixture properly vaporized. This is done with exhaust manifold heat, engine coolant heat, or both, depending on engine type.

V-type engines have the intake manifold in the valley between the cylinder banks, figure 8-2. An exhaust crossover passage inside the manifold carries hot exhaust gases near the base of the carburetor, figure 8-3. In a few manifolds, engine coolant is routed near the carburetor to heat the air-fuel mixture.

Most inline engines have both manifolds on the same side of the engine, with the intake manifold on top of the exhaust manifold, figure 8-4. A chamber between the two fills with exhaust gases and creates a hot spot to improve fuel vaporization in the intake manifold passages. Some inline engines have the two manifolds mounted on opposite sides of the cylinder head, figure 8-5. This is called a crossflow design. Coolant passages and a heat jacket on the intake manifold supply the heat.

Even with the help of heat from the exhaust manifold and engine coolant, the air-fuel mixture usually does not completely vaporize in the intake manifold. This results in an unequal mixture distribution among the cylinders, with some cylinders receiving more fuel and devel-

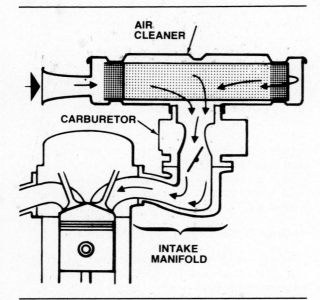

Figure 8-1. The intake manifold has passages that route the air-fuel mixture from the carburetor to the intake valve ports.

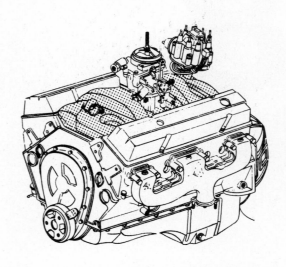

Figure 8-2. V-type engines normally have the intake manifold between the two banks of cylinders, shaded area in the illustration. (Chevrolet)

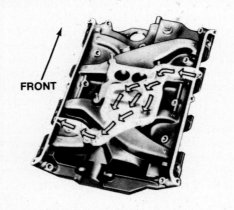

Figure 8-3. Exhaust gases are routed from ports in the cylinder heads through separate passages in the intake manifold to form the manifold hot spot. (Ford)

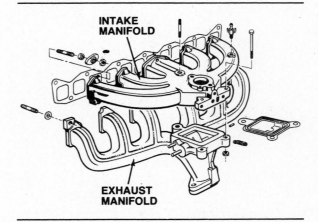

Figure 8-4. Most inline engines have both manifolds on one side of the engine. (Chrysler)

oping more power than others. This problem is greater during engine warmup, when less than normal heat is available to vaporize the fuel.

Causes of Unequal Distribution

The air-fuel mixture reaching the engine cylinders may vary in amount and ratio for several reasons:
• The mixture flow is directed against one side of the manifold by the carburetor throttle valve, figure 8-6. To some extent, the choke plate influences flow in a similar manner.
• Lighter particles of the air-fuel mixture will turn corners in the manifold easily, while heavier particles tend to continue in one direction, figure 8-7.
• Cylinders closer to the carburetor will receive a leaner mixture than will those farther from the carburetor. Carburetor and manifold design can minimize this, though.

Depending upon manifold design, the use of a 2- or 4-barrel carburetor can improve mixture distribution, because each carburetor barrel supplies fewer cylinders.

The efficiency of a manifold is determined by the shape, surface, and size of its passages. They should be as short as possible, and with

Intake and Exhaust Manifolds

109

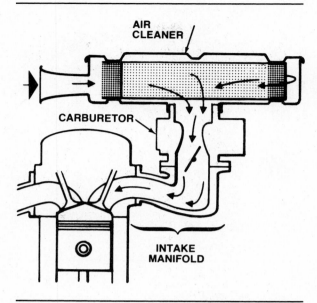

Figure 8-5. A few inline engines have the two manifolds on opposite sides of the engine. (Chrysler)

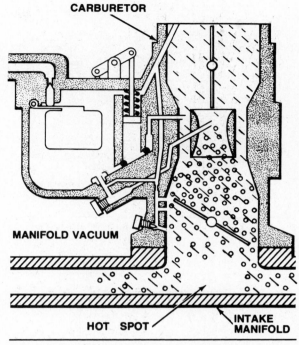

Figure 8-6. The angle of the carburetor throttle plate can affect the flow of the air-fuel mixture above the hotspot.

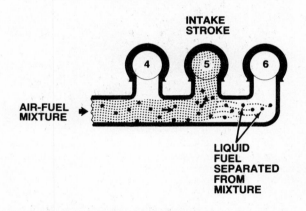

Figure 8-7. An intake manifold with large passages and sharp angles will cause liquid fuel to separate out of the air-fuel mixture. (Chevrolet)

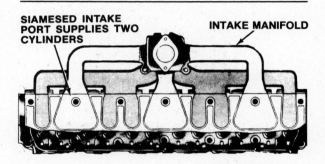

Figure 8-8. By using siamesed ports, this 6-cylinder engine needs only three intake manifold runners. (Pontiac)

no sharp corners, bends, or turns to interfere with mixture flow. Smooth passage surfaces speed mixture flow. Rough surfaces aid vaporization by slowing down and breaking up mixture flow. Passages must be large enough in diameter to supply all cylinders with equal amounts of mixture. Passages that are too large or too small will slow down flow.

BASIC INTAKE MANIFOLD TYPES

Passenger car intake manifolds are made of cast iron or aluminum. The exact design and number of outlets to the engine depends on the engine type, number of cylinders, carburetion, and valve port arrangement. Individual intake and exhaust ports may be used, or two cylinders can be supplied by a single port. This is called a **siamesed** port, figure 8-8. Siamesed intake ports are common on inline engines but rare in V-type engines.

Siamesed: Joined together. A siamesed port on an intake manifold is a single port that supplies the air-fuel mixture to two cylinders.

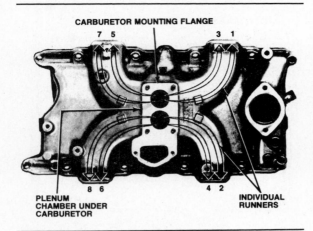

Figure 8-9. This single-plane manifold feeds eight cylinders from the single plenum chamber. (Chrysler)

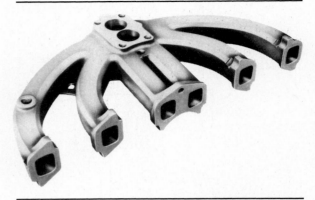

Figure 8-10. This Chrysler intake manifold has curved runners for better mixture distribution. (Chrysler)

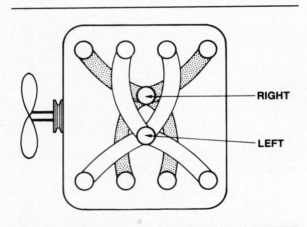

Figure 8-11. The two chambers of a 2-plane manifold are of different heights. (AC Delco)

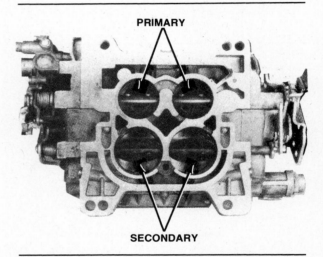

Figure 8-12. A spread-bore carburetor, showing the size difference between the primaries and the secondaries.

Manifold Planes

Production intake manifolds for passenger cars are classified as either single-plane or 2-plane designs. A single-plane manifold, figure 8-9, uses short branches, called **runners**, to connect all of the engine's inlet ports to a single common chamber. This chamber, called the **plenum chamber**, is simply a storage area for the air-fuel mixture. The mixture accumulates within the chamber until it is drawn out by one of the cylinders. Cylinder-to-cylinder mixture distribution is more equal with the single-plane design, but it usually allows less airflow in low-to-intermediate speed ranges. Short runner length causes a drop in mixture velocity. The single-plane manifold design can produce more horsepower at high rpm than can the 2-plane design.

Inline engines use a single-plane manifold. The carburetor usually is placed in the middle of the manifold, leading to the air-fuel flow condition illustrated in figure 8-7. Cylinders closer to the carburetor will receive too much air in the mixture, while those farther away will receive too much fuel. By using siamesed ports, an inline engine needs fewer manifold runners. Chrysler inline 6-cylinder engines do not have siamesed intake ports, but the intake manifolds have long, curved runners with no sharp turns, figure 8-10.

V-8 engine intake manifolds may be either a single-plane or a 2-plane design. A 2-plane intake manifold has two separate plenum chambers connected to the intake ports of the engine, figure 8-11. Each chamber feeds two central and two end cylinders. Mixture velocity is greater at low-to-intermediate engine speeds than with the single-plane manifold, but mixture distribution is usually less even. When a 4-barrel carburetor is used with a 2-plane manifold, each side

Intake and Exhaust Manifolds

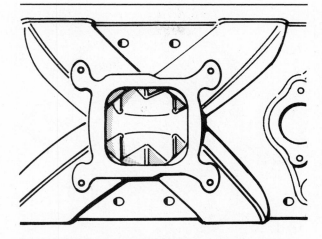

Figure 8-13. The floor of this single-plane manifold is shaped to equalize the mixture flow to all cylinders.

of the carburetor (one primary and one secondary barrel) feeds one plenum chamber. When a 2-barrel carburetor is used, each barrel feeds one chamber.

Manifolds for Spread-Bore Carburetors

Until emission control became an important factor in automotive design, all four throats of a 4-barrel carburetor were the same size. Most late-model 4-barrel carburetors are a spread-bore design, with smaller primaries and larger secondaries, figure 8-12. The way the flow goes through spread-bore designs makes it hard to adapt these carburetors to manifolds not designed for them. It is generally true that 2-plane manifolds work best in the low-to-middle speed ranges, and single-plane designs work better at higher speeds. Engineers have discovered, however, that a single-plane manifold can be designed to use the best features of both types. The design, length, and arrangement of the runners are all important.

A spread-bore carburetor and a 2-plane manifold distribute a fairly even mixture to all eight cylinders. In effect, each group of four cylinders is fed by similar but independent 2-barrel carburetors. When the same carburetor is fitted to a single-plane manifold, mixture flow tends to seek the shortest path. Each carburetor throat ends up feeding the two closest manifold runners. This means that four cylinders are fed by the larger secondaries, and the other four are fed by the smaller primaries. A lean mixture reaches the front four cylinders, and a rich mixture is fed to the rear four cylinders.

To use a spread-bore carburetor on a single-plane manifold and equalize the air-fuel flow, the plenum chamber floor is grooved and ridged, figure 8-13. This speeds up the flow to the front cylinders, closest to the primaries, and slows down the flow to the rear cylinders, closest to the secondaries. This is the most efficient manifold design to use with a spread-bore carburetor.

EXHAUST MANIFOLDS

As the piston moves upward during the exhaust stroke, it forces combustion gas through the open exhaust valve and out of the cylinder, figure 8-14. Whenever gas is pushed through a passageway, turbulence and friction along the sides of the passage cause a resistance, called **backpressure**. A piston encounters backpressure each time it comes up on the exhaust

---

Runners: The short branches of an intake manifold that connect the manifold's plenum chamber to the engine's inlet ports.

Plenum Chamber: The area of an intake manifold that the air-fuel mixture from the carburetor enters before it is distributed to the cylinders.

Backpressure: The resistance, caused by turbulence and friction, that is created as a gas or liquid is forced through a passage.

■ The Torque Box

Designing manifolds for production engines is a series of compromises dictated by engine design, emission requirements, driveability, manufacturing costs, and various other factors. In general, the dual plane manifold provides better throttle response, a broader torque curve, and higher maximum torque than a single plane design. The single plane manifold is responsible for high maximum horsepower (more efficiency) at high rpm, but is not good at idle, and can contribute to icing during cold weather.

An attempt to combine the best features of both designs resulted in what engineers call a "torque box." This design uses the long runners from the dual plane manifold with the plenum chamber of the single plane design. The result is good torque characteristics and higher maximum horsepower.

The best known example of a torque box used on production cars is the dual 4-bbl. manifold fitted to early Z/28 Camaros.

The torque box became one of the most common racing manifold designs. Its use on production engines was limited by stricter emission requirements, the use of smaller and less powerful engines, and the emphasis on fuel economy.

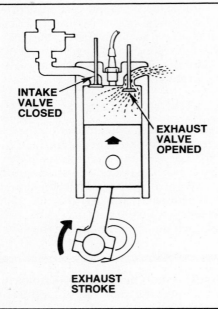

Figure 8-14. Burned gases are pushed out of the combustion chamber by the piston's upward exhaust stroke.

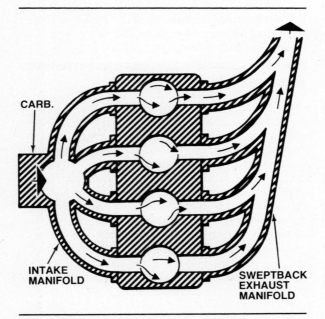

Figure 8-16. Streamlining the exhaust manifold can reduce backpressure.

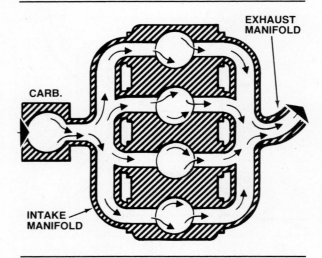

Figure 8-15. Exhaust backpressure is caused, in part, by the many turns of the exhaust manifold passages.

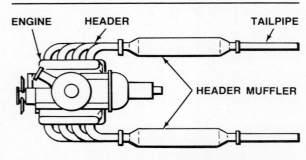

Figure 8-17. Headers used with V-type engines usually have separate muffler systems.

stroke. This resistance causes a loss of power which would otherwise go to the engine flywheel. A direct and unrestricted flow of exhaust gas causes less backpressure and prevents this unnecessary power loss.

Backpressure in the exhaust manifold and exhaust system can contaminate the intake manifold's fresh air-fuel mixture. A brief **camshaft overlap** period combined with backpressure in an exhaust manifold can cause one cylinder's intake stroke to draw exhaust gases from a nearby cylinder.

Dual exhaust systems reduce exhaust backpressure by splitting the exhaust gas flow into two outlet lines. Since the exhaust manifold is at the beginning of the flow, manifold design is the most important factor in reducing backpressure. However, good muffler design can also minimize restrictions.

Sharp turns and narrow passages will slow the flow of gases from the exhaust ports, increasing the amount of backpressure in the system, figure 8-15. Many original equipment manifold designs have duplicated high performance manifolding by using the sweptback manifold design, figure 8-16. The high-performance exhaust manifolds are called **headers**, figure 8-17. They can do more than just reduce backpressure. They can also create a slight suction

Intake and Exhaust Manifolds

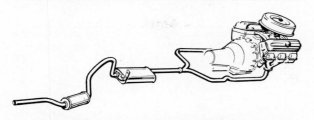

Figure 8-18. This V-type engine uses a Y-pipe to connect its two exhaust manifolds to a single muffler system.

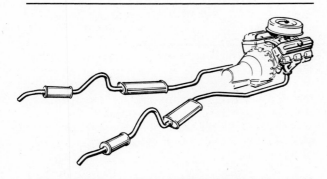

Figure 8-20. V-type engines can have a separate muffler system for each exhaust manifold.

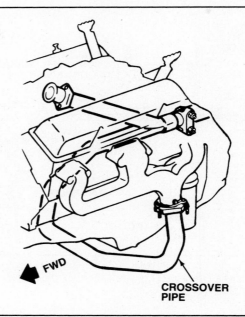

Figure 8-19. A crossover pipe can also connect the two exhaust manifolds to a single muffler system. (Chevrolet)

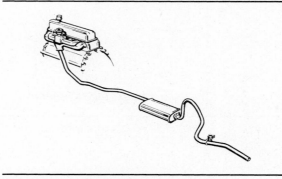

Figure 8-21. Inline engines have a single muffler system attached to the exhaust manifold.

called **scavenging**, which helps pull the exhaust out of the cylinder.

Each cylinder bank of a V-type engine has its own exhaust manifold. When a single exhaust system is used, the two manifolds will connect with a Y-pipe, figure 8-18, or a crossover pipe, figure 8-19. With dual exhaust systems, each manifold is connected to its own exhaust pipe, muffler, and tailpipe, figure 8-20.

Since an inline engine has a single exhaust manifold, a single exhaust system is generally used, with the exhaust pipe connecting directly to the exhaust manifold flange, figure 8-21.

INTAKE MANIFOLD HEAT CONTROL

When the engine is cold, the incoming air-fuel mixture must be heated for complete vaporization. Hot exhaust gases are the most efficient heat source. A heat control valve, or heat riser, in the exhaust system routes a small part of the exhaust through passages in the intake manifold, in order to provide the heat source.

Heat control valves can be operated by:
- A thermostatic spring
- A vacuum diaphragm.

Each of these reacts to engine heat to control the valve operation, as we will see.

Camshaft Overlap: The brief period of time during which a camshaft lobe may be forcing one valve open before another lobe has allowed another valve to close.

Headers: Exhaust manifolds on high-performance engines that reduce backpressure by using larger passages with gentle curves.

Scavenging: A slight suction caused by a vacuum drop through a well designed header system. Scavenging helps pull exhaust gases out of an engine cylinder.

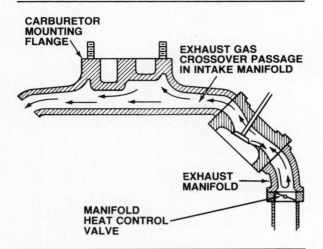

Figure 8-22. A thermostatic heat control valve or heat riser, can be installed between the manifold and the exhaust pipe.

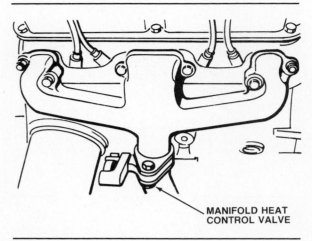

Figure 8-23. The heat control valve can also be within the manifold itself.

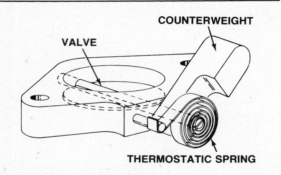

Figure 8-24. The manifold heat control valve can be operated by a thermostatic spring and a counterweight.

Thermostatic Heat Control Valve

Used by automakers for decades, this type of valve is located between the exhaust manifold and the exhaust pipe, figure 8-22, or in the manifold itself, figure 8-23. The valve is held closed by a thermostatic spring when the engine is cold, figure 8-24. This directs the hot exhaust gases around the intake manifold to preheat and help vaporize the air-fuel mixture, figure 8-25, position A. As the engine warms up, the thermostatic spring unwinds. The counterweight and the pressure of exhaust gas open the valve. Exhaust gases now pass directly out through the exhaust system, figure 8-25, position B.

The thermostatic heat control valve has a tendency to stick, because the exhaust gases rust and corrode it. If the valve sticks open, it can cause increased fuel consumption, poor performance during warmup, and too much emissions because the choke stays on. If the valve sticks closed, it can cause poor acceleration, a lack of power, and poor high-speed performance.

Vacuum-Operated Heat Control Valve

GM and Ford introduced vacuum-operated manifold heat control valves, figure 8-26, on some 1975 engines. Chrysler followed suit in 1977 with vacuum-operated valves on its California V-8 engines. AMC continues to use the thermostatic type on all engines. GM calls its device an early fuel evaporation (EFE) valve; Ford's is called a vacuum-operated heat control valve (HCV); Chrysler's is a power heat control valve. All work in the same way to provide more precise control of the manifold heat and to reduce emissions while improving driveability.

A rotating flapper valve is contained in a cast-iron body, figure 8-27. This valve body is installed between the manifold and exhaust pipe. The valve shaft extends through the valve body and is linked to a diaphragm in a vacuum motor. Intake manifold vacuum operates the vacuum diaphragm. The manifold vacuum source is controlled by a switch that reacts to either coolant temperature or oil temperature. Ford, Chrysler, and some GM engines have a thermal vacuum switch (TVS) installed in the cooling system; GM's inline 6-cylinder engines use an oil temperature switch.

When an engine is cold, vacuum is applied to the heat control valve diaphragm, closing the valve. This directs exhaust gas through the intake manifold passage. As the engine reaches normal operating temperature, vacuum is shut off from the valve. In systems using a thermostatic vacuum switch, the TVS closes when cool-

Intake and Exhaust Manifolds

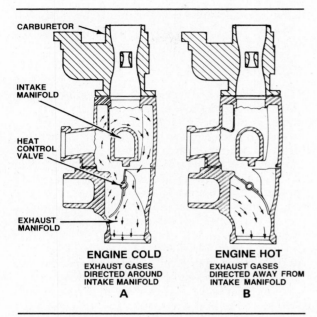

Figure 8-25. In this cross section of an inline engine manifold, the heat control valve forces the exhaust gases to flow either around the intake manifold (Position A), or directly out the exhaust system (Position B).

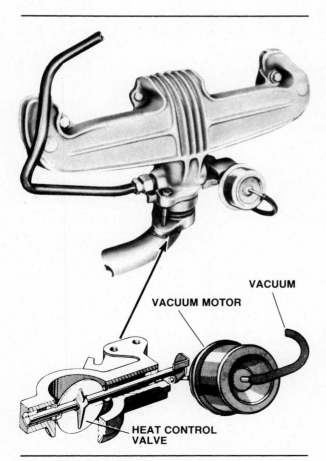

Figure 8-27. A vacuum-operated heat control valve installed in the exhaust manifold. (Chrysler)

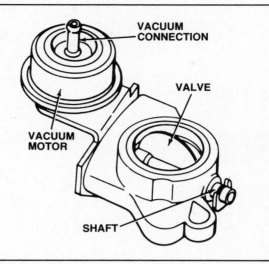

Figure 8-26. A vacuum-operated heat control valve.

■ What's Next In Emission Controls?

Emission control devices and systems now in use have been refined about as much as possible. Further control of emissions will have to be made by new approaches to the problem. One device that looks promising for use in the near future is the exhaust manifold thermal reactor.

Thermal reactors have been under study for several years. In its basic form, a thermal reactor is an oversized exhaust manifold that has sheet metal tubing and a series of port liners. The tubing is centrally located in the cast iron reactor housing, but is insulated from the housing. The port liners are located in the cylinder head's exhaust ports. They are aligned with the reactor inlet ports, but do not touch them.

The thermal reactor is basically an oxidation unit. Air is injected into it to maintain the high (1,600-1,800° F) temperature needed to completely oxidize the HC and CO in exhaust gas. The thermal reactor is a "holding area" to get rid of the unwanted emissions by thermally oxidizing them to form water and carbon dioxide.

The thermal reactor will not replace the catalytic converter, but will work with it to reduce emission levels below those now considered acceptable. So far, the main drawback to the thermal reactor is the higher underhood temperatures it causes. But engineers are studying how to move the battery and other temperature-sensitive parts around so that the heat will not affect them.

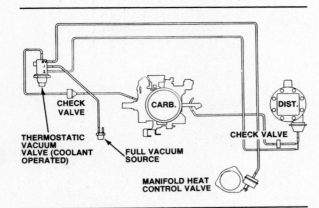

Figure 8-28. A vacuum diagram of the heat control valve installation. This valve responds to engine coolant temperature. (Buick)

ant temperature reaches a specified level, figure 8-28. This blocks vacuum from the valve. Where an oil temperature switch is used, the rising oil temperature opens a normally closed thermostatic switch. This deenergizes the vacuum solenoid, and a spring in the diaphragm opens the valve. Exhaust now flows through the exhaust system instead of bypassing to the intake manifold.

SUMMARY

Good manifold design is critical for smooth performance and low emissions required of today's engines. The intake manifold must evenly distribute the air-fuel mixture to each cylinder, and the exhaust manifold must quickly remove the exhaust from the cylinders so the next incoming charge of air and fuel will not be contaminated. Runner length, the smoothness of the inside surfaces, and arrangement of headers and exhaust pipes are all important for smooth flow of gases.

Intake manifold designs favor good low-to-middle speed operation, good high-speed operation, or a compromise of the two. The type of carburetor used and its placement on the manifold has a relationship to manifold efficiency. To promote better vaporization, a heat control valve is used to route exhaust gases through the intake manifold while a cold engine warms up.

Review Questions

Choose the single most correct answer.
Compare your answers to the correct answers on page 248.

1. The intake manifold is heated by:
   a. Exhaust manifold heat
   b. Engine coolant heat
   c. Both a and b
   d. Neither a nor b

2. Unequal distribution of the air-fuel mixture occurs because of:
   a. Fuel flow from the throttle
   b. Sharp turns in the manifold
   c. Distance of travel to the cylinders
   d. All of the above

3. The most efficient manifold design to use with a spread-bore carburetor is a:
   a. Single-plane
   b. Single-plane with siamesed intake ports
   c. Two-plane
   d. Single-plane with grooved plenum chamber floor

4. Exhaust backpressure:
   a. Causes power loss
   b. Can contaminate air-fuel mixture in the intake manifold
   c. Can be minimized by exhaust manifold streamlining
   d. All of the above

5. Headers:
   a. Are made of aluminum
   b. Cause engines to operate at higher temperatures
   c. Produce a slight suction called scavenging
   d. Increase horsepower by 30%

6. Manifold heat control valves:
   a. Can be located between the carburetor and the intake manifold
   b. Can be located between the exhaust manifold and exhaust pipe
   c. Are open when the engine is cold
   d. None of the above

7. GM's thermostatic vacuum switch reacts to:
   a. Oil temperature
   b. Coolant temperature
   c. Exhaust gases
   d. The vacuum solenoid

Chapter 9
Electronic Fuel Injection

A carburetor is a mechanical device. Like most such devices, it is neither totally accurate nor particularly fast at responding to changing engine needs. The intake manifold is also a mechanical device, and usually is a design compromise between the best performance and the available space in the engine compartment. The result, when these two devices are used together, is an engine that does not run efficiently at low and high speeds.

If the air-fuel mixture can be precisely controlled, and the distribution to the cylinders can be made more efficient, then today's strict emissions standards can be met, while at the same time the engine will run better and consume less fuel. Fuel injection is one of the best ways to get this precise control.

In this chapter, we will discuss:
• The advantages of fuel injection over carburetion
• The differences between various mechanical fuel injection systems
• The fundamentals of electronic fuel injection (EFI)
• The subsystems and components of typical EFI systems.

FUEL INJECTION ADVANTAGES

Fuel injection is not a new development. It has been commonly used with diesel and racing engines for many years. Chevrolet, Pontiac, and Chrysler offered a fuel injection option during the late 1950's and early 1960's. Some imported cars have been equipped with fuel injection since 1968. In 1975, Cadillac made electronic fuel injection standard on the Seville and optional on other Cadillac models.

Fuel injection systems offer several major advantages over carburetors:
• Injection can precisely match fuel delivery to engine requirements under all load and speed conditions. This reduces fuel consumption with no loss of engine performance.
• Because intake air and fuel are mixed at the engine port or in the combustion chamber, keeping an even mixture temperature is not as much of a problem with fuel injection systems. There is no need for manifold heat valves.
• The manifold in an injection system carries only air, so there is no problem of the air and fuel separating.
• By maintaining a precise air-fuel ratio according to engine requirements, exhaust emissions are lowered. The improved air-fuel flow in an injection system also helps to reduce emission levels.

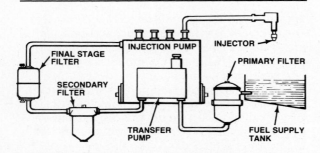

Figure 9-1. A typical diesel fuel injection system. (Fram)

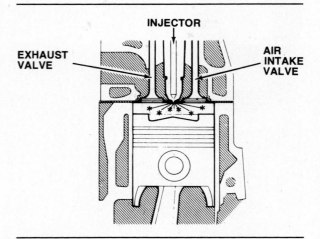

Figure 9-2. Diesel combustion occurs when fuel is injected into the hot, highly compressed air in the cylinder. (Cummins)

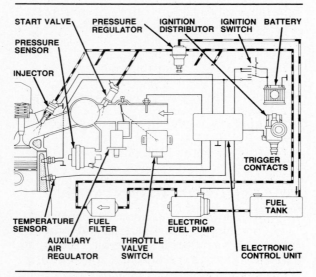

Figure 9-3. A schematic diagram of Bosch's EFI-D systems. (Bosch)

MECHANICAL FUEL INJECTION SYSTEMS

Mechanical fuel injection is used with all diesel engines. Before solid-state electronic research made electronic fuel injection possible, mechanical systems were used with some gasoline engines. There are two general types of mechanical fuel injection systems:
- Combustion chamber injection
- Continuous port injection.

Combustion Chamber Injection

This type of fuel injection is used with diesel engines, figure 9-1. When air is compressed, it gets very hot very fast. With the high compression ratios used by diesel engines, air temperature in the cylinder can range from 1,000° to 1,650° F. This is hot enough to ignite a fine spray of fuel injected into the combustion chamber.

Fuel is drawn from the tank by a transfer pump and then sent to the injection pump. The injection pump boosts system pressure before delivering the fuel through separate lines to the injectors.

During the compression stroke, air temperature increases greatly. At the start of the power stroke, fuel is injected under pressure directly into the hot, highly compressed air in the cylinder, and combustion takes place, figure 9-2. Combustion chamber injection can also be used with gasoline engines, but today's emission and fuel economy requirements make it impractical.

Continuous Port Injection

Most mechanical injection systems for gasoline engines use continuous port injection. A mechanical pump delivers fuel to a manifold near the intake ports. The air and fuel are drawn into each cylinder with each intake stroke. Because the system depends upon engine vacuum, it is difficult to control the fuel flow at idle and low-speed operation.

Two common continuous port mechanical injection systems are the Rochester Ramjet and the Bosch Continuous Injection System (CIS).

ELECTRONIC FUEL INJECTION

Mechanical injection systems are suitable for diesel engines, but not for most gasoline engines. Until recently, the injection systems were not accurate or reliable enough, and were too expensive, to be used widely.

In the early 1950's, the Bendix Corporation began work on electronic fuel injection systems. The first system to enter production was offered as an option on 1958 Chryslers. The system used vacuum-tube electronics, and again

Electronic Fuel Injection

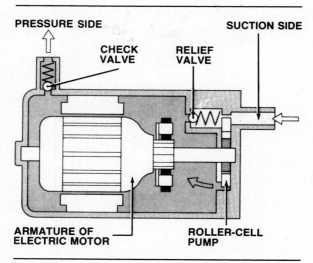

Figure 9-4. The Bosch EFI-D system uses an electric fuel pump, shown in cross-section. (Bosch)

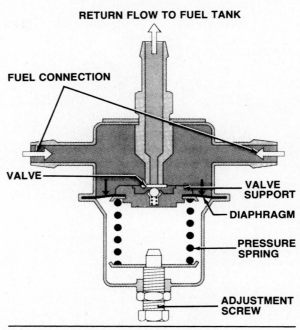

Figure 9-5. This adjustable pressure regulator assures the same fuel pressure at each EFI-D injector. (Bosch)

proved unsuitable for mass production.

With the invention of solid-state semiconductors in the early 1960's, EFI became practical. Bendix holds many patents covering basic EFI technology, and all EFI systems made today are produced under license from Bendix. Three commonly used systems are Bosch's EFI-D and EFI-L, and the Bendix system used by Cadillac.

Each of these uses continuous port injection, and each has the same four basic subsystems: the fuel delivery, the air induction, the sensor, and the electronic control unit subsystems.

Instead of a mechanical metering valve controlling the amount of fuel entering the manifold, the electronic systems have solenoid-operated injectors controlled by computer. Engine sensors feed information on operating conditions to the computer, which processes the information to find the exact amount of fuel needed.

Bosch EFI-D System

This is an intermittent (start-and-stop) low-pressure injection system, figure 9-3. It is used on 1968-73 Volkswagens, as well as some Saab, Mercedes, and Volvo models. It is controlled by intake manifold pressure and engine speed.

Fuel delivery subsystem

An electric fuel pump, figure 9-4, delivers fuel from the tank to the solenoid-operated injection valves under pressure. Fuel pressure at the injectors is maintained by an adjustable pressure regulator, figure 9-5. Excess fuel is returned to the tank at atmospheric pressure. This recirculation results in the delivery of cool fuel, and prevents vapor bubbles in the fuel system. A

■ Smoke Signals Mean A Lot

Even in these days of tough emission standards and a maze of electronics under the hood, the presence of visible smoke in the exhaust can still tell you a lot about the condition of the car's engine.

Experienced mechanics can still eyeball a car and give a pretty good analysis of what ails it, simply by looking at the exhaust. These are generally good indicators for gasoline-fueled engines:

- White smoke — moisture has entered the combustion chambers. Check for a blown head gasket or a cracked head.
- Black smoke — usually caused by an overly rich fuel mixture. Check for a closed choke, a flooded carburetor, or a float level that is too high. Also, if the carburetor power valve is a vacuum operated type, a stuck vacuum piston or a ruptured vacuum diaphragm can cause the power valve to stay open when it shouldn't.
- Blue or blue-gray smoke — oil has entered the combustion chambers or the exhaust system. If the blue smoke is seen during wide-open throttle operation, worn or stuck piston rings may be the cause. If the smoke is seen during deceleration, failed or worn valve stem seals or worn valve guides are most often the culprits.

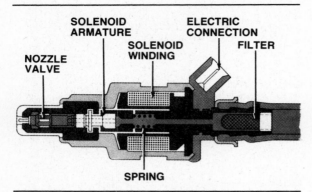

Figure 9-6. A cross-sectional view of the injector used by the Bosch EFI-D system. (Bosch)

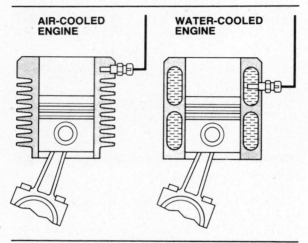

Figure 9-8. Temperature sensors installed in an air-cooled and in a water-cooled engine. (Bosch)

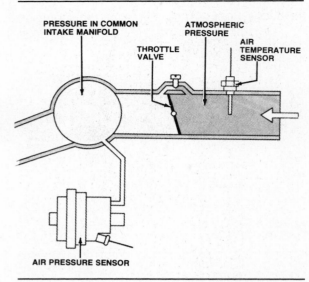

Figure 9-7. In the EFI-D system, the air distributor serves as a common intake manifold for each cylinder. (Bosch)

filter between the pump and the injectors removes any dirt or contamination which might plug the injector nozzle valves.

The injectors are located in the intake manifold and spray fuel in front of each intake valve. Each solenoid-operated injector, figure 9-6, consists of a valve body, a nozzle valve, a solenoid armature, and a filter. Electric signals from the control unit generate a magnetic field in the solenoid winding. The movable solenoid armature is drawn back, lifting the nozzle valve from its seat. Fuel under pressure passes through the nozzle valve channel until the solenoid valve is deenergized by the control unit. The armature spring then pushes the nozzle valve tightly against the valve seat to shut off fuel flow.

Injectors used in the EFI-D system are divided into two groups and each group is connected electrically in parallel and open at the same time. Only one group operates at a time.

Air induction subsystem

Combustion air is supplied to an air distributor through an air filter and throttle valve. This air distributor is a common intake manifold, figure 9-7. Individual equal-length intake manifold branches connect the air distributor with each cylinder. The throttle valve switch signals idle, acceleration, and full-load conditions to the control unit.. A start valve sprays extra fuel into the intake manifold when the engine is started at low temperatures.

A second air regulator meters extra air for mixing with the additional fuel provided by the start valve. As engine temperature increases, the airflow through this second air regulator decreases. When engine coolant reaches normal operating temperature, the regulator is completely closed.

Sensor subsystem

A temperature sensor controls fuel enrichment for cold starting and engine warm-up. When the EFI-D system is used with an air-cooled engine, an engine temperature sensor is installed in the cylinder head, figure 9-8, and an air temperature sensor is installed in the intake manifold, figure 9-7. Water-cooled engines have the engine temperature sensor in the thermostat housing or water jacket to sense coolant temperature, figure 9-8.

The air pressure sensor is located in a sealed housing connected to the air distributor, figure 9-7. As engine load changes the manifold pressure changes, and the pressure sensor sends an electrical signal to the control unit. The length of this signal determines how long the injectors will remain open.

Electronic Fuel Injection

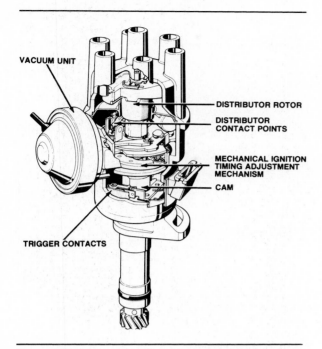

Figure 9-9. In Bosch's EFI-D system, fuel injection is triggered by a special set of contact points in the ignition distributor. (Bosch)

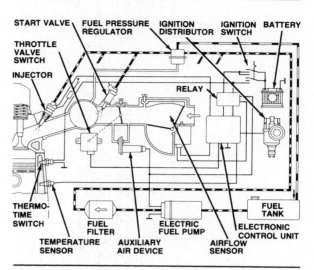

Figure 9-10. Bosch's EFI-L system responds to airflow instead of manifold pressure, as the EFI-D does. (Bosch)

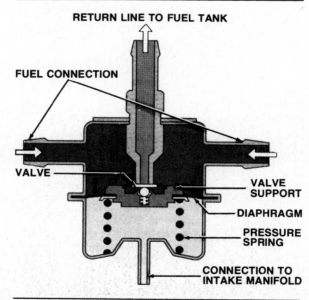

Figure 9-11. The EFI-L fuel pressure regulator is not adjustable. (Bosch)

Exactly when the injectors will open is determined by a pair of trigger contacts located in the ignition distributor, figure 9-9, under the centrifugal advance mechanism. These contacts are operated by a cam on the distributor shaft, and send an electrical signal to the control unit to begin injection.

Electronic control unit

This programmed analog computer receives data about intake manifold pressure, intake air temperature, coolant and cylinder head temperature, movement and position of the throttle valve, the starting cycle, engine speed, and the start of injection. The ECU processes this information to switch the fuel pump on and to control the length of injection.

Bosch EFI-L System

This is also an intermittently (start-and-stop) operating, low-pressure injection system, figure 9-10. EFI-L is a refinement of the EFI-D system and was developed by Bosch to reduce exhaust emissions. It first was used in 1974, when it replaced EFI-D on many foreign cars. With the EFI-L system, injection is controlled by the amount of air drawn into the engine, instead of engine manifold pressure as in EFI-D. This reduces emissions, and results in a simpler system with fewer parts. The EFI-L system is also known as the Jetronic-L or airflow-controlled system.

EFI-L uses the same basic subsystems as EFI-D, with some operating changes described in the following paragraphs.

Fuel delivery subsystem

The fuel pressure regulator, figure 9-11, is not adjustable. Its spring chamber is connected to the intake manifold. This keeps a constant difference between the intake manifold pressure and the fuel pressure. The pressure drop across the injectors is equal under all load conditions.

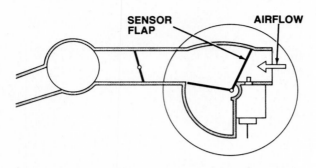

Figure 9-12. This airflow sensor generates an electric signal in proportion to the air drawn into the engine. (Bosch)

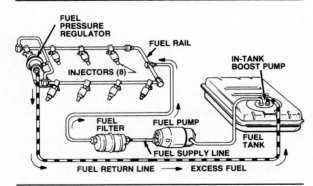

Figure 9-14. The Cadillac EFI fuel delivery subsystem. (Cadillac)

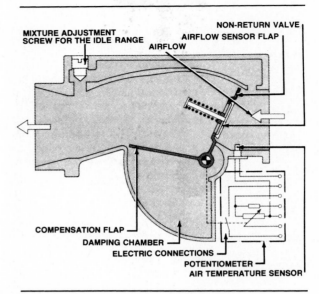

Figure 9-13. A schematic diagram of the airflow sensor used with Bosch's EFI-L system. (Bosch)

EFI-L injectors are about the same as those used with EFI-D but they have a smaller nozzle valve opening. This smaller opening is necessary since the injectors are combined into a single group. All injectors operate at the same time. This means that one-half of the required fuel for an operating cycle is injected twice during each camshaft rotation. With this system, half the cylinders receive fuel for combustion on or near their intake cycle. The other half briefly stores the injected fuel in the intake manifold until their intake stroke occurs.

Air induction subsystem
The major difference in the EFI-L air intake system is the use of an airflow sensor, figure 9-12. This sensor generates an electric current in proportion to the air drawn into the engine. The air movement pushes against a spring-loaded sensor flap. A damping flap attached to the sensor flap reduces backpressure vibrations. The position of the sensor flap is measured by a **potentiometer**, figure 9-13. Since the airflow sensor measures only the fresh air drawn into the system, exhaust gas can be recirculated in the EFI-L system to lower the combustion chamber temperatures.

Sensor subsystem
Since all injectors operate at the same time, the trigger contacts in the ignition distributor are not necessary. Injector opening is triggered instead by every other opening of the distributor breaker points. The breaker points also send signals to the control unit, in order for it to count engine rpm. The temperature sensor in the EFI-L system is a temperature-time switch that controls the length of time the cold-enrichment is activated for starting.

Electronic control unit
The control unit receives battery voltage from a relay when the ignition is switched on. The same relay turns on the electric fuel pump, the start valve, the temperature-time switch, and the auxiliary air device. In the EFI-D system, these relay jobs are done by the electronic control unit. In most other ways, the control units of the EFI-D system and the EFI-L system operate the same. Because the EFI-L system uses **integrated circuits**, fewer parts are required.

Cadillac EFI System

The EFI system made by Bendix and used by Cadillac since 1975 is similar to the Bosch EFI-D system. Intake manifold pressure and engine speed control its operation. The Cadillac EFI version also uses the four major subsystems, but the parts are different from those found in the Bosch system.

Fuel delivery subsystem
This consists of the fuel tank, two electric fuel pumps, a filter, the fuel rails (or tubes) with eight injectors, a fuel pressure regulator, supply

Electronic Fuel Injection

Figure 9-15. On the Cadillac Seville, the fuel filter is mounted near the electric fuel pump under the rear of the car.

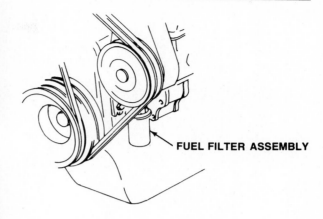

Figure 9-16. On Cadillac models other than the Seville, the fuel filter is mounted in place of the mechanical fuel pump.

lines, and return lines, figure 9-14. The fuel tank contains a special "bathtub" reservoir under the in-tank fuel booster pump. This ensures enough fuel pickup at all times.

The in-tank pump supplies fuel to the chassis-mounted main pump and prevents vapor lock on the suction side of the system. The chassis-mounted pump is a constant displacement type. A check valve in this pump prevents backflow and maintains fuel pressure when the pump is off. An internal relief valve protects against too much pressure. The pump is located under the car, ahead of the rear wheels. Both pumps are controlled by, and wired in parallel with, the electronic control unit (ECU). They are protected by a 10-ampere inline fuse, located in the wiring harness near the ECU.

The fuel filter has a casing which contains a throw-away paper filter element. The filter is mounted near the fuel pump on the Seville, figure 9-15, and in place of the mechanical fuel pump, figure 9-16, on other Cadillac models.

The pressure regulator is mounted on the fuel rail at the front of the engine. It has air and fuel chambers separated by a spring-loaded diaphragm. The air chamber is connected to the throttle body by a hose. Pressure in the regulator is the same as that in the intake manifold, similar to the pressure regulator used in the more recent Bosch EFI-L system.

Manifold pressure changes cause the diaphragm valve in the pressure regulator to open or close the fuel chamber orifice. This keeps a constant 39 psi fuel pressure differential across the injectors. Excess fuel is routed back to the bathtub reservoir in the fuel tank by a return line.

The injectors are solenoid-operated valves controlled by the ECU. Like the EFI-D system, the eight injectors are divided into two groups. All four injectors in a group open at the same time. The two groups operate alternately. Each of the injector groups operates once for every camshaft revolution.

Potentiometer: A variable resistor.

Integrated Circuits: Electronic circuits in which the parts have been miniaturized and assembled together to reduce the number of parts in the circuit.

■ Bosch's Mechanical Injection

The Bosch Continuous Injection System (CIS) is a hydraulically controlled, mechanical injection system. In the CIS system, air enters the engine through the air venturi and is measured by the airflow sensor. The fuel distributor receives fuel from an electric fuel pump. It meters fuel to each injector according to the measured airflow. In this way, the correct amount of fuel is delivered as required.

A cold-start valve injects extra fuel into the intake manifold for cold engine starts. A pressure regulator stabilizes fuel flow at engine operating temperature. A fuel warm-up regulator makes extra fuel available during engine warmup. The additional air required by engine warmup is supplied by the auxiliary air regulator.

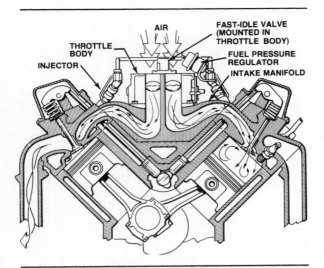

Figure 9-17. Cadillac's EFI intake manifold resembles the 2-plane manifolds used with carbureted engines. (Cadillac)

Figure 9-18. In the Cadillac EFI system, injection timing is controlled by a speed sensor mounted near the ignition distributor shaft. (Delco-Remy)

Air induction subsystem

This has a throttle body, a fast-idle valve, and an intake manifold, figure 9-17. Air enters the throttle body and is distributed to each cylinder through the intake manifold. The throttle body contains two bores with a throttle valve in each one. The valves are connected to the accelerator pedal linkage and are used to control primary airflow rate. A screw on the front of the throttle body adjusts warm engine idle speed.

An electrically operated fast-idle valve on top of the throttle body is connected to the fuel pump circuit of the control unit. It allows extra air to bypass the throttle when a cold engine is started. A heater in the valve warms a thermostatic element, which expands as it is heated. This closes a plunger in the air orifice. The flow of extra air is reduced in this way, and engine rpm drops to the normal idle speed. The warmer the air temperature, the faster the valve will close to reduce engine speed.

The intake manifold looks like the two-plane manifold used with a carbureted engine. Injectors are located in a port above each cylinder. This allows the fuel to be injected directly toward the top of the cylinder intake valve.

Sensor subsystem

Just as in the Bosch systems, sensors monitor various engine operating conditions, and transmit electrical signals to the control unit. The manifold absolute pressure (MAP) sensor is located inside the control unit and is connected to the throttle body by a vacuum line. It monitors manifold pressure changes which result from changes in engine speed, load, and atmospheric pressure. As manifold pressure increases, the sensor signals the control unit to increase the length of injection. A decrease in pressure causes a shorter injection time.

The throttle position switch on the throttle body is connected to the throttle shaft. This switch signals the control unit about the throttle shaft position. Two sensors provide the ECU with air and coolant temperatures. The air temperature sensor is located on the intake manifold. The coolant temperature sensor either is in the intake manifold or below the thermostat. The sensors are identical and interchangeable.

A speed sensor is built into the distributor, figure 9-18. This provides the same information as the cam-operated trigger contacts in the EFI-D system. The speed sensor is a combination external-internal unit. The external part is a plastic housing that contains two reed switches. This is attached to the distributor shaft housing. The internal unit is a rotor with two magnets attached to the distributor shaft. As the shaft rotates, the rotor with its magnets causes the reed switches to open and close, creating signals that time the correct injector group with intake valve timing. It also provides engine rpm information for fuel delivery.

Electronic control unit

The ECU is a programmed analog computer housed in a metal case and located inside the passenger compartment. Control units are adjusted differently according to car model, and

Electronic Fuel Injection

cannot be interchanged. The ECU receives its power from the battery and its signals from the engine sensors. It processes these signals to determine the air-fuel requirements for all driving conditions.

When the ignition is turned on, the ECU turns on the fuel pumps, all the engine sensors, and the fast-idle valve. If the engine is not cranked or started, the pumps and fast-idle valve will shut down after about one second. As the engine is cranked, the ECU receives signals from the sensors and begins to activate the right systems. The EGR solenoid is activated to block the vacuum signal to the EGR valve until coolant temperature reaches a certain level. On some California engines, a vacuum retard solenoid is activated by the same signal used to control the EGR valve operation. This sends vacuum to the retard side of the distributor diaphragm. Once the engine is started, the ECU continues to receive and process sensor signals to control injector duration.

SUMMARY

Fuel injection offers several advantages over carburetion, such as more precise mixture control, better mixture distribution, and lower emissions. Mechanical injection systems have seldom been used on mass-production gasoline engines because of cost and complexity. However, solid-state electronics have made electronic fuel injection practical.

EFI systems are controlled by either manifold pressure or airflow. All contain the same basic subsystems, but may use different components. Engine sensors feed basic information on engine speed, load, and manifold pressure or airflow to an electronic control unit. The ECU determines air-fuel requirements for the engine's operating conditions. It controls the time during which the injectors remain open. The instantaneous response of electronics in an EFI system makes it far more responsive to the engine's needs than a carbureted system can be.

Review Questions

Choose the single most correct answer.
Compare your answers to the correct answers on page 248.

1. Which is *not* true of a fuel injection system:
   a. It precisely meters fuel to meet engine requirements
   b. Fuel consumption is high
   c. Mixture temperature is not a significant problem
   d. Exhaust emissions are lower

2. Diesel engines use:
   a. Rochester Ramjet injection systems
   b. Bosch continuous injection systems
   c. Combustion chamber injection
   d. Intake manifold injection

3. All electronic fuel injection systems are presently under license from:
   a. Chrysler
   b. Bosch
   c. Cadillac
   d. Bendix

4. This illustration is:
   a. A vacuum breaker switch
   b. A throttle valve switch
   c. A pressure sensor
   d. An adjustable pressure regulator

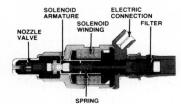

5. On most EFI systems fuel metering is done by:
   a. A metering valve
   b. Solenoid-operated injectors controlled by computer
   c. A metering rod
   d. A pressure sensor

6. In an EFI system:
   a. The fuel injectors are located in each cylinder
   b. The fuel injectors are vacuum operated
   c. Fuel distribution varies from cylinder to cylinder
   d. None of the above

7. The air pressure sensor in an EFI is located in:
   a. The cylinder head
   b. A sealed housing connected to the air distributor
   c. The intake manifold
   d. The water jacket

8. The Bosch EFI-L was introduced in:
   a. 1972
   b. 1973
   c. 1974
   d. 1975

9. In EFI-L fuel systems:
   a. The fuel pressure regulator is adjustable
   b. Pressure drop across the injectors is constant
   c. The injectors operate sequentially (one right after another)
   d. Injection is controlled by manifold pressure

10. In the EFI-L system:
    a. Trigger contacts are not necessary in the distributor
    b. The control unit turns on the temperature-time switch
    c. There is no air sensor
    d. All of the above

11. The pressure regulator in the Cadillac EFI maintains a constant fuel pressure differential across the injectors of:
    a. 79 psi
    b. 59 psi
    c. 39 psi
    d. 19 psi

12. The throttle body in the Cadillac EFI:
    a. Has two primary bores and two secondary bores
    b. Has two bores and two throttle valves
    c. Has a vacuum-operated idle valve
    d. Has four venturis

13. The MAP sensor feeds the ECU signals about:
    a. Throttle shaft position
    b. Valve timing
    c. Manifold pressure
    d. Air and coolant temperature

14. The electronic control unit in the Cadillac EFI:
    a. Is a small analog computer
    b. Is attached to the 12-volt battery
    c. Shuts off when engine reaches normal operating temperature
    d. Is activated by the fuel pump and the fast-idle valve

Chapter 10
Supercharging and Turbochargers

Engines with carburetors rely upon atmospheric pressure to push an air-fuel mixture into the combustion chamber vacuum created by a piston downstroke. The mixture is then compressed before ignition to increase the force of the burning, expanding gases.

The greater the mixture compression, the greater the power resulting from combustion. One way to increase mixture compression is with a high engine compression ratio. In the late 1960's, compression ratios on high-performance auto engines reached 11:1. High compression ratios have two major benefits. First, volumetric efficiency is improved because the piston displaces a larger percentage of the total cylinder volume on each intake stroke. Second, because temperature increases as pressure increases, thermal efficiency is higher with a high-compression engine.

Since 1971, emission control requirements have brought compression ratios down to the range of 8:1 or 8.5:1, because higher-compression engines tend to emit too much NO_x. Lower compression ratios have also been required because of the reduced lead content in gasoline.

SUPERCHARGING

Another way to increase mixture compression is to deliver the air-fuel charge to the cylinders at greater than atmospheric pressure. This is called **supercharging** the engine. The result is much the same as a high compression ratio, but the effect can be controlled during idle and deceleration to avoid high emissions. These control methods will be explained later in the chapter.

Boost

In a supercharger, air is pressurized to greater than atmospheric pressure. This amount of pressurization above atmospheric pressure is called **boost**. Boost can be measured in the same way as atmospheric pressure, which we studied in Chapter 7. Normal sea-level atmospheric pressure is 14.7 psi, or 760 mm Hg. Some superchargers almost double that pressure, providing about 12 psi or 600 mm Hg boost *above* atmospheric pressure. Superchargers used on racing engines can double or even triple the horsepower of a **normally aspirated** engine. On passenger cars, where reliability and driveability are of more concern, the horsepower increases by about 50 percent. For example, the 1978 Buick V-6 engine with a 2-barrel carburetor has a 40 percent increase in horsepower when turbocharged (another form of supercharging). The

Supercharging and Turbochargers

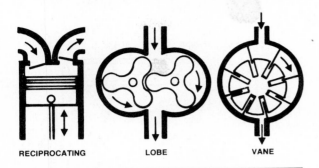

Figure 10-1. These are three common positive-displacement pump designs.

same engine, with a 4-barrel carburetor, gains about 60 percent horsepower when turbocharged.

Superchargers

The term supercharger usually applies to an air pressurizing pump that is mechanically driven by the engine itself. Gears, shafts, chains or belts from the engine crankshaft turn the pump. These mechanical linkages rob a lot of power from the engine. Some supercharger designs must often be driven at speeds of 50,000 to 90,000 rpm. Engine-driven superchargers were never extensively used for passenger car engines.

There are two general types of supercharger pumps:
- Positive displacement
- Centrifugal.

Positive displacement pumps include the reciprocating, the lobe (roots), and the vane designs, figure 10-1. These are called positive-displacement pumps because every revolution pumps the same volume of air. They compress air by taking in a large volume and forcing it into a smaller area. Some of these designs are very inefficient, and the efficient designs are usually too expensive for passenger car use. They more often can be found on diesel truck or racing engines.

The centrifugal pump, figure 10-2, contains an impeller that looks like a many-bladed fan. As the impeller turns at high speed, air pulled into the center of the impeller is speeded up and then thrown outward from the blades by centrifugal force. This fast-moving air enters a diffuser passage. As the volume of the passage increases, the air diffuses, or slows down. The air's deceleration causes an increase in its pressure. The diffuser passage routes the air to the engine intake manifold.

Engine-driven centrifugal pumps have been tried, but the mechanical drive is always a weak link. Also, the mechanical drive can require up

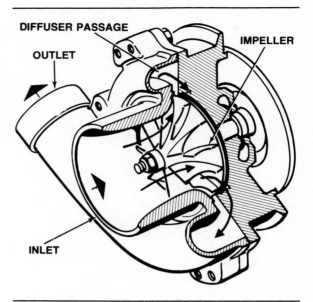

Figure 10-2. The centrifugal pump design accelerates and then slows the air in order to compress it.

to 20 percent of the engine's horsepower output. However, the development of an exhaust-gas turbine-driven supercharger (usually called a turbocharger) made supercharging practical.

TURBOCHARGERS

A **turbocharger** turbine, figure 10-3, looks much like the centrifugal pump. Hot exhaust gases flow from the combustion chamber to the turbine wheel. The gases are heated and expanded as they leave the engine. It is not the speed or force of the exhaust gases that forces the turbine wheel to turn, as is commonly thought, but the expansion of the hot gases against the turbine wheel's blades.

The turbine's main advantage over a mechanically driven supercharger is that the turbine

---

Supercharging: Use of an air pump to deliver an air-fuel mixture to the engine cylinders at a pressure greater than atmospheric pressure.

Boost: A measure of the amount of air pressurization, above atmospheric, that a supercharger can deliver.

Normally Aspirated: An engine that uses normal vacuum to draw in its air-fuel mixture. Not supercharged.

Turbocharger: A supercharging device that uses exhaust gases to turn a turbine that forces extra air into the cylinders.

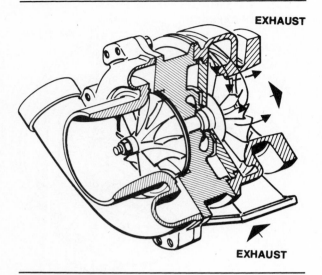

Figure 10-3. A turbine wheel is turned by the expansion of gases against its blades.

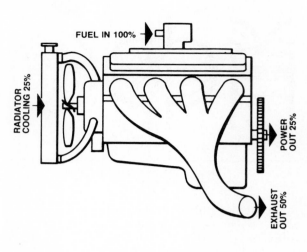

Figure 10-4. A turbine uses some of the heat energy that would normally be wasted.

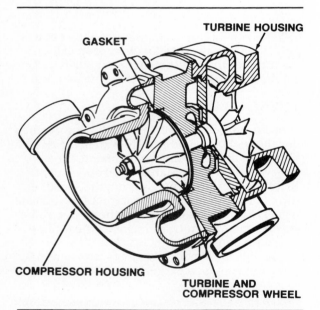

Figure 10-5. A cutaway view of a typical automotive turbocharger.

does not drain power from the engine. In a normally aspirated engine, about half of the heat energy contained in the fuel goes out the exhaust system, figure 10-4. Another 25 percent is lost through radiator cooling. Only about 25 percent is actually converted to mechanical power. A mechanically driven pump uses up some of this mechanical output, but a turbine gets its energy from the exhaust gases, converting more of the fuel's heat energy into mechanical energy.

Turbocharger Design

Most automobile turbochargers use a centrifugal pump because of cost, size, reliability, and efficiency. Figure 10-5 is a cutaway view of a typical automobile turbocharger. The turbine wheel and pump impeller are on the same shaft, so that they turn at the same speed. Because they also turn in the same direction, air flows as shown in figure 10-6.

Turbocharger Installation

A turbocharger can be installed in one of two ways:
- Between the air cleaner and the carburetor, figure 10-7
- Between the carburetor and the engine, figure 10-8.

Each of these has its disadvantages. If compressed air is being forced through the carburetor, as in figure 10-7, the fuel pressure must be great enough so that fuel can enter the airflow. This also means that the carburetor and all fuel lines must be able to hold the increased pressure with no leaks.

If the air and fuel are mixed and then compressed, figure 10-8, the mixture can easily separate. Also, the carburetor must be mounted farther away from the engine, which can cause some problems in cold weather.

Diesel truck engines are often turbocharged.

Supercharging and Turbochargers

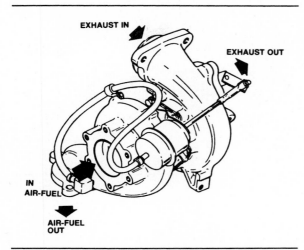

Figure 10-6. Airflow through a turbocharger. (Buick)

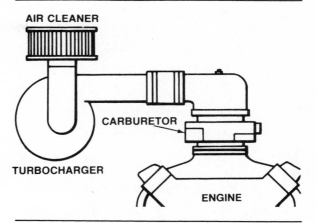

Figure 10-7. The turbocharger can force air into the carburetor.

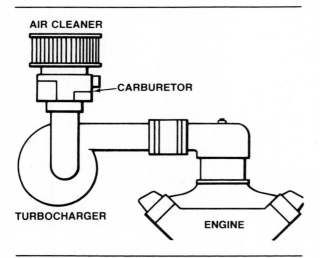

Figure 10-8. The turbocharger can force the air-fuel mixture directly into the intake manifold.

With a diesel fuel injection system, the installation problems just mentioned do not exist. The turbocharger delivers air to the induction system independently of the fuel injection system. The only modern gasoline engine using both a turbocharger and fuel injection is the German-built Porsche Turbo Carrera and the Swedish Saab turbo, which use the Bosch CIS injection mentioned in Chapter 9.

Engine Emissions

Turbochargers have just about the same effect as an increased compression ratio. Why, then, can turbochargers be used on emission-controlled engines when increased compression ratios cannot be used?

The advantage of a turbocharger is the way it controls compression. An engine built with a certain compression ratio will always have that ratio. As we said before, high compression engines tend to create high exhaust emissions during idle, deceleration, and choked operation. Turbocharger boost can be controlled so that the engine runs as a normally aspirated unit during these operating modes. Under acceleration or heavy load conditions, turbocharger boost is applied to the intake manifold. The resulting increase in compression improves the engine's volumetric efficiency because the piston displaces a greater percentage of the total possible cylinder volume.

■ Turbocharging Vs. Emissions

Does turbocharging a conventional automotive engine reduce emission levels from that engine? Turbocharger manufacturers think so, and point to several closely controlled tests to prove their point. These tests show a 14 percent reduction of HC, 13 percent CO and 8 percent NO_x by simply bolting a turbocharger on a standard production engine. Four reasons explain this effect on exhaust emissions.
1. A low-compression engine that runs satisfactorily on low lead, low octane gasoline is ideal for turbocharging.
2. The smaller displacement engine is more efficient, burning less fuel and giving off a smaller amount of combustion byproducts under normal use.
3. Turbocharging causes backpressure on the engine's exhaust system during periods of full speed and power. This increases the temperature and pressure of exhaust gases, leading to a more complete burning in the exhaust manifold.
4. Because the intake manifold and carburetor work under pressure instead of vacuum, the turbocharged engine maintains more precise air-fuel mixtures in each engine cylinder.

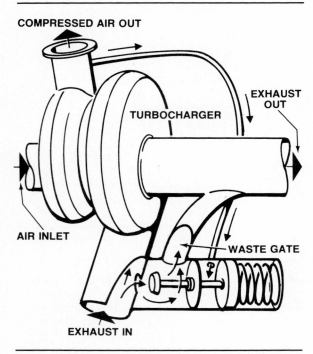

Figure 10-9. The waste gate reacts to intake manifold pressure and controls the amount of exhaust gas reaching the turbine wheel.

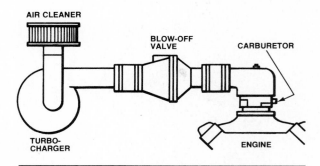

Figure 10-10. A blowoff valve can open to relieve intake manifold pressure.

The few turbocharged cars currently in production use essentially the same emission control devices as do nonturbocharged versions of the same engines.

Engine Detonation

Excessive heating of the air-fuel mixture can cause an explosion in the combustion chamber known as detonation. This wastes power and can damage the engine, as we learned in Chapter 3. Because a turbocharger compresses and therefore heats the air-fuel mixture, detonation is a common problem. Various ways have been devised to control or avoid detonation.

TURBOCHARGER CONTROLS

The action of a turbocharger must be controlled to avoid high exhaust emissions and detonation. The maximum boost pressure of a turbocharger must also be controlled, or the higher compression and power potential could damage the engine. There are three general ways of controlling a turbocharger system:
1. Changing the amount of boost
2. Cooling the compressed mixture
3. Altering the spark timing.

Each of these methods controls several aspects of a turbocharger's operation. The following paragraphs explain how the controls are achieved and what effects they have upon turbocharger operation.

Boost Control

Boost control systems can either limit the amount of exhaust gas reaching the turbine, or vent off some of the compressed air mixture or air-fuel mixture before it reaches the combustion chamber. The mechanisms involved are waste gates and blowoff valves.

Waste gate

A waste gate, figure 10-9, controls the flow of exhaust gas to the turbine. It can allow all of the exhaust into the turbine, or it can route the exhaust directly to the muffler system.

Waste gates are normally controlled by a diaphragm linkage. The diaphragm is exposed to intake manifold pressure. When this pressure reaches a certain level, the diaphragm moves far enough so that the waste gate routes exhaust gases directly to the muffler and bypasses the turbine. Because waste gates are constantly exposed to corrosive exhaust gases, they must be made of high-quality alloys.

The waste gate action controls maximum boost pressure by limiting the speed of the exhaust turbine. It also controls detonation, to some extent, because the chances of detonation increase as the boost pressure increases.

Blowoff valve

A blowoff valve, or pressure control valve, affects the flow of compressed air between the turbocharger and the engine, figure 10-10. The valve can be operated in different ways depending upon what it is meant to control.

If the valve is to control the maximum boost pressure, then it is a simple spring-loaded unit. When boost pressure reaches a certain level, the spring tension is overcome, and some of the boost pressure is allowed to escape. If the turbocharger is compressing only air, as shown in figure 10-10, then the air can be vented to the atmosphere. If the turbocharger is compressing the combustible air-fuel mixture, then the mixture must be vented back into the turbocharger inlet.

Supercharging and Turbochargers

Figure 10-11. The Buick turbocharged V-6. (Buick)

If the blowoff valve is meant to control exhaust emissions, then it is operated by intake manifold vacuum. During idle, closed-throttle deceleration, and choked operation, vacuum in the intake manifold causes a diaphragm to move and open the valve. The engine operates as a normally-aspirated unit, avoiding the excessive emissions that the turbocharger compression would cause. Some blowoff valves are operated by vacuum routed to one side of the diaphragm and boost pressure routed to the other.

Mixture Cooling

Two methods are used to cool the compressed mixture before it reaches the combustion chamber: intercoolers and water injection systems. The main aim of these systems is to control detonation.

Intercoolers

Between the turbocharger and the engine, the compressed mixture flows through a radiator core which can contain water, air, or special chemicals. Although intercoolers require extra space and special plumbing connections, they are very efficient. Intercooling is often used on racing engines and on turbocharged diesel trucks.

Water injection

Water injection systems cool the compressed mixture by spraying a mist of water, or water and special chemicals, into the intake manifold. The injection can be powered by an electric pump or by boost pressure.

Spark Timing

Retarding the spark timing can control detonation by lowering the peak temperature of combustion. Spark control systems can be mechanically or electronically operated.

Mechanical operation

A special diaphragm at the ignition distributor is exposed to intake manifold pressure. As the boost increases, the diaphragm moves to retard the ignition timing.

Electronic operation

A sensor in the engine detects detonation and signals an electronic control module. The module's solid-state circuitry then adjusts ignition timing.

SPECIFIC TURBOCHARGERS

As we have said, few manufacturers have produced supercharged engines for everyday use. In the 1950's, some Ford, Packard, Studebaker, and Kaiser models were supercharged. In the early 1960's, some Oldsmobile F-85 Jetfires and Corvair Monzas and Corsas were turbocharged. The only turbocharged units in production today are the Porsche Turbo Carrera, the Saab turbo, and some Buick V-6 engines. Because the Buick is based on a conventional domestic engine and is produced in greater numbers than the Porsche or Saab, we will concentrate on examining the Buick turbocharger system.

The turbocharged Buick V-6, introduced in 1978, is based on a normally aspirated production engine, figure 10-11. A few internal changes were made, such as strengthening the pistons and using a different piston ring design. The turbocharged and nonturbocharged versions share the same compression ratio and emission control devices.

The Buick turbocharger is mounted between the carburetor and the engine. The carburetor air cleaner is remotely mounted in front of the engine to keep the engine as vertically short as possible. The turbocharger is used with either a 2-barrel or a 4-barrel carburetor. The 2-barrel version, with the M2ME-210 Dualjet carburetor, adds 45 horsepower to the 105 horsepower of the standard V-6. The 4-barrel version, with the M4ME Quadrajet, adds an additional 60 horsepower.

The 2-barrel turbo operates with a maximum boost of 7 to 8 psi. The 4-barrel version provides a boost of 8 to 9 psi. Maximum boost is controlled by a manifold-pressure-operated wastegate. An actuator with a pressure-operated,

Figure 10-12. The cutaway turbocharger shows how air flows in the system. (Buick)

Figure 10-13. Some of the special parts used with Buick's turbocharger system. (Buick)

■ Switching The Emphasis On Turbos

Most drivers think of turbochargers as exotic trappings for souped-up street cars and racers. That's natural when you consider that automotive engineers normally bring a turbo into play high up on the engine's performance curve to deliver maximum power at top rpm. The result is additional peak power that has won many races, and further convinced the general driving public that turbine-operated superchargers are not for them.

The reappearance of turbocharging as a factory option on some 1978 V-6 engines approaches the subject from the opposite viewpoint. Engineers wanted 6-cylinder fuel economy with V-8 performance. The new turbo is used for low- and medium-speed passing, and acceleration from about 1,200 rpm up. As engine speed climbs, the waste gate begins to open, preventing an overload of the engine. This means that the turbocharger will be used for only about five percent of the time the engine is operating.

spring-loaded diaphragm is tapped into the compressor outlet. Through mechanical linkage, the diaphragm opens and closes the exhaust wastegate in the turbine housing. When the wastegate is open, part of the exhaust bypasses the turbine to lower the turbocharger speed and limit boost pressure.

Exhaust gases are routed from the top front of the right-hand exhaust manifold to the turbine. The left-hand exhaust manifold is connected to the right-hand manifold by a crossover pipe, figure 10-12. Exhaust leaves the turbine through a single, large-diameter exhaust pipe that runs to the catalytic converter and muffler.

The turbocharger is mounted directly on top of a special intake manifold. Buick says this location reduces "turbo lag" (the delay between a rapid throttle opening and the delivery of increased boost). Lubrication is provided to the turbo by an external oil line that runs from the front of the engine to the center of the turbo housing. Figure 10-13 shows some of the external engine parts used with the Buick turbocharger installation.

The carburetor is mounted on a special plenum chamber casting that attaches to the compressor inlet. By placing the carburetor on the inlet side of the compressor, the need for a

Supercharging and Turbochargers

Figure 10-14. The solid-state turbo control center helps prevent detonation by adjusting the spark timing. (Buick)

pressurized carburetor and fuel delivery system was eliminated.

The air cleaner is forward of the left cylinder bank, near the left side of the car's grille. It is attached to the carburetor by flexible tubing and a metal shroud. Heated air is provided to the thermostatic air cleaner by a heat stove and tube on the left exhaust manifold. The air cleaner damper is controlled by a conventional vacuum motor.

Buick is using an electronic spark timing control system, which can sense detonation when it starts to occur and retard the ignition timing approximately 20 degrees to control these explosions. The heart of the system is the solid-state turbo control module, or electronic spark control (ESC) controller, which is mounted on the top of the engine fan shroud, figure 10-14.

Battery voltage is applied to the controller through a wire from the ignition switch. Operation is controlled by a relay on the ground side of the controller. The relay is operated by voltage from the switch side of the starter solenoid.

The ESC controller receives signals from a detonation sensor screwed into the intake manifold. The sensor is an accelerometer that responds to the explosions caused by detonation.

The ESC controller is also connected, in parallel with the distributor pickup coil connections, to the ignition control module. The controller processes the signals from the detonation sensor and compares them to the timing signals from the pickup coil. It then sends a timing delay signal back to the ignition module when detonation occurs. Timing returns to its normal setting for speed and load conditions within 20 seconds after detonation stops.

SUMMARY

Like fuel injection, supercharging and turbochargers are not new. The concept of increasing mixture compression by providing the engine with an air-fuel mixture at greater than atmospheric pressure has been used on diesel and racing engines for years, and occasionally on passenger car engines.

Superchargers and turbochargers differ in the way they operate. The supercharger is an air pump mechanically driven by the engine. Turbochargers are driven by exhaust gas. They do not consume engine power for their operation.

As the automotive industry moves into the 1980's, the emphasis is on smaller cars with more fuel-efficient engines. Used as a boost device, turbocharging helps improve engine performance without affecting fuel economy or emission levels. This is possible because the engine will run normally under idle, deceleration, and choked conditions. Turbocharger boost is applied only during acceleration or heavy load conditions. The use of electronic spark timing control prevents detonation, a common problem with turbochargers.

Review Questions
Choose the single most correct answer.
Compare your answers to the correct answers on page 248.

1. Supercharging delivers the air-fuel mixture to the cylinder at:
   a. Lower than atmospheric pressure
   b. Atmospheric pressure
   c. Higher than atmospheric pressure
   d. Three times the atmospheric pressure

2. Which is *not* true of superchargers:
   a. They can operate at very low rpm
   b. They are mechanically driven
   c. There are two types
   d. Mechanical superchargers consume a lot of engine power

3. Positive displacement pumps:
   a. Contain impellers
   b. Pump the same volume of air each revolution
   c. Increase air pressure by decelerating it
   d. Are highly efficient

4. A turbine wheel turns because of:
   a. Centrifugal force
   b. Exhaust gas speed
   c. Manifold vacuum
   d. Expansion of hot exhaust gases

5. Of the heat energy contained in gasoline:
   a. 50 percent is converted to mechanical power
   b. 50 percent is lost to cooling
   c. 25 percent is converted to engine power
   d. 50 percent goes out the exhaust system

6. Despite high compression pressures achieved, turbochargers can be used on emission-controlled engines because:
   a. They have fixed compression ratios
   b. Turbocharger boost can be varied to meet engine needs
   c. They are placed between the air cleaner and the carburetor
   d. They are placed between the carburetor and the engine

7. Which is *not* a method of controlling a turbocharger system:
   a. Changing the amount of boost
   b. Cooling the compressed mixture
   c. Readjusting the carburetor idle
   d. Altering spark timing

8. The waste gate controls boost by:
   a. Controlling flow of exhaust gas
   b. Controlling compressed air
   c. Controlling the air-fuel mixture
   d. Controlling exhaust emissions

9. Retarding spark time will control detonation by:
   a. Cooling the compressed mixture
   b. Cooling the exhaust manifold
   c. Reducing compression pressure
   d. Lowering peak temperature of combustion

10. The Buick turbocharged V-6 was introduced in:
    a. 1950
    b. 1960
    c. 1977
    d. 1978

11. The Buick V-6:
    a. Has its turbocharger mounted between the air cleaner and carburetor
    b. Has a waste gate control
    c. Uses mechanical spark timing control
    d. Uses a blowoff valve

PART THREE

Specific Carburetors

Chapter Eleven
Autolite-Motorcraft Carburetors

Chapter Twelve
Carter Carburetors

Chapter Thirteen
Holley Carburetors

Chapter Fourteen
Rochester Carburetors

11
Autolite-Motorcraft Carburetors

Autolite-Motorcraft carburetors are used on Ford Motor Company and American Motors vehicles. When using the *Shop Manual* to overhaul specific carburetors, you may find the exploded drawings and carburetor descriptions in this chapter useful. The following carburetors are included:
- Model 1100 One-Barrel
- Model 1250 One-Barrel
- Models 2100 and 2100-D Two-Barrel (AMC Model 6200)
- Model 2150 Two-Barrel
- Model 4100 Four-Barrel
- Models 4300, 4300-A and 4300-D Four-Barrel
- Model 4350 Four-Barrel
- Model 2700 VV Two-Barrel
- Model 740 Two-Barrel

AUTOLITE 1100 ONE-BARREL

The Autolite 1100, figure 11-1, has five fuel metering circuits: idle, main, accelerator pump, power enrichment, and choke. The upper body houses the idle, main, and power enrichment circuits. The float chamber vent and the fuel inlet system are also in the upper body. The lower body contains the fuel bowl and accelerator pump circuit. A hydraulic dashpot may also be included in the lower body. Cold starting is helped by a manifold-heated piston-type choke attached to the lower body.

AUTOLITE 1250 ONE-BARREL

The Model 1250, figure 11-2, is a single-barrel carburetor used only on the Pinto 1600-cc engine. It has five fuel metering circuits: idle, main, accelerator pump, choke, and power enrichment. The upper body contains the fuel inlet and bowl vent, the thermostatic choke, and the main metering system. The throttle body contains the fuel bowl and the idle and accelerator pump circuits. Cold starting is aided by a water-heated thermostatic choke attached to the upper body.

AUTOLITE 2100 TWO-BARREL

Early versions were designated Model 6200 when used by American Motors. The Autolite 2100 or 6200, figure 11-3, has five fuel metering circuits: idle, main, accelerator pump, choke, and power enrichment. The airhorn is the main body cover and contains the choke valve and the internal and external fuel vents. All fuel metering circuits are in the main body, to which the thermostatic choke and dashpot are attached. Model 2100-D is vented to a charcoal canister and has a choke diaphragm assembly in

Autolite-Motorcraft Carburetors

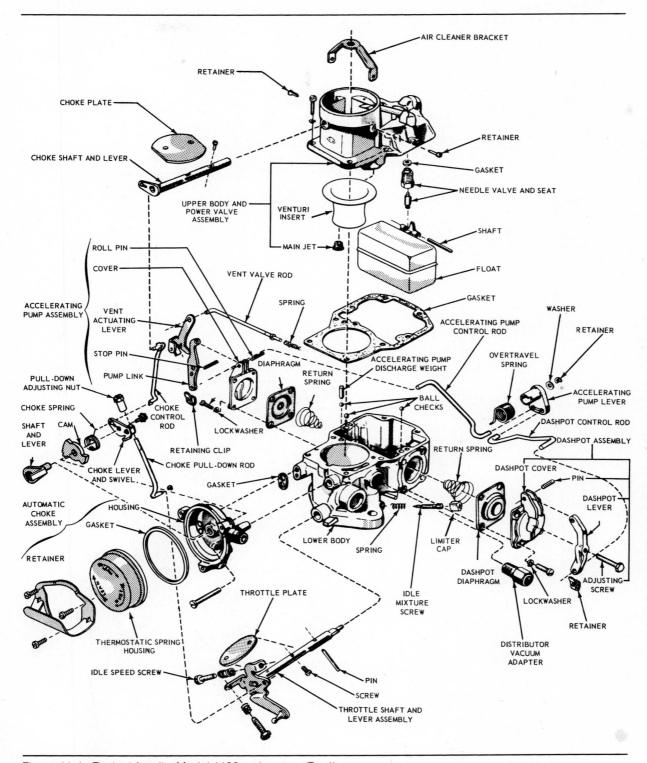

Figure 11-1. Typical Autolite Model 1100 carburetor. (Ford)

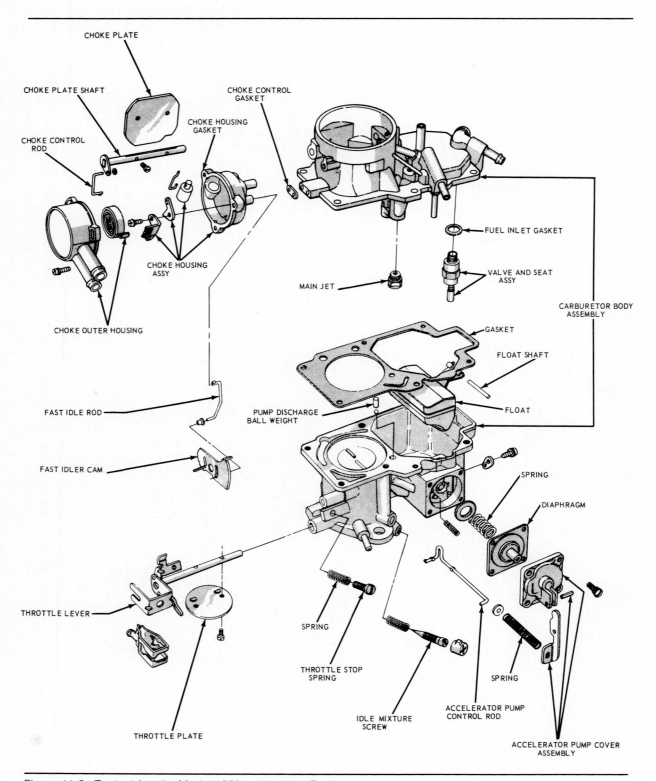

Figure 11-2. Typical Autolite Model 1250 carburetor. (Ford)

Autolite-Motorcraft Carburetors

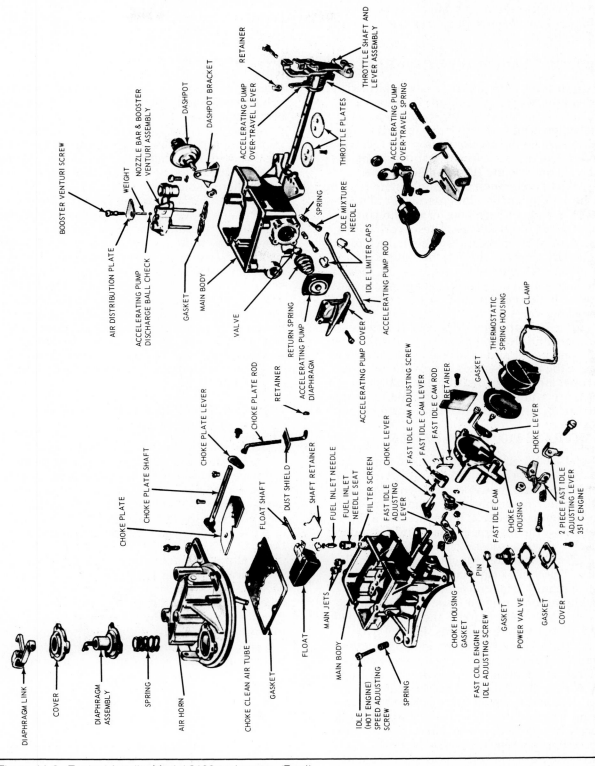

Figure 11-3. Typical Autolite Model 2100 carburetor. (Ford)

the airhorn. Some Model 2100 carburetors use a staged choke pulldown system, explained in Chapter 7, while later versions have an electric-assist choke.

MOTORCRAFT 2150 TWO-BARREL

Introduced in 1975, this more sophisticated version of the Model 2100, figure 11-4, uses a pullover enrichment system to provide extra fuel when airflow through the airhorn is high. Fuel is drawn through orifices in the fuel bowl and passes into the airhorn through air bleeds. All versions use the electric-assist choke.

A fuel deceleration metering system is in the airhorn of the carburetors used on Ford's 2.8-liter V-6 engine. Each booster venturi contains high-speed bleed orifices. These work with a mechanical bleed control system, which is controlled by a cam on the throttle shaft. This system uses reverse-tapered metering rods in the high-speed bleed orifices to provide more precise high-speed operation and better low-speed response.

Some 1976 and later Model 2150 carburetors have an altitude compensator. Intake air enters a bypass intake and is metered into the airflow above the throttle valves to provide the leaner air-fuel mixture needed at higher altitudes. The airflow is controlled by a valve operated by an aneroid bellows at the rear of the carburetor.

AUTOLITE 4100 FOUR-BARREL

Model 4100, figure 11-5, uses six fuel metering circuits: idle, primary and secondary main, accelerator pump, power enrichment, and choke. The airhorn assembly is the main body cover and contains the choke plate, the fuel inlet, a hot-idle compensator, primary and secondary fuel bowl vents, secondary throttle vacuum tubes, and an automatic choke clean air pickup tube. The choke heat chamber in the right exhaust manifold is connected to the clean air pickup tube by a rubber hose and steel tube.

The main body contains the twin fuel bowls and all fuel metering circuits. Each of two primary (front) barrels has a main and a booster venturi, main fuel discharge, idle fuel discharge, accelerator pump discharge, and a primary throttle valve. The two secondary (rear) barrels contain a main fuel discharge and vacuum-operated throttle valves. The primary and secondary barrels are equal in size. A manifold-heated piston-type thermostatic choke, attached to the main body, assists in cold starts.

AUTOLITE 4300 FOUR-BARREL

Nearly a complete revision of the original 4100, the Model 4300, figure 11-6, uses a 3-piece body rather than the 2-piece body of the 4100. Other major differences include a single float bowl with a pontoon float, internal balance and mechanical atmospheric vents, secondary throttle valves mechanically linked to the primaries, and secondary barrels with larger throats than the primaries. The manifold-heated, piston-type thermostatic choke housing is a built-in part of the throttle body. Many Model 4300 carburetors have an electric-assist heater built into the choke thermostatic cap.

Model 4300-A carburetors have auxiliary air valve plates above the secondary main venturis. An integral hydraulic dashpot is used to dampen any sudden movements of the air valve plates, preventing flutter and erratic engine operation. On 4300-D carburetors, figure 11-7, the secondary air valve plates are located above the secondary bore. On some 4300-D carburetors, the choke is heated by both manifold air and engine coolant flowing through a housing held against the choke cap by a spring clip. Other 4300-D carburetors have electric-assist chokes. A staged choke release is operated by a vacuum diaphragm.

MOTORCRAFT 4350 FOUR-BARREL

The 4350, figure 11-8, was introduced in 1975 and differs from the 4300 chiefly in its use of vacuum piston rods in the primary fuel metering system. These provide more precise control over enrichment. A vacuum-operated delayed choke pulldown diaphragm controls the top-step pulloff of the fast-idle cam. Some versions also use a mechanical vent valve for the fuel bowl.

An altitude compensator was added to the 1976 and later Model 4350 carburetors. Like the Model 2150, intake air entering the bypass intake is metered into the airflow above the throttle valves to provide the leaner air-fuel mixture required at higher altitudes. The airflow is controlled by a valve operated by an aneroid bellows in the rear of the carburetor housing.

Autolite-Motorcraft Carburetors

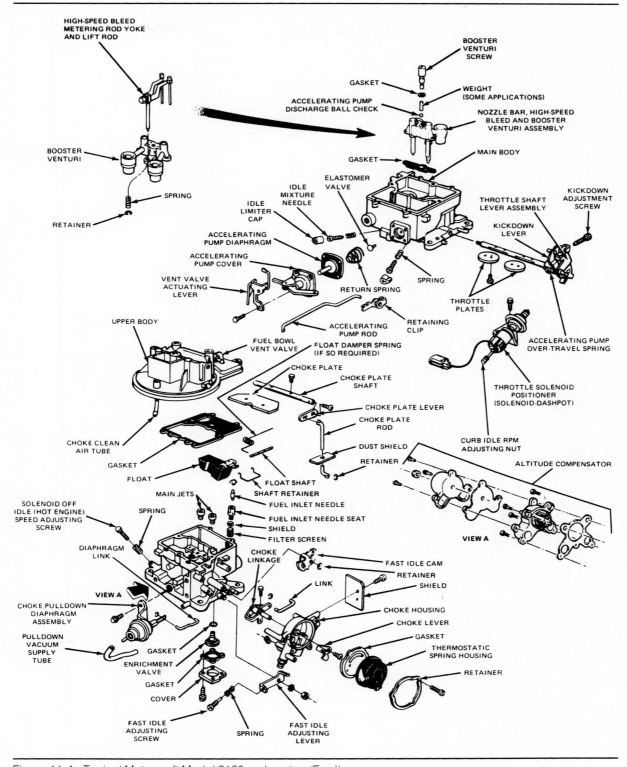

Figure 11-4. Typical Motorcraft Model 2150 carburetor. (Ford)

Chapter Eleven

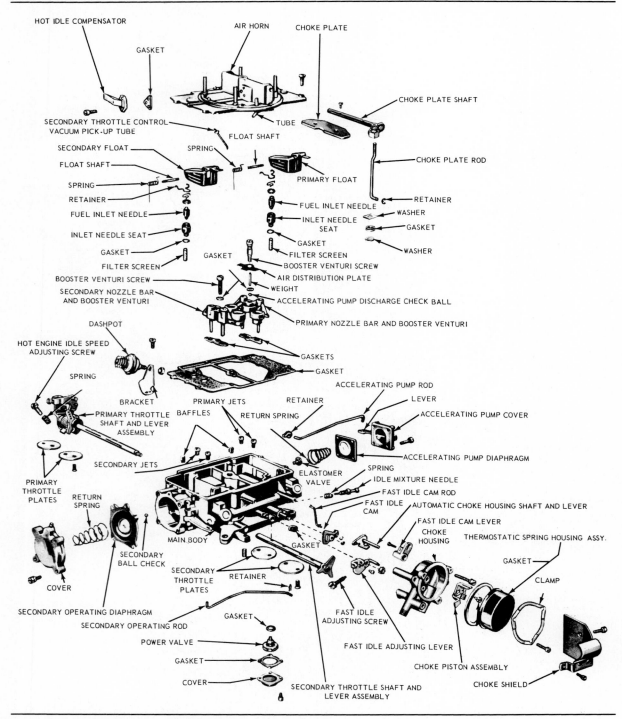

Figure 11-5. Typical Autolite Model 4100 carburetor. (Ford)

Autolite-Motorcraft Carburetors

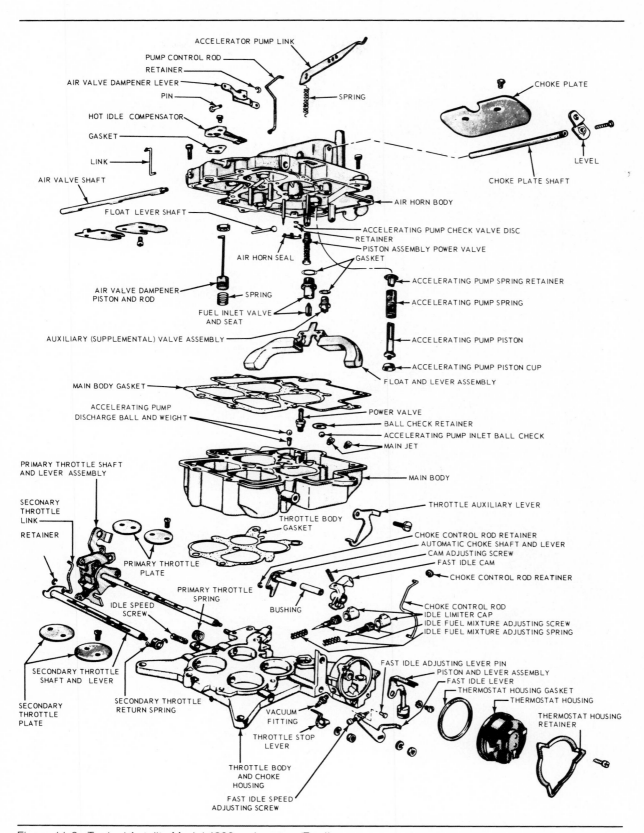

Figure 11-6. Typical Autolite Model 4300 carburetor. (Ford)

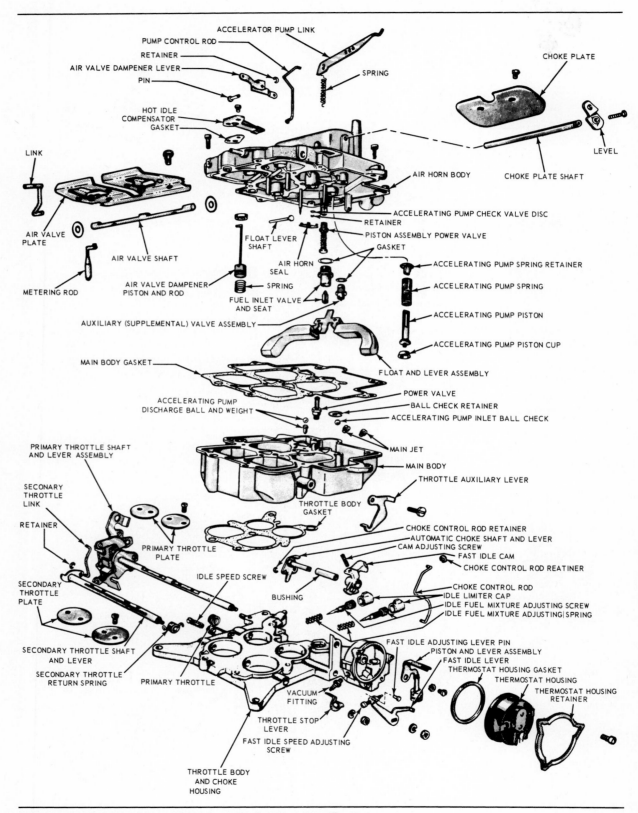

Figure 11-7. Typical Motorcraft Model 4300-D carburetor. (Ford)

Autolite-Motorcraft Carburetors

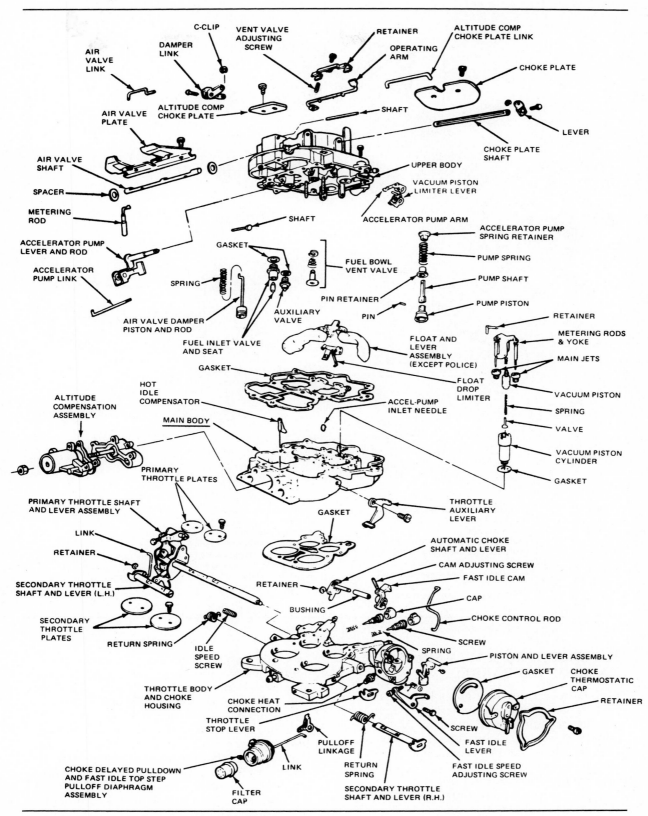

Figure 11-8. Typical Motorcraft Model 4350 carburetor. (Ford)

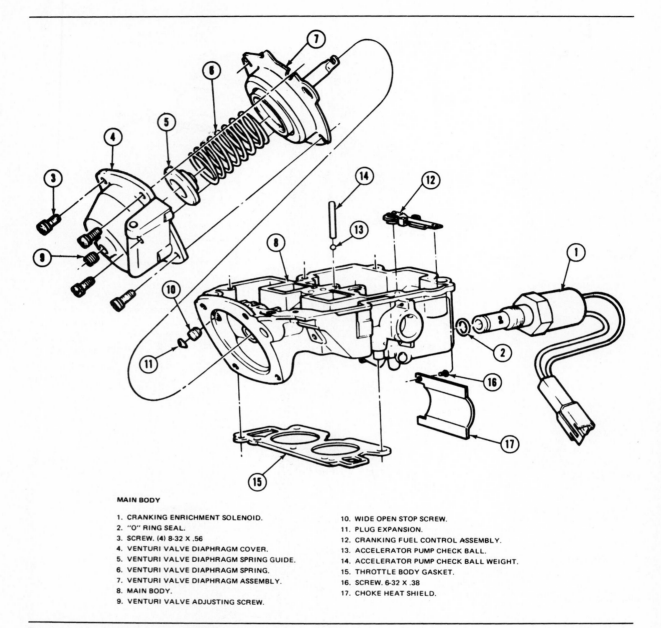

MAIN BODY

1. CRANKING ENRICHMENT SOLENOID.
2. "O" RING SEAL.
3. SCREW. (4) 8-32 X .56
4. VENTURI VALVE DIAPHRAGM COVER.
5. VENTURI VALVE DIAPHRAGM SPRING GUIDE.
6. VENTURI VALVE DIAPHRAGM SPRING.
7. VENTURI VALVE DIAPHRAGM ASSEMBLY.
8. MAIN BODY.
9. VENTURI VALVE ADJUSTING SCREW.
10. WIDE OPEN STOP SCREW.
11. PLUG EXPANSION.
12. CRANKING FUEL CONTROL ASSEMBLY.
13. ACCELERATOR PUMP CHECK BALL.
14. ACCELERATOR PUMP CHECK BALL WEIGHT.
15. THROTTLE BODY GASKET.
16. SCREW. 6-32 X .38
17. CHOKE HEAT SHIELD.

Figure 11-9. Motorcraft Model 2700 carburetor, main body. (Ford)

MOTORCRAFT 2700 VV VARIABLE-VENTURI TWO-BARREL

This 2-barrel carburetor, figures 11-9, 11-10, and 11-11, varies its venturi area according to the demands of engine speed and load. All other Autolite-Motorcraft carburetors have fixed venturis. The Model 2700 VV uses a movable venturi valve, which moves in and out of the airflow through the carburetor throats.

Engine vacuum and throttle position control the venturi valve, which is connected to two main metering rods to control fuel flow. Air speed through the carburetor remains just about constant, instead of varying as in fixed-venturi designs. Additional circuits such as the idle, power enrichment, and choke systems are not required.

Other features include a system which trims the fuel flow at idle, an accelerator pump system similar to that used on the Model 4350 carburetor, cranking and cold enrichment systems, a control vacuum regulator, and an external vent. More information on this carburetor is in Chapter 7.

Autolite-Motorcraft Carburetors

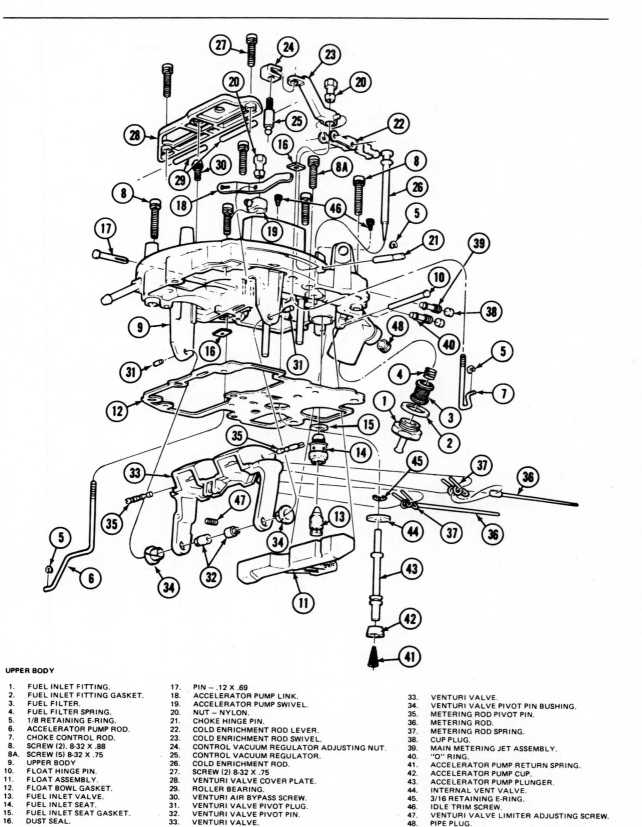

UPPER BODY

1. FUEL INLET FITTING.
2. FUEL INLET FITTING GASKET.
3. FUEL FILTER.
4. FUEL FILTER SPRING.
5. 1/8 RETAINING E-RING.
6. ACCELERATOR PUMP ROD.
7. CHOKE CONTROL ROD.
8. SCREW (2). 8-32 X .88
8A. SCREW (5) 8-32 X .75
9. UPPER BODY
10. FLOAT HINGE PIN.
11. FLOAT ASSEMBLY.
12. FLOAT BOWL GASKET.
13. FUEL INLET VALVE.
14. FUEL INLET SEAT.
15. FUEL INLET SEAT GASKET.
16. DUST SEAL.
17. PIN — .12 X .69
18. ACCELERATOR PUMP LINK.
19. ACCELERATOR PUMP SWIVEL.
20. NUT — NYLON.
21. CHOKE HINGE PIN.
22. COLD ENRICHMENT ROD LEVER.
23. COLD ENRICHMENT ROD SWIVEL.
24. CONTROL VACUUM REGULATOR ADJUSTING NUT.
25. CONTROL VACUUM REGULATOR.
26. COLD ENRICHMENT ROD.
27. SCREW (2) 8-32 X .75
28. VENTURI VALVE COVER PLATE.
29. ROLLER BEARING.
30. VENTURI AIR BYPASS SCREW.
31. VENTURI VALVE PIVOT PLUG.
32. VENTURI VALVE PIVOT PIN.
33. VENTURI VALVE.
33. VENTURI VALVE.
34. VENTURI VALVE PIVOT PIN BUSHING.
35. METERING ROD PIVOT PIN.
36. METERING ROD.
37. METERING ROD SPRING.
38. CUP PLUG.
39. MAIN METERING JET ASSEMBLY.
40. "O" RING.
41. ACCELERATOR PUMP RETURN SPRING.
42. ACCELERATOR PUMP CUP.
43. ACCELERATOR PUMP PLUNGER.
44. INTERNAL VENT VALVE.
45. 3/16 RETAINING E-RING.
46. IDLE TRIM SCREW.
47. VENTURI VALVE LIMITER ADJUSTING SCREW.
48. PIPE PLUG.

Figure 11-10. Motorcraft Model 2700 carburetor, upper body. (Ford)

148   Chapter Eleven

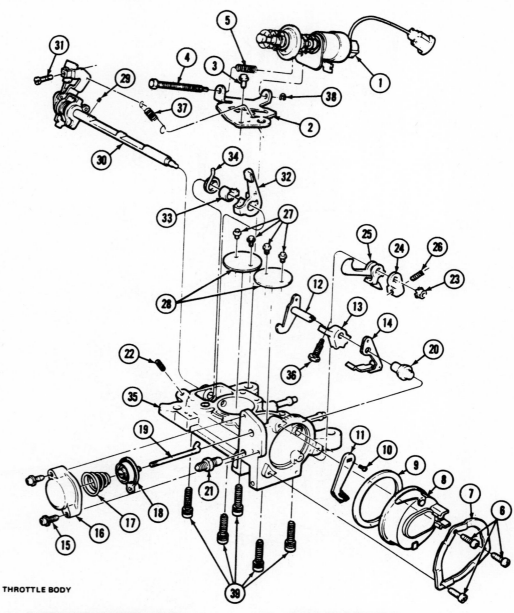

THROTTLE BODY

1. THROTTLE RETURN CONTROL DEVICE.
2. THROTTLE RETURN CONTROL DEVICE BRACKET.
3. MOUNTING SCREW. 10-32 X .50
4. ADJUSTING SCREW (TSP ON)
5. ADJUSTING SCREW SPRING.
6. SCREW (3). 8-32 X .50
7. CHOKE THERMOSTATIC HOUSING RETAINER.
8. CHOKE THERMOSTATIC HOUSING.
9. CHOKE THERMOSTATIC HOUSING GASKET.
10. SCREW. 6-32 X .50
11. CHOKE THERMOSTATIC LEVER.
12. CHOKE LEVER AND SHAFT ASSEMBLY.
13. FAST IDLE CAM.
14. HIGH CAM SPEED POSITIONER ASSEMBLY
15. SCREW (2). 8-32 X .75
16. HIGH CAM SPEED POSITIONER DIAPHRAGM COVER.
17. HIGH CAM SPEED POSITIONER DIAPHRAGM SPRING.
18. HIGH CAM SPEED POSITIONER DIAPHRAGM ASSEMBLY.
19. HIGH CAM SPEED POSITIONER ROD.
20. CHOKE HOUSING BUSHING.
21. CHOKE HEAT TUBE FITTING.
22. CURB IDLE ADJUSTING SCREW (TSP OFF).
23. RETAINING NUT. 10-32
24. FAST IDLE ADJUSTING LEVER.
25. FAST IDLE LEVER.
26. FAST IDLE ADJUSTING SCREW.
27. THROTTLE PLATE SCREWS (4).
28. THROTTLE PLATES.
29. VENTURI VALVE LIMITER STOP PIN.
30. THROTTLE SHAFT ASSEMBLY.
31. TRANSMISSION KICKDOWN ADJUSTING SCREW.
32. VENTURI VALVE LIMITER LEVER.
33. VENTURI VALVE LIMITER BUSHING.
34. VENTURI VALVE LIMITER SPRING.
35. THROTTLE BODY.
36. FAST IDLE CAM ADJUSTING SCREW.
37. TRANSMISSION KICKDOWN LEVER RETURN SPRING.
38. 3/16 RETAINING E-RING.
39. SCREW (5) 8-32 X .75

Figure 11-11. Motorcraft Model 2700 VV carburetor, throttle body. (Ford)

Autolite-Motorcraft Carburetors

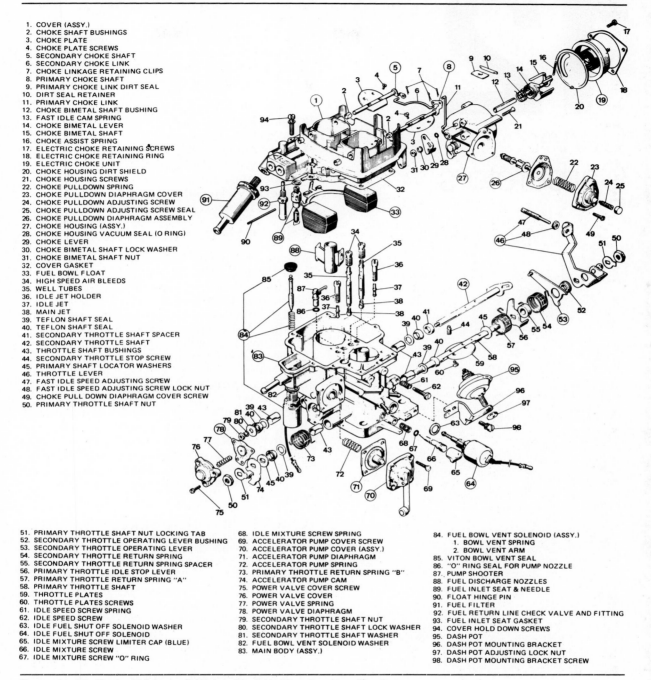

1. COVER (ASSY.)
2. CHOKE SHAFT BUSHINGS
3. CHOKE PLATE
4. CHOKE PLATE SCREWS
5. SECONDARY CHOKE SHAFT
6. SECONDARY CHOKE LINK
7. CHOKE LINKAGE RETAINING CLIPS
8. PRIMARY CHOKE SHAFT
9. PRIMARY CHOKE LINK DIRT SEAL
10. DIRT SEAL RETAINER
11. PRIMARY CHOKE LINK
12. CHOKE BIMETAL SHAFT BUSHING
13. FAST IDLE CAM SPRING
14. CHOKE BIMETAL LEVER
15. CHOKE BIMETAL SHAFT
16. CHOKE ASSIST SPRING
17. ELECTRIC CHOKE RETAINING SCREWS
18. ELECTRIC CHOKE RETAINING RING
19. ELECTRIC CHOKE UNIT
20. CHOKE HOUSING DIRT SHIELD
21. CHOKE HOUSING SCREWS
22. CHOKE PULLDOWN SPRING
23. CHOKE PULLDOWN DIAPHRAGM COVER
24. CHOKE PULLDOWN ADJUSTING SCREW
25. CHOKE PULLDOWN ADJUSTING SCREW SEAL
26. CHOKE PULLDOWN DIAPHRAGM ASSEMBLY
27. CHOKE HOUSING (ASSY.)
28. CHOKE HOUSING VACUUM SEAL (O RING)
29. CHOKE LEVER
30. CHOKE BIMETAL SHAFT LOCK WASHER
31. CHOKE BIMETAL SHAFT NUT
32. COVER GASKET
33. FUEL BOWL FLOAT
34. HIGH SPEED AIR BLEEDS
35. WELL TUBES
36. IDLE JET HOLDER
37. IDLE JET
38. MAIN JET
39. TEFLON SHAFT SEAL
40. TEFLON SHAFT SEAL
41. SECONDARY THROTTLE SHAFT SPACER
42. SECONDARY THROTTLE SHAFT
43. THROTTLE SHAFT BUSHINGS
44. SECONDARY THROTTLE STOP SCREW
45. PRIMARY SHAFT LOCATOR WASHERS
46. THROTTLE LEVER
47. FAST IDLE SPEED ADJUSTING SCREW
48. FAST IDLE SPEED ADJUSTING SCREW LOCK NUT
49. CHOKE PULL DOWN DIAPHRAGM COVER SCREW
50. PRIMARY THROTTLE SHAFT NUT
51. PRIMARY THROTTLE SHAFT NUT LOCKING TAB
52. SECONDARY THROTTLE OPERATING LEVER BUSHING
53. SECONDARY THROTTLE OPERATING LEVER
54. SECONDARY THROTTLE RETURN SPRING
55. SECONDARY THROTTLE RETURN SPRING SPACER
56. PRIMARY THROTTLE IDLE STOP LEVER
57. PRIMARY THROTTLE RETURN SPRING "A"
58. PRIMARY THROTTLE SHAFT
59. THROTTLE PLATES
60. THROTTLE PLATES SCREWS
61. IDLE SPEED SCREW SPRING
62. IDLE SPEED SCREW
63. IDLE FUEL SHUT OFF SOLENOID WASHER
64. IDLE FUEL SHUT OFF SOLENOID
65. IDLE MIXTURE SCREW LIMITER CAP (BLUE)
66. IDLE MIXTURE SCREW
67. IDLE MIXTURE SCREW "O" RING
68. IDLE MIXTURE SCREW SPRING
69. ACCELERATOR PUMP COVER SCREW
70. ACCELERATOR PUMP COVER (ASSY.)
71. ACCELERATOR PUMP DIAPHRAGM
72. ACCELERATOR PUMP SPRING
73. PRIMARY THROTTLE RETURN SPRING "B"
74. ACCELERATOR PUMP CAM
75. POWER VALVE COVER SCREW
76. POWER VALVE COVER
77. POWER VALVE SPRING
78. POWER VALVE DIAPHRAGM
79. SECONDARY THROTTLE SHAFT NUT
80. SECONDARY THROTTLE SHAFT LOCK WASHER
81. SECONDARY THROTTLE SHAFT WASHER
82. FUEL BOWL VENT SOLENOID WASHER
83. MAIN BODY (ASSY.)
84. FUEL BOWL VENT SOLENOID (ASSY.)
   1. BOWL VENT SPRING
   2. BOWL VENT ARM
85. VITON BOWL VENT SEAL
86. "O" RING SEAL FOR PUMP NOZZLE
87. PUMP SHOOTER
88. FUEL DISCHARGE NOZZLES
89. FUEL INLET SEAT & NEEDLE
90. FLOAT HINGE PIN
91. FUEL FILTER
92. FUEL RETURN LINE CHECK VALVE AND FITTING
93. FUEL INLET SEAT GASKET
94. COVER HOLD DOWN SCREWS
95. DASH POT
96. DASH POT MOUNTING BRACKET
97. DASH POT ADJUSTING LOCK NUT
98. DASH POT MOUNTING BRACKET SCREW

Figure 11-12. Typical Motorcraft-Weber Model 740 carburetor. (Ford)

MOTORCRAFT 740 TWO-BARREL

The Motorcraft-Weber 740, figure 11-12, is a 2-stage, 2-barrel design with equal size primary and secondary barrels. It was first used on Ford's Fiesta in 1977. The secondary throttle is mechanically operated by linkage from the primary throttle. Five fuel metering circuits are used: idle, main, accelerator pump, power enrichment, and choke systems. A fuel shutoff solenoid is used to prevent dieseling. The bowl vent is also solenoid operated.

Chapter 12
Carter Carburetors

Carter carburetors are used on American Motors, Chrysler, and Ford Motor Company cars. Some models have also been used on a few GM cars. This chapter contains descriptions and exploded drawings of the following eight Carter models:
- Model YF and YFA One-Barrel
- Model RBS One-Barrel
- Model BBS One-Barrel
- Model BBD Two-Barrel
- Model WCD Two-Barrel
- Model AFB Four-Barrel
- Model AVS Four-Barrel
- Model Thermo-Quad (TQ) Four-Barrel

CARTER YF AND YFA ONE-BARREL

Models YF, figure 12-1, and YFA, figure 12-2, have three sections: the airhorn, the main body, and the throttle body. The airhorn is the main body cover. It contains the choke plate and choke control assembly, the fuel inlet fitting, the needle valve and seat, and the float and lever. The main body contains the idle, main, and accelerator pump systems. The throttle body holds the throttle valve and shaft assembly and the idle mixture adjusting screw.

Some versions have an antistall dashpot or solenoid throttle positioner attached to the airhorn. Model YFA uses a choke hot air inlet adapter and is vented to a charcoal canister by a mechanical external bowl vent. Some 1977 versions also have altitude compensators to automatically lean the air-fuel mixture at higher elevations.

CARTER RBS ONE-BARREL

A single aluminum casting, the Model RBS, figure 12-3, has a detachable fuel bowl of stamped steel. All calibration points, including two internal vapor vents, are within the casting. An additional mechanical vent is provided on California carburetors. The RBS uses four fuel metering circuits: idle, main, accelerator pump, and choke. Fuel supply is controlled by a diaphragm-operated, step-up metering rod and a spring-operated accelerator pump. The thermostatic choke on California models has an electric assist.

CARTER BBS ONE-BARREL

The Model BBS, figure 12-4, has three sections: the airhorn, the main body, and the throttle body. The airhorn is the main body cover and contains the choke plate and accelerator pump plunger. The main body contains the fuel bowl, fuel inlet, hot-idle compensator, and the idle, main, and accelerator metering systems. The

Carter Carburetors

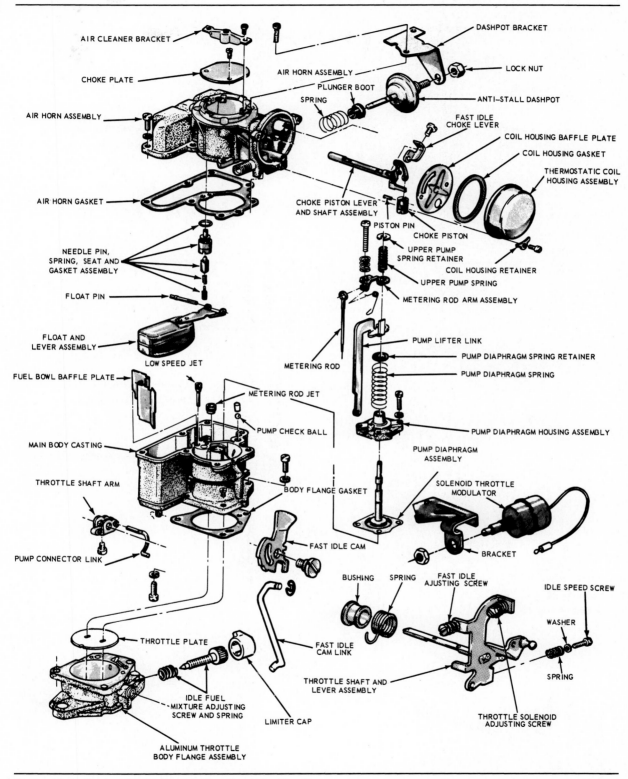

Figure 12-1. Typical Carter Model YF carburetor. (Ford)

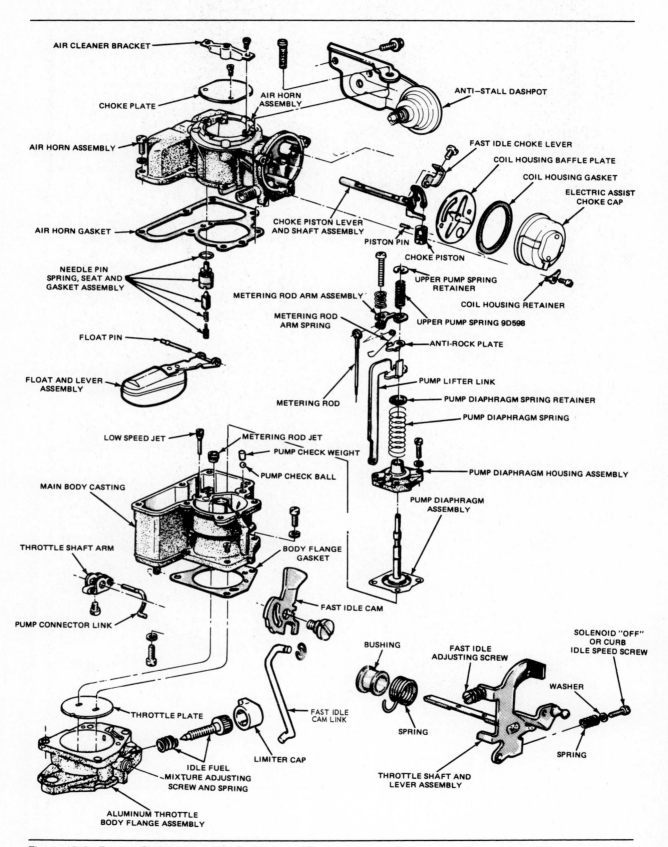

Figure 12-2. Typical Carter Model YFA carburetor. (Ford)

Carter Carburetors

153

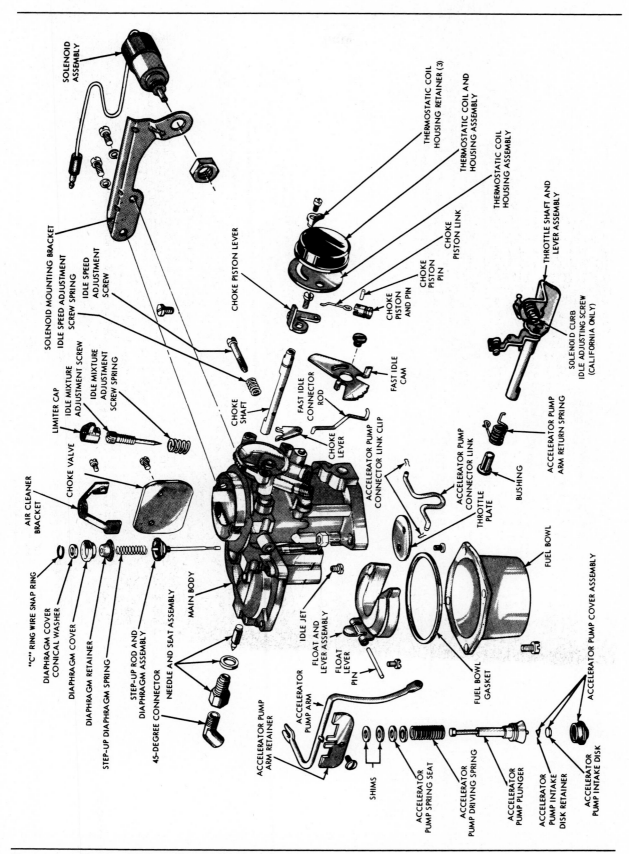

Figure 12-3. Typical Carter Model RBS carburetor. (Ford)

Chapter Twelve

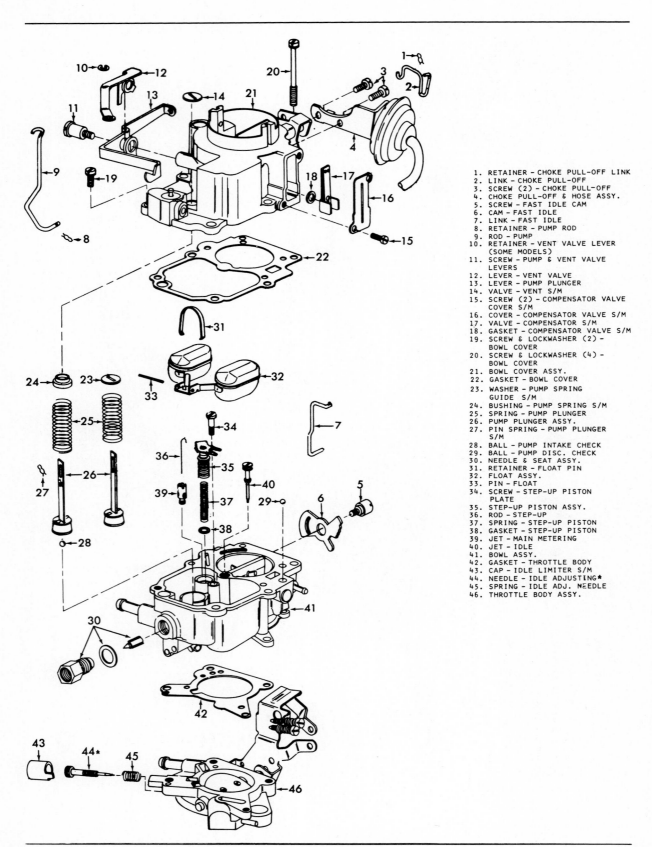

Figure 12-4. Typical Carter Model BBS carburetor. (Borg-Warner)

Carter Carburetors

throttle body holds the throttle valve and shaft assembly and the idle mixture screw. Fuel supply is controlled by a step-up metering rod and a spring-operated accelerator pump.

CARTER BBD TWO-BARREL

The Model BBD, figure 12-5, is basically a 2-barrel version of the Model BBS. It has the same three main sections: the airhorn, the main body, and the throttle body. The airhorn is the main body cover and contains the choke plate and choke control assembly, the bowl vent, and the accelerator pump plunger. The main body contains twin fuel bowls, the fuel inlet and needle valve assembly, two pontoon floats, and a hot-idle compensator. Also located here are the five fuel metering circuits: idle, main, accelerator, power, and choke. The throttle body contains the throttle valve and shaft assembly and two idle mixture screws. Some BBD versions may use a dashpot attached to the airhorn. Others have an idle speed solenoid. Some BBD models used by Chrysler have a distributor ground switch to retard the distributor advance when the engine is at idle.

CARTER WCD TWO-BARREL

The Model WCD, figure 12-6, has four sections: the airhorn, the main body cover, the main body, and the throttle body. Four fuel metering circuits are used: idle, main, accelerator, and choke. These are contained in the main body, along with twin fuel bowls and floats. The airhorn contains the choke valve and shaft assembly. The fuel inlet, metering rods, and accelerator pump piston are all housed in the main body cover. The throttle body holds the throttle valve and shaft assembly and two idle mixture adjusting screws.

CARTER AFB FOUR-BARREL

A 3-piece carburetor of aluminum castings, figure 12-7, the Model AFB uses four fuel metering circuits located in the main body: idle, main, accelerator, and choke. Twin fuel bowls each feed a primary and secondary nozzle on one side of the carburetor. Step-up rods, pistons, and springs can be serviced without removing the airhorn from the carburetor or the carburetor from the engine. Both the primary and secondary barrels have removable venturi clusters with high-speed and low-speed circuit parts. Mechanical secondaries linked to the primaries work with offset air velocity valves.

CARTER AVS FOUR-BARREL

Quite similar in appearance to the AFB, the Model AVS, figure 12-8, shares the same features. The AVS, however, has no venturis in the secondary barrels. Airflow is controlled, instead, by air velocity valves at the tops of the barrels.

CARTER THERMO-QUAD (TQ) FOUR-BARREL

A 3-piece carburetor, the Thermo-Quad, figure 12-9, differs somewhat from other Carter designs. A molded plastic fuel bowl and suspended metering system result in lower fuel temperatures. This in turn means more precise fuel metering and better emission control. Twin metering rods are operated by a combination mechanical-vacuum system, according to engine load. A cam on the primary throttle shaft raises the rods as the throttle starts to open; then vacuum control takes over.

The airhorn, or upper body, contains all fuel metering circuits: the choke valve and shaft assembly; secondary air valves and controls; the fuel inlet system, the low-speed, high-speed, and accelerator pump circuits; a vacuum-controlled step-up piston; fuel discharge nozzles; and all air bleeds. The throttle body contains the hot-idle compensator, the throttle valve and shaft assemblies, and the idle mixture adjusting screws. The main body is simply the molded plastic fuel bowl.

Fuel mixes with air *after* leaving the nozzle, instead of *before*. Mechanical secondaries are linked to the primaries. The vacuum-operated air valve is controlled by a spring on the mounting shaft and a diaphragm with two purposes. It holds back initial air valve operation when the throttle is opened quickly, and provides choke pulldown when a cold engine is started.

Chapter Twelve

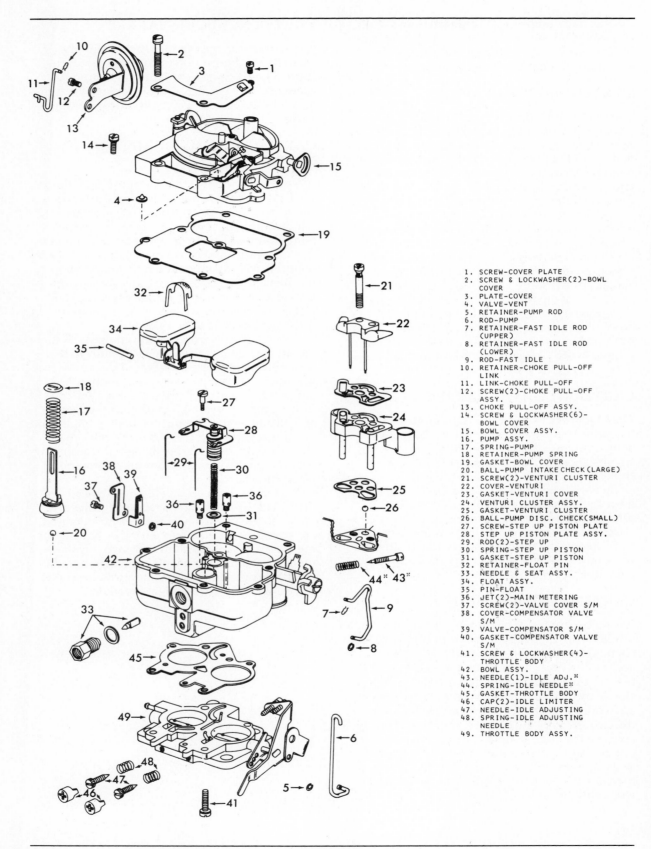

Figure 12-5. Typical Carter Model BBD carburetor. (Borg-Warner)

Carter Carburetors

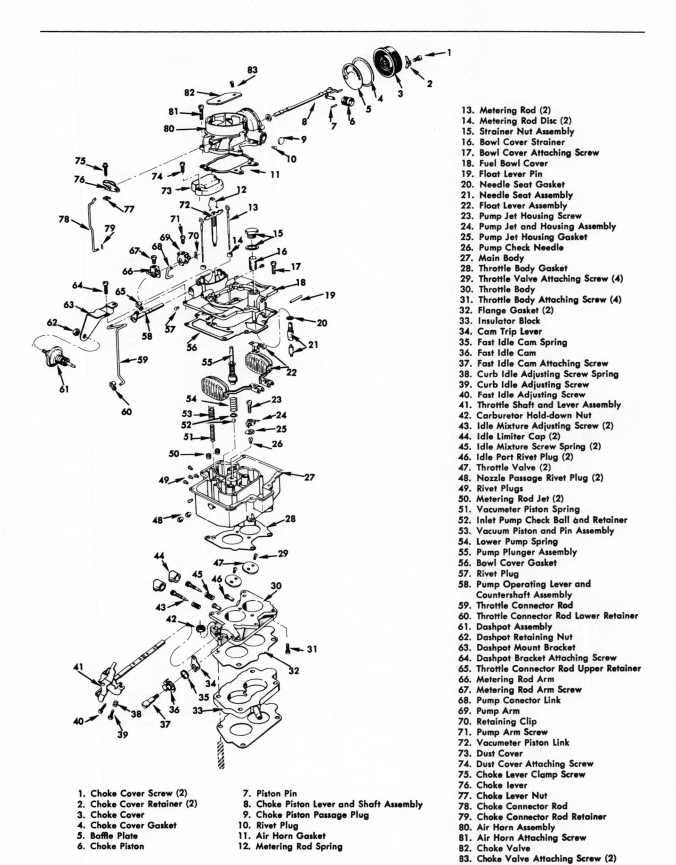

13. Metering Rod (2)
14. Metering Rod Disc (2)
15. Strainer Nut Assembly
16. Bowl Cover Strainer
17. Bowl Cover Attaching Screw
18. Fuel Bowl Cover
19. Float Lever Pin
20. Needle Seat Gasket
21. Needle Seat Assembly
22. Float Lever Assembly
23. Pump Jet Housing Screw
24. Pump Jet and Housing Assembly
25. Pump Jet Housing Gasket
26. Pump Check Needle
27. Main Body
28. Throttle Body Gasket
29. Throttle Valve Attaching Screw (4)
30. Throttle Body
31. Throttle Body Attaching Screw (4)
32. Flange Gasket (2)
33. Insulator Block
34. Cam Trip Lever
35. Fast Idle Cam Spring
36. Fast Idle Cam
37. Fast Idle Cam Attaching Screw
38. Curb Idle Adjusting Screw Spring
39. Curb Idle Adjusting Screw
40. Fast Idle Adjusting Screw
41. Throttle Shaft and Lever Assembly
42. Carburetor Hold-down Nut
43. Idle Mixture Adjusting Screw (2)
44. Idle Limiter Cap (2)
45. Idle Mixture Screw Spring (2)
46. Idle Port Rivet Plug (2)
47. Throttle Valve (2)
48. Nozzle Passage Rivet Plug (2)
49. Rivet Plugs
50. Metering Rod Jet (2)
51. Vacumeter Piston Spring
52. Inlet Pump Check Ball and Retainer
53. Vacuum Piston and Pin Assembly
54. Lower Pump Spring
55. Pump Plunger Assembly
56. Bowl Cover Gasket
57. Rivet Plug
58. Pump Operating Lever and Countershaft Assembly
59. Throttle Connector Rod
60. Throttle Connector Rod Lower Retainer
61. Dashpot Assembly
62. Dashpot Retaining Nut
63. Dashpot Mount Bracket
64. Dashpot Bracket Attaching Screw
65. Throttle Connector Rod Upper Retainer
66. Metering Rod Arm
67. Metering Rod Arm Screw
68. Pump Conector Link
69. Pump Arm
70. Retaining Clip
71. Pump Arm Screw
72. Vacumeter Piston Link
73. Dust Cover
74. Dust Cover Attaching Screw
75. Choke Lever Clamp Screw
76. Choke lever
77. Choke Lever Nut
78. Choke Connector Rod
79. Choke Connector Rod Retainer
80. Air Horn Assembly
81. Air Horn Attaching Screw
82. Choke Valve
83. Choke Valve Attaching Screw (2)

1. Choke Cover Screw (2)
2. Choke Cover Retainer (2)
3. Choke Cover
4. Choke Cover Gasket
5. Baffle Plate
6. Choke Piston
7. Piston Pin
8. Choke Piston Lever and Shaft Assembly
9. Choke Piston Passage Plug
10. Rivet Plug
11. Air Horn Gasket
12. Metering Rod Spring

Figure 12-6. Typical Carter Model WCD carburetor. (Borg-Warner)

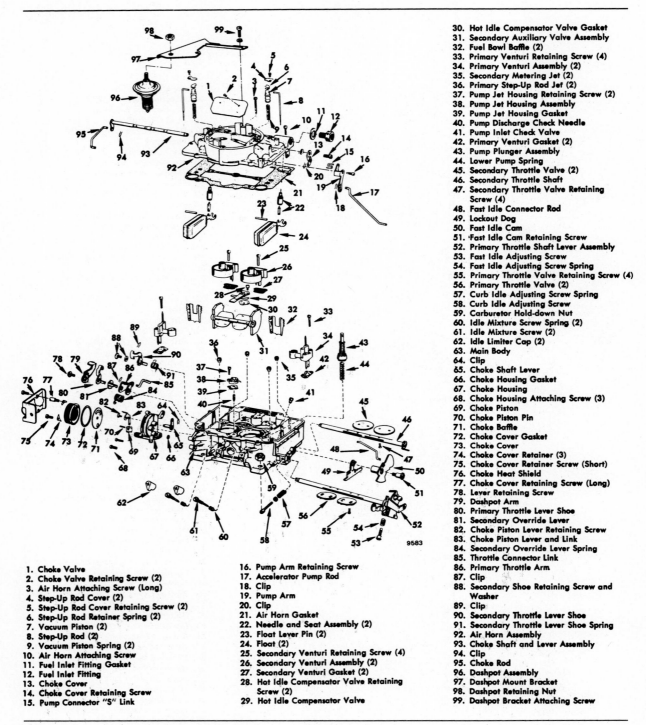

Figure 12-7. Typical Carter Model AFB carburetor. (Borg-Warner)

Carter Carburetors

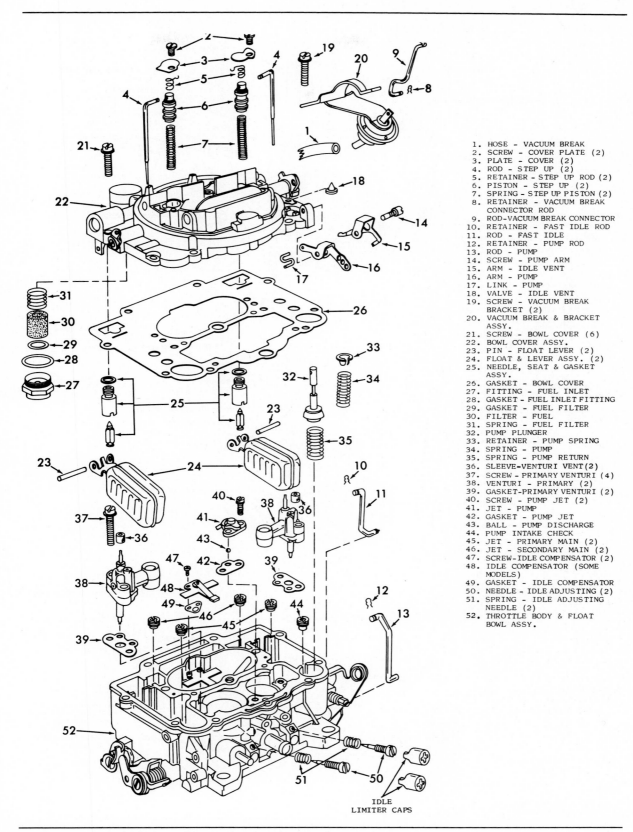

Figure 12-8. Typical Carter Model AVS carburetor. (Borg-Warner)

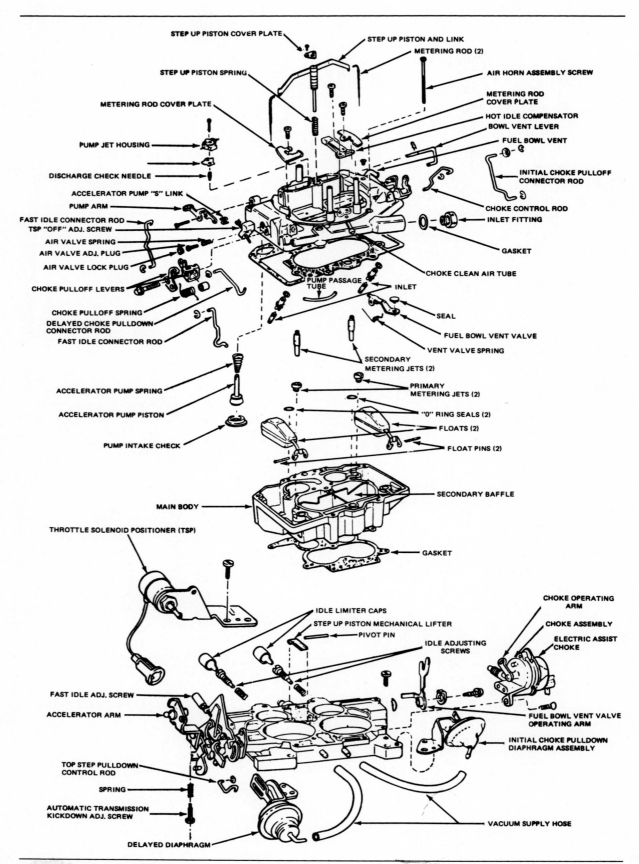

Figure 12-9. Typical Carter Thermo-Quad carburetor. (Chrysler)

Chapter 13
Holley Carburetors

Holley carburetors have been used by all domestic automakers. Basic carburetor features are described in this chapter, and a typical exploded drawing of each is provided to help you understand the design. The following carburetors are included:
- Model 1920 One-Barrel
- Model 1931 One-Barrel
- Models 1940 and 1945 One-Barrel
- Model 2209 Two-Barrel
- Models 2210 and 2245 Two-Barrel
- Model 2300 Two-Barrel
- Models 5200 and 5210-C Two-Barrel
- Models 4150 and 4160 Four-Barrel
- Models 4165 and 4175 Four-Barrel
- Model 4360 Four-Barrel

HOLLEY 1920 ONE-BARREL

The Model 1920, figure 13-1, is a single casting with detachable fuel bowl, main well, and power valve (economizer) assemblies. Four fuel metering circuits are used: idle, main, power enrichment, and accelerator pump. The vacuum-operated, spring-staged choke valve is contained within the carburetor bore and connected by linkage to a remote choke device in the intake manifold. A power valve in the metering body is operated by manifold vacuum. Some versions use a dashpot mounted on the body assembly to retard throttle return to the idle position.

HOLLEY 1931 ONE-BARREL

The Model 1931, figure 13-2, has a 1-piece casting with detachable fuel bowl and integral choke. Four fuel metering circuits are used: idle, main, power enrichment, and accelerator pump. The choke valve is contained within the carburetor bore and is connected to the thermostatic spring in the choke housing by a shaft and lever assembly. Manifold vacuum is used to operate the power valve in the metering body. Fuel bowl venting is by a fixed internal vent tube to the airhorn. An external vent opens at idle speeds.

HOLLEY 1940 AND 1945 TWO-BARREL

This 3-piece design, figure 13-3, has an airhorn, a main body, and a throttle body. Models 1940 and 1945 include four basic fuel metering circuits: idle-transfer, main, accelerator pump, and power enrichment. An insulating gasket between the main body and throttle body prevents heat transfer to the fuel bowl. The fuel bowl completely surrounds the venturi and uses dual floats. The main body contains the fuel inlet and all metering systems. The choke system is built into the Model 1940 and remote

162   **Chapter Thirteen**

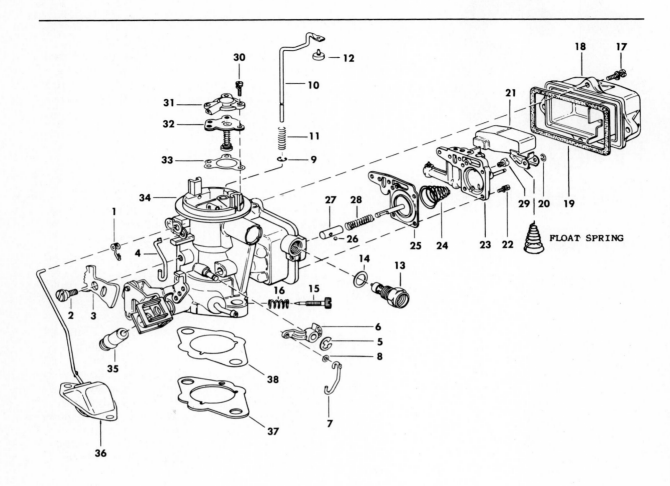

1. Choke rod retainer clip
2. Fast-idle cam screw
3. Fast-idle cam
4. Fast-idle rod
5. Pump lever retainer ("E" washer)
6. Pump lever
7. Pump link
8. Pump-link washer
9. Bowl vent rod retainer ("E" washer)
10. Bowl vent rod
11. Bowl vent rod spring
12. Bowl vent valve
13. Needle and seat assembly
14. Needle seat gasket
15. Idle-mixture adjusting needle
16. Idle adjusting needle spring
17. Screw and lockwasher assembly
18. Fuel bowl
19. Fuel-bowl gasket

20. Float lever retainer ("E" washer)
21. Float assembly
22. Screw and lockwasher assembly
23. Main well and economizer body assy.
24. Pump diaphragm spring
25. Pump diaphragm assembly
26. Pump push-rod sleeve ball
27. Pump push-rod sleeve
28. Pump push-rod spring
29. Main-metering jet
30. Screw and lockwasher assembly
31. Economizer diaphragm cover
32. Economizer diaphragm assembly
33. Economizer diaphragm gasket
34. Body assembly
35. Throttle-rod insulator bushing
36. Automatic choke assy. (well type)
37. Manifold flange gasket, thick
38. Manifold flange gasket, thin

Figure 13-1. Typical Holley Model 1920 carburetor. (Borg-Warner)

Holley Carburetors

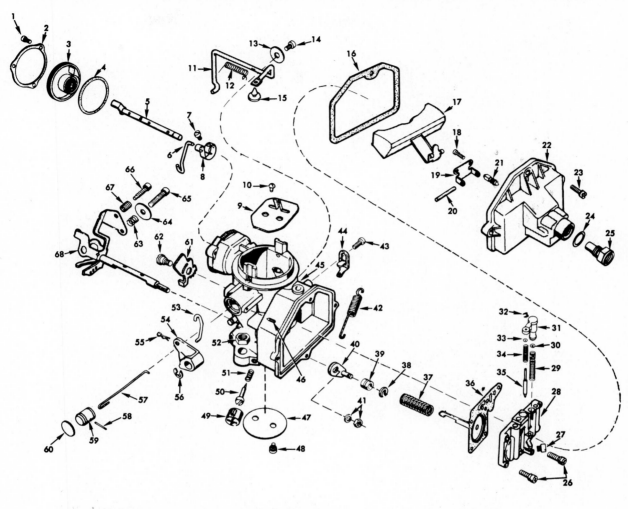

1. Choke Cover Screw (3)
2. Choke Cover Retainer
3. Choke Cover Assembly
4. Choke Cover Gasket
5. Choke Shaft Assembly
6. Fast Idle Connector Rod
7. Choke Lever Screw
8. Choke Lever
9. Choke Valve
10. Choke Valve Screw (2)
11. Bowl Vent Rod
12. Bowl Vent Rod Spring
13. Vent Rod Retainer Washer (2)
14. Vent Rod Retainer Screw (2)
15. Bowl Vent Valve
16. Fuel Bowl Gasket
17. Float Assembly
18. Float Shaft Bracket Screw (2)
19. Float Shaft Retaining Bracket
20. Float Shaft
21. Fuel Inlet Needle
22. Fuel Bowl
23. Fuel Bowl Retaining Screw (4)
24. Fuel Inlet Gasket
25. Fuel Inlet Fitting
26. Metering Block Screw (5)
27. Main Jet
28. Metering Block
29. Power Valve Piston Spring
30. Power Valve Shim
31. Power Valve Piston
32. Power Valve Piston Retainer
33. Power Valve Spring Washer
34. Power Valve Spring
35. Power Valve
36. Pump Diaphragm Assembly
37. Pump Diaphragm Spring
38. Retaining Washer
39. Lever Pin Collar
40. Return Spring Lever
41. Return Lever Lock Washer And Nut
42. Throttle Return Spring
43. Perch Bracket Retaining Screw
44. Perch Bracket
45. Main Body
46. Choke Piston Stop Screw
47. Throttle Valve
48. Throttle Valve Screw (2)
49. Idle Limiter Cap
50. Idle Mixture Adjustment Screw
51. Idle Mixture Screw Spring
52. Carburetor Hold-down Nut
53. Pump Link
54. Pump Operating Lever
55. Pump Link Retainer
56. Pump Operating Lever Retainer
57. Choke Piston Link
58. Choke Piston Pin
59. Choke Piston
60. Choke Piston Plug
61. Fast Idle Cam
62. Fast Idle Cam Screw
63. Curb Idle Adjusting Screw Spring
64. Vent Rod Operating Washer
65. Curb Idle Adjusting Screw
66. Fast Idle Adjusting Screw
67. Fast Idle Adjusting Screw Spring
68. Throttle Shaft Assembly

Figure 13-2. Typical Holley Model 1931 carburetor. (Holley)

164

Chapter Thirteen

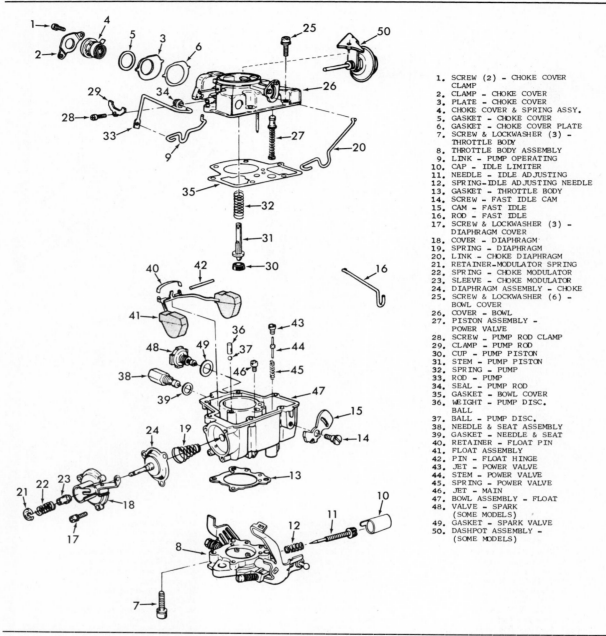

Figure 13-3. Typical Holley Model 1940 carburetor. (Borg-Warner)

for the Model 1945. This later, more sophisticated Model 1945 design is vented to a charcoal canister. It has an idle enrichment valve on some versions.

HOLLEY 2209 TWO-BARREL

Figure 13-4 illustrates this 3-piece design that uses an airhorn, a main body, and a throttle body. The Model 2209 contains four basic fuel metering cirucits: idle, main, accelerator pump, and power enrichment. The integral choke is connected by linkage to the choke valve in the airhorn. A thermostatic spring in the choke housing and an external diaphragm assembly control choke operation. Fuel inlet is through the airhorn, which also contains the power valve piston. The throttle body houses the throttle valve and shaft assembly and two idle mixture adjusting screws. A dashpot is used on some models to retard throttle closing at idle.

Holley Carburetors

HOLLEY 2210 AND 2245 TWO-BARREL

The Models 2210 and 2245 are a 3-piece design, figure 13-5, using an airhorn, a main body, and a throttle body. This design has five basic fuel metering circuits: basic idle, idle enrichment, main, accelerator pump, and power enrichment. Fuel inlet is through the airhorn, with a single fuel bowl and float feeding both barrels. Both versions use a remote choke device linked to the choke valve in the airhorn. The throttle body contains the throttle valve and shaft assembly and two idle mixture adjusting screws. Some versions include a hot-idle compensator.

HOLLEY 2300 TWO-BARREL

This model, figure 13-6, has a horizontally mounted fuel bowl and metering block assembly attached to the main body, resting on the throttle body. Four fuel metering circuits are used: idle, main, accelerator pump, and power enrichment. These are helped by the fuel inlet and choke systems. When used in 2-nount and 3-mount versions, only the center carburetor contains the choke, idle, accelerator pump, and power enrichment circuits. The outboard carburetors use a throttle control vacuum diaphragm and are mechanically connected to the center carburetor's slotted throttle lever by adjustable connector rods. Outboard carburetor throttle valves open by vacuum and close mechanically.

HOLLEY 5200 AND 5210-C TWO-BARREL

A 2-piece design built under license from Weber, Models 5200 and 5210-C, figure 13-7, are 2-stage carburetors with primary bores smaller than the secondaries. The secondary is mechanically linked to the primary. The primary stage uses five basic fuel metering circuits: idle, transfer and low-speed, main, accelerator pump, and power enrichment. The secondary stage uses a transfer circuit, a main metering circuit, and a power enrichment circuit. The primary and secondary stages are fed by a single fuel bowl vented to a charcoal canister.

Ford designates the Model as 5200. GM applications are designated 5210-C. Early 5200 versions use a hot-water-heated integral choke, but later models have an electric assist built into the choke housing cap. The Model 5210-C uses a hot-water-heated integral choke with vacuum break diaphragms. Both carburetors may have various throttle stop solenoids and dashpots, according to model year and application. Those used on 1977 Ford vehicles have a mechanical altitude compensator as explained in Chapter 7.

HOLLEY 4150 AND 4160 FOUR-BARREL

This Holley design uses a main body attached to a throttle body and fitted with primary and secondary fuel bowls and metering blocks, figure 13-8. The fuel inlet system uses an external distribution tube to route fuel from the primary inlet to the secondary inlet. The primary fuel bowl is vented with a vent valve operated by a throttle shaft lever.

The primary stage has a fuel bowl and vent, a metering block, and an accelerator pump circuit. The power system is also contained within the primary metering block. Each primary bore has a primary and a booster venturi, a main fuel discharge nozzle, an idle fuel passage, and a throttle valve.

The secondary stage of the 4150 has a fuel bowl, a metering block, and a secondary throttle operating diaphragm. Each secondary bore contains a primary and a booster venturi, idle fuel passages, a transfer system, a main secondary fuel discharge nozzle, and a throttle valve. The primary and secondary throats are the same size.

The Model 4160 is similar to the 4150, except that a simpler secondary metering plate is used instead of a metering block.

HOLLEY 4165 AND 4175 FOUR-BARREL

The Models 4165 and 4175, figure 13-9, are 4-barrel spread-bore carburetors designed to replace the Rochester Quadrajet 4M and Carter Thermo-Quad carburetors. Both models are the same basic design as the 4150, and work in the same way. While they appear very similar to the Models 4150 and 4160, parts and gaskets are not interchangeable. The 4165 has mechanically operated secondaries and an accelerator pump in both the primary and secondary sides. The 4175 uses a vacuum-operated secondary system and does not need an accelerator pump on the secondary side.

HOLLEY MODEL 4360 FOUR-BARREL

This spread-bore model, figure 13-10, was introduced in 1976 as a replacement for the Rochester Quadrajet. Unlike other Holley four-barrel carburetors, it has a single fuel bowl built into the main body casting. The carburetor consists of an airhorn, a main body, and a throttle body. The main metering circuit for the primary side is similar to the one used in the Models 5200 and 5210-C. Both the primary and the secondary sides have idle systems, but only the primary side is adjustable. A spring-driven, piston-type accelerator pump is used. The secondary throttles are mechanically operated. The main jets have metric threads.

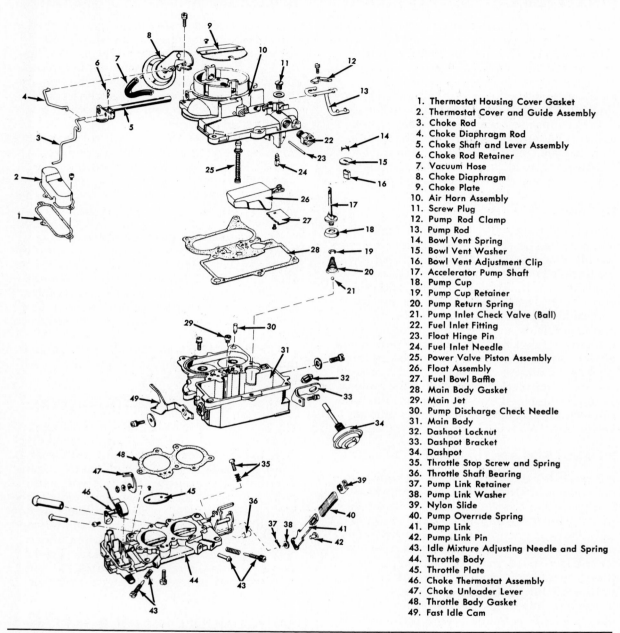

Figure 13-4. Typical Holley Model 2209 carburetor. (Holley)

Holley Carburetors

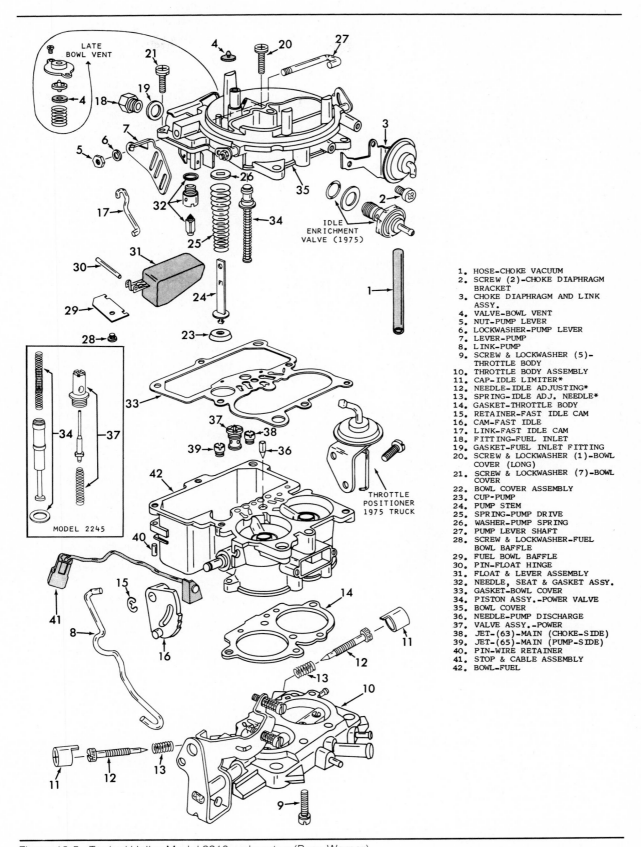

Figure 13-5. Typical Holley Model 2210 carburetor. (Borg-Warner)

168

Chapter Thirteen

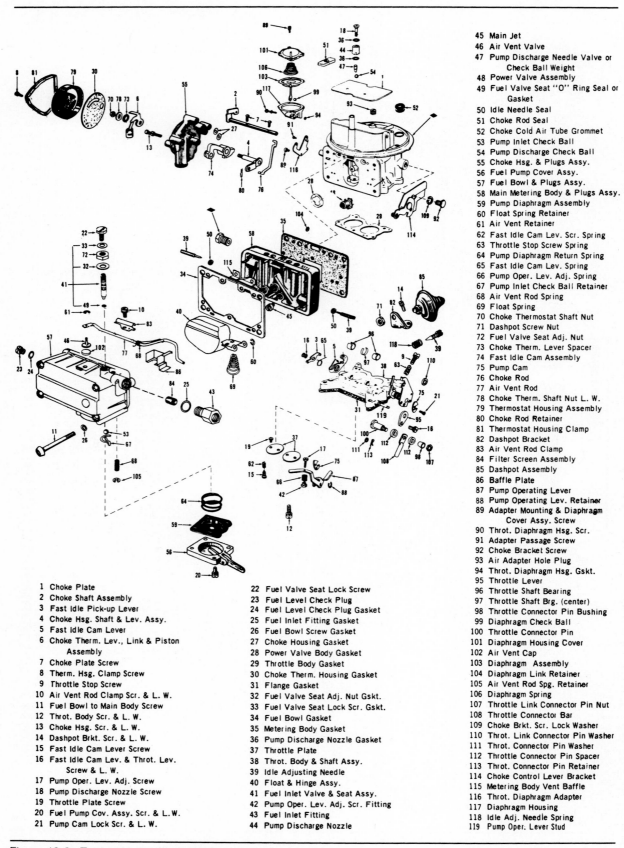

Figure 13-6. Typical Holley Model 2300 carburetor. (Holley)

Holley Carburetors

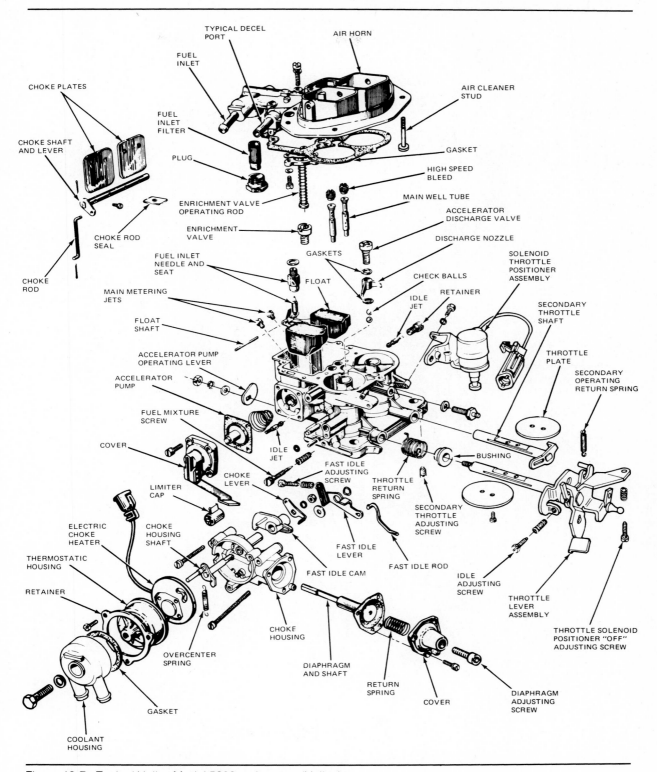

Figure 13-7. Typical Holley Model 5200 carburetor. (Holley)

Figure 13-8. Typical Holley Model 4150 carburetor. (Holley)

Holley Carburetors

Figure 13-9. Typical Holley Model 4165 carburetor. (Holley)

1. Choke Plate
2. Choke Shaft Assembly
3. Fast Idle Pick Up Lever
4. Choke Control Lever
5. Fast Idle Cam Lever
6. Choke Lever & Swivel Assembly
7. Choke Rod Lever & Bushing Assembly
8. Choke Plate Screw
9. Choke Lever Swivel Screw
10. Choke Diaphragm Bracket Screw & L.W.
11. Fuel Pump Cover Screw & L.W. Primary
12. Fuel Pump Cover Screw & L.W. Secondary
13. Fuel Bowl Screw (Long) Pri. & Sec.
14. Pump Lever Adjusting Screw Secondary
15. Throttle Body Screw & L.W.
16. Fast Idle Cam Lever Adj. Screw
17. Fast Idle Cam Lever Screw & L.W.
18. Pump Cam Lever Screw & L.W.
19. Pump Lever Adj. Screw Primary
20. Fast Idle Cam Plate Screw & L.W.
21. Throttle Plate Screw Pri. & Sec.
22. Choke Wire Bracket Clamp Screw
23. Pump Cam Screw
24. Pump Discharge Nozzle Screw
25. Fuel Valve Seat Lock Screw
26. Pump Operating Lever Adj. Screw
27. Float Shaft Bracket Screw & L.W.
28. Throttle Stop Screw
29. Fuel Bowl Screw (Short) Pri. & Sec.
30. Throttle Lever Extension Screw No. 12
30A. Throttle Body Channel Plug
31. Fuel Level Check Plug
32. Vacuum Tuba Plug
33. Fuel Level Check Plug Gasket
34. Fuel Bowl Screw Gasket
35. Power Valve Gasket
36. Fuel Valve Seat Adj. Nut Gasket
37. Fuel Valve Seat Lock Screw Gasket
38. Pump Discharge Nozzle Gasket
39. Metering Body Gasket Pri. & Sec.
40. Fuel Inlet Filter Gasket
41. Fuel Inlet Fitting Gasket
42. Fuel Valve Gasket Pri. & Sec.
43. Throttle Body Gasket
44. Flange Gasket
45. Fuel Bowl Gasket Pri. & Sec.
46. Throttle Plate Secondary
47. Throttle Plate Primary
48. Throttle Lever Extension
49. Cam Follower Lever Assy.
50. Throttle Body & Shaft Assy.
51. Fuel Line Tubing
52. Idle Adjusting Needle
53. Float & Hinge Assy. Primary
54. Float & Hinge Assy. Secondary
55. Float Shaft
56. Fuel Inlet Valve & Seat Assy.
57. Pump Lever Adj. Screw Fitting
58. Fuel Inlet Fitting
59. Pump Discharge Nozzle Pri.
60. Pump Discharge Nozzle Sec.
61. Main Jet - Primary
62. Main Jet - Secondary
63. Pump Check Valve
64. Power Valve Assy. Primary
65. Power Valve Assy. Secondary
66. Fuel Line Tube "o" Ring Seal
67. Fuel Valve Seat "o" Ring Seal
68. Idle Needle Seal
69. Choke Rod Seal
70. Choke Diaphragm Link
71. Back-Up Plate & Stud Assy.
72. Fast Idle Cam Plate
73. Fuel Pump Cover Assy. Pri.
74. Fuel Pump Cover Assy. Sec.
75. Fuel Bowl & Plugs Assy. Pri.
76. Fuel Bowl & Plugs Assy. Sec.
77. Metering Body & Plugs Assy. Primary
78. Metering Body & Plugs Assy. Secondary
79. Pump Diaphragm Assy. Pri.
80. Pump Diaphragm Assy. Sec.
81. Choke Diaphragm Assy.
82. Float Hinge Retainer
83. Cam Follower Lever Assy. Retainer
84. Choke Control Lever Retainer
85. Pump Lever Stud
86. Fast Idle Cam Plunger Spring
87. Fast Idle Cam Lever Screw Spring
88. Throttle Stop Screw Spring
89. Diaphragm Return Spring Pri.
90. Choke Spring
91. Float Spring Secondary
92. Float Spring Primary
93. Pump Lever Adj. Screw Spring Primary
94. Pump Lever Adj. Screw Spring Secondary
95. Fuel Inlet Filter Spring
96. Diaphragm Return Spring Sec.
97. Throttle Return Spring Sec.
98. Fast Idle Cam Lever Spring
99. Choke Wire Bracket Clamp Screw Nut
100. Back-Up Plate Stud Nut
101. Choke Lever Nut
102. Fuel Valve Seat Adj. Nut
103. Pump Operating Lever Adj. Nut
104. Throttle Lever Ext. Screw Nut
105. Pump Cam-Primary
106. Fast Idle Cam & Shaft Assy.
107. Pump Cam-Secondary
108. Fast Idle Cam Assy.
109. Pump Operating Lever Screw Sleeve
110. Choke Rod
111. Secondary Connecting Rod
112. Back-Up Plate Stud Nut L.W.
113. Choke Control Lever Nut L.W.
114. Secondary Connecting Rod Washer
115. Throttle Seal Washer
116. Choke Spring Washer
117. Secondary Connecting Rod Cotter Pin
118. Choke Wire Bracket Clamp
119. Choke Wire Bracket
120. Fast Idle Cam Plunger
121. Choke Vacuum Hose
122. Metering Body Vent Baffle Pri. & Sec.
123. Float Shaft Retaining Bracket
124. Baffle Plate - Primary
125. Baffle Plate - Secondary
126. Fuel Inlet Filter
127. Pump Operating Lever Primary
128. Pump Operating Lever & Guide Assy. Secondary
129. Pump Cam Lever Secondary
130. Pump Operating Lever Retainer Pri. & Sec.
131. Choke Therm. Lever
132. Choke Therm. Cover Screw
133. Choke Housing Screw
134. Choke Housing Gasket
135. Choke Therm. Cover Gasket
136. Tube & "O" Ring Assy.
137. Idle Adj. Needle Limiter Cap
138. Choke Housing & Plugs Assy.
139. Choke Therm. Cover Retainer
140. Choke Therm. Shaft Nut
141. Choke Shaft Nut Lock Washer
142. Choke Housing Screw & L.W.
143. Choke Therm. Cover Assy.
144. Choke Shaft Spacer

172 Chapter Thirteen

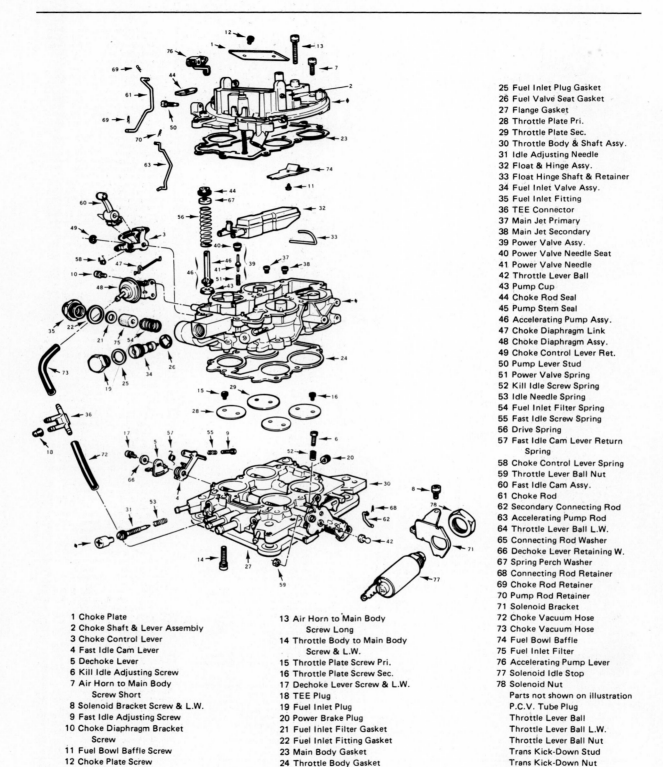

Figure 13-10. Typical Holley Model 4360 carburetor. (Holley)

Chapter 14
Rochester Carburetors

Although primarily used by General Motors, Rochester carburetors have been used by American Motors, Chrysler, and Ford Motor Company. The following carburetors are described in this chapter:
- Models B, BC, and BV One-Barrel
- Models H and HV One-Barrel
- Models M, MV, and 1MV One-Barrel
- Models 2G, 2GV and 2GC Two-Barrel
- Models 2MC and M2MC Two-Barrel
- Model 4GC Four-Barrel
- The Quadrajet Four-Barrel (Models 4M, 4MC, 4MV, M4MC, M4ME, and M4MEA).

ROCHESTER B, BC, AND BV ONE-BARREL

Model B carburetors, Figure 14-1, are a 3-piece design, consisting of an airhorn, a main body, and a throttle body. Letter designations indicate the type of choke used: Model B (manual choke), Model BC (integral automatic choke), and Model BV (remote automatic choke). Four conventional fuel metering circuits are used: idle, main, accelerator pump, and power enrichment.

The fuel bowl completely surrounds the main bore and has a centrally located main nozzle to prevent fuel spillage, despite engine angle. The main metering jet and power valve are contained within the main well, which is attached to the airhorn. Suspending the main well assembly in this manner prevents engine heat from being transmitted to the main well. This minimizes percolation spillover during hot engine operation.

ROCHESTER H AND HV ONE-BARREL

These single-barrel carburetors, figure 14-2, are used in pairs on Corvairs only. They vary considerably in internal design despite uniform external appearance. The 1960 version (Model H) has no choke, fast-idle, or choke unloader mechanism. These are part of the separate airhorn assembly located between the carburetors. A manual choke and fast-idle cam are fitted to 1961 versions. A remote automatic choke is used on 1962-64 carburetors, along with a vacuum break diaphragm. A nonadjustable power enrichment circuit was added to 1965 carburetors. With 4-carburetor installations, the secondary carburetors are Model H, and use no choke, idle, low-speed, or power enrichment systems.

ROCHESTER M, MV, AND 1MV ONE-BARREL

Model M carburetors, figure 14-3, have a 3-piece design consisting of an airhorn, a main body, and a throttle body. The throttle body is

Chapter Fourteen

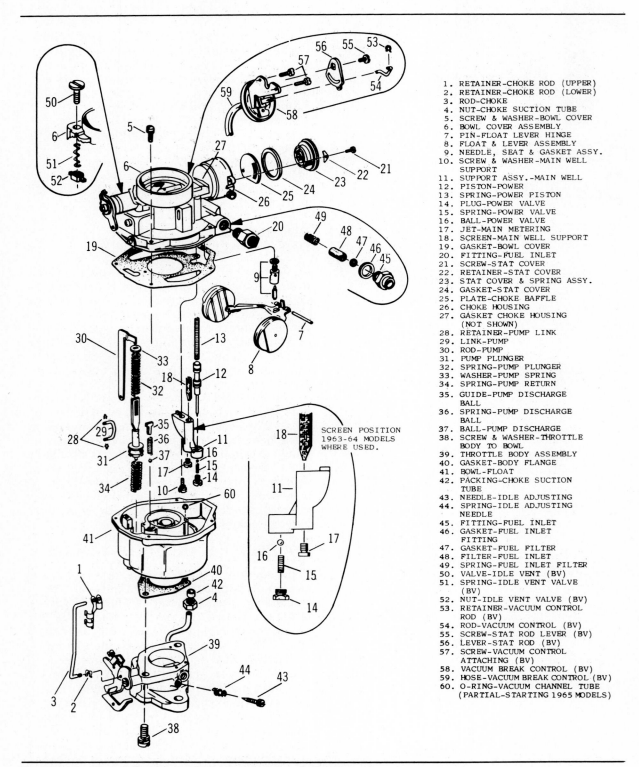

Figure 14-1. Typical Rochester Model B carburetor. (Borg-Warner)

1. RETAINER-CHOKE ROD (UPPER)
2. RETAINER-CHOKE ROD (LOWER)
3. ROD-CHOKE
4. NUT-CHOKE SUCTION TUBE
5. SCREW & WASHER-BOWL COVER
6. BOWL COVER ASSEMBLY
7. PIN-FLOAT LEVER HINGE
8. FLOAT & LEVER ASSEMBLY
9. NEEDLE, SEAT & GASKET ASSY.
10. SCREW & WASHER-MAIN WELL SUPPORT
11. SUPPORT ASSY.-MAIN WELL
12. PISTON-POWER
13. SPRING-POWER PISTON
14. PLUG-POWER VALVE
15. SPRING-POWER VALVE
16. BALL-POWER VALVE
17. JET-MAIN METERING
18. SCREEN-MAIN WELL SUPPORT
19. GASKET-BOWL COVER
20. FITTING-FUEL INLET
21. SCREW-STAT COVER
22. RETAINER-STAT COVER
23. STAT COVER & SPRING ASSY.
24. GASKET-STAT COVER
25. PLATE-CHOKE BAFFLE
26. CHOKE HOUSING
27. GASKET CHOKE HOUSING (NOT SHOWN)
28. RETAINER-PUMP LINK
29. LINK-PUMP
30. ROD-PUMP
31. PUMP PLUNGER
32. SPRING-PUMP PLUNGER
33. WASHER-PUMP SPRING
34. SPRING-PUMP RETURN
35. GUIDE-PUMP DISCHARGE BALL
36. SPRING-PUMP DISCHARGE BALL
37. BALL-PUMP DISCHARGE
38. SCREW & WASHER-THROTTLE BODY TO BOWL
39. THROTTLE BODY ASSEMBLY
40. GASKET-BODY FLANGE
41. BOWL-FLOAT
42. PACKING-CHOKE SUCTION TUBE
43. NEEDLE-IDLE ADJUSTING
44. SPRING-IDLE ADJUSTING NEEDLE
45. FITTING-FUEL INLET
46. GASKET-FUEL INLET FITTING
47. GASKET-FUEL FILTER
48. FILTER-FUEL INLET
49. SPRING-FUEL INLET FILTER
50. VALVE-IDLE VENT (BV)
51. SPRING-IDLE VENT VALVE (BV)
52. NUT-IDLE VENT VALVE (BV)
53. RETAINER-VACUUM CONTROL ROD (BV)
54. ROD-VACUUM CONTROL (BV)
55. SCREW-STAT ROD LEVER (BV)
56. LEVER-STAT ROD (BV)
57. SCREW-VACUUM CONTROL ATTACHING (BV)
58. VACUUM BREAK CONTROL (BV)
59. HOSE-VACUUM BREAK CONTROL (BV)
60. O-RING-VACUUM CHANNEL TUBE (PARTIAL-STARTING 1965 MODELS)

Rochester Carburetors

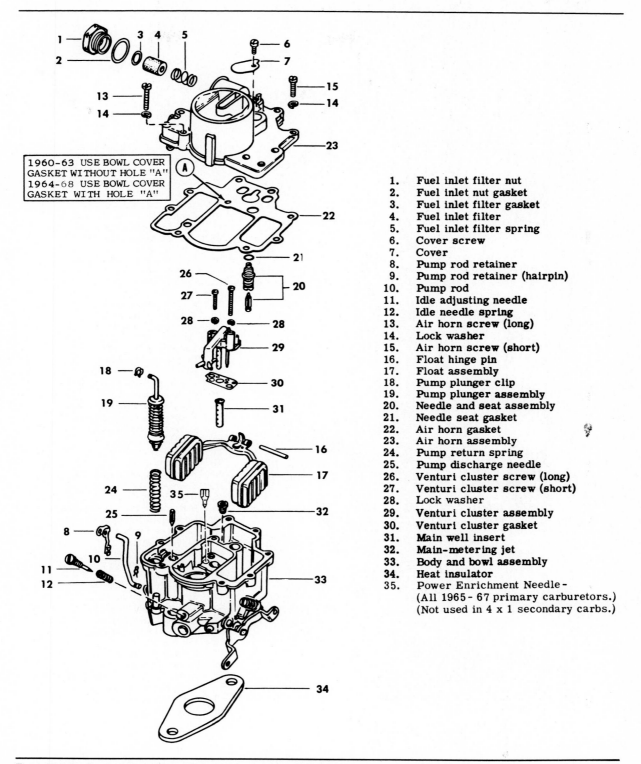

1960-63 USE BOWL COVER GASKET WITHOUT HOLE "A"
1964-68 USE BOWL COVER GASKET WITH HOLE "A"

1. Fuel inlet filter nut
2. Fuel inlet nut gasket
3. Fuel inlet filter gasket
4. Fuel inlet filter
5. Fuel inlet filter spring
6. Cover screw
7. Cover
8. Pump rod retainer
9. Pump rod retainer (hairpin)
10. Pump rod
11. Idle adjusting needle
12. Idle needle spring
13. Air horn screw (long)
14. Lock washer
15. Air horn screw (short)
16. Float hinge pin
17. Float assembly
18. Pump plunger clip
19. Pump plunger assembly
20. Needle and seat assembly
21. Needle seat gasket
22. Air horn gasket
23. Air horn assembly
24. Pump return spring
25. Pump discharge needle
26. Venturi cluster screw (long)
27. Venturi cluster screw (short)
28. Lock washer
29. Venturi cluster assembly
30. Venturi cluster gasket
31. Main well insert
32. Main-metering jet
33. Body and bowl assembly
34. Heat insulator
35. Power Enrichment Needle - (All 1965-67 primary carburetors.) (Not used in 4 x 1 secondary carbs.)

Figure 14-2. Typical Rochester Model H carburetor. (Borg-Warner)

176 Chapter Fourteen

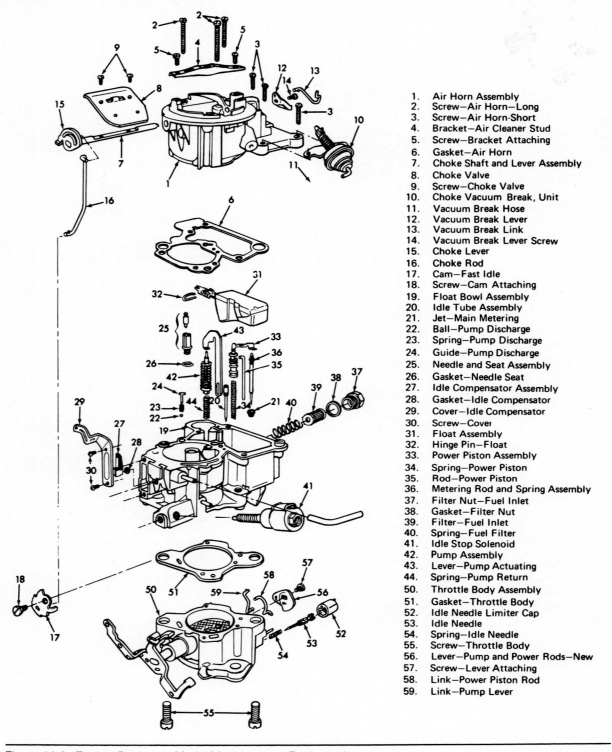

1. Air Horn Assembly
2. Screw—Air Horn—Long
3. Screw—Air Horn-Short
4. Bracket—Air Cleaner Stud
5. Screw—Bracket Attaching
6. Gasket—Air Horn
7. Choke Shaft and Lever Assembly
8. Choke Valve
9. Screw—Choke Valve
10. Choke Vacuum Break, Unit
11. Vacuum Break Hose
12. Vacuum Break Lever
13. Vacuum Break Link
14. Vacuum Break Lever Screw
15. Choke Lever
16. Choke Rod
17. Cam—Fast Idle
18. Screw—Cam Attaching
19. Float Bowl Assembly
20. Idle Tube Assembly
21. Jet—Main Metering
22. Ball—Pump Discharge
23. Spring—Pump Discharge
24. Guide—Pump Discharge
25. Needle and Seat Assembly
26. Gasket—Needle Seat
27. Idle Compensator Assembly
28. Gasket—Idle Compensator
29. Cover—Idle Compensator
30. Screw—Cover
31. Float Assembly
32. Hinge Pin—Float
33. Power Piston Assembly
34. Spring—Power Piston
35. Rod—Power Piston
36. Metering Rod and Spring Assembly
37. Filter Nut—Fuel Inlet
38. Gasket—Filter Nut
39. Filter—Fuel Inlet
40. Spring—Fuel Filter
41. Idle Stop Solenoid
42. Pump Assembly
43. Lever—Pump Actuating
44. Spring—Pump Return
50. Throttle Body Assembly
51. Gasket—Throttle Body
52. Idle Needle Limiter Cap
53. Idle Needle
54. Spring—Idle Needle
55. Screw—Throttle Body
56. Lever—Pump and Power Rods—New
57. Screw—Lever Attaching
58. Link—Power Piston Rod
59. Link—Pump Lever

Figure 14-3. Typical Rochester Model M carburetor. (Rochester)

an aluminum casting for improved heat distribution. It is used with an insulating body-to-bowl gasket to prevent too much heat from reaching the float bowl. Model M uses a manual choke, Model MV has a remote automatic choke and Model 1MV uses an electrically operated idle stop solenoid.

Four conventional fuel metering circuits are used: idle, main, accelerator pump, and power enrichment. A triple venturi is used with a plain tube nozzle. Fuel flow through the main metering system is controlled by mechanical and vacuum devices. A main well air bleed and variable orifice are also used. The fuel bowl is internally vented to the airhorn, with an external idle vent on some versions. A hot idle compensator may also be used.

ROCHESTER 2G, 2GV, AND 2GC TWO-BARREL

These 3-piece carburetors, figure 14-4, consist of an airhorn, a fuel bowl, and a throttle body. The 2G uses a manual choke, the 2GV a remote automatic choke, and the 2GC an integral automatic choke. Four fuel metering circuits are used: idle, main, accelerator pump, and power enrichment. Each barrel has two venturis and a separate fuel feed, with centrally located metering. A removable venturi cluster contains the main nozzle, idle tubes, mixture passages, air bleeds, and pump jets. The main nozzle and idle tubes are part of the air horn and are suspended in the float bowl main wells. This helps prevent percolation spillover during hot engine operation. Some versions use a hot-idle compensator.

ROCHESTER 2MC AND M2MC TWO-BARREL

A single-stage carburetor introduced in 1975, the Dualjet, figure 14-5, uses the design features of the primary side of the Quadrajet four-barrel carburetor. The 2MC uses the same casting as the Quadrajet but contains no secondary throttles or system. Each bore of the 2MC uses a separate and independent idle system. All primary circuits are the same as the Quadrajet. A revised version of the 2MC was introduced in 1977. Called the M2MC200 series, figure 14-6, it retains most of the features of the 2MC but does not use the same Quadrajet body casting.

■ Alternative Fuels

Since the 1974 energy crisis, intense effort has been spent to develop a replacement fuel for gasoline. Three of the most promising alternatives are hydrogen, LNG or liquid natural gas, and methanol.

Hydrogen is present in water (it's the H in H_2O) as well as in all fossil fuels. It's lightweight, odorless, nontoxic, and is easily converted into energy when mixed with pure oxygen and ignited. This combination produces water as a byproduct of combustion, instead of harmful pollutants.

There are some disadvantages to hydrogen. It burns fast and creates large amounts of NO_x. Hydrogen also produces about 15 percent less power than the same amount of gasoline. But these and other such problems can be largely overcome by the engines.

LNG is a liquid form of the same clean-burning natural gas used in the home. Liquified by chilling to −258° F, it is stored in thermos-type containers. Extensive testing of LNG in automobiles shows that exhaust pollutants are practically eliminated.

Methanol was used as a substitute for gasoline during World War II, and is still used by auto racers. Highly volatile, it can be used full strength, or in a blend with gasoline or water. How it is used depends on the desired results.

Why is methanol so attractive to fuel-conscious researchers? Because it is made by reducing wood or straw through a heat process — one ton of wood chips will yield about 112 gallons of pure methanol. Plenty of such raw material exists as byproducts of the farming and lumbering industries. Unlike petroleum, this makes it a constantly self-renewing resource.

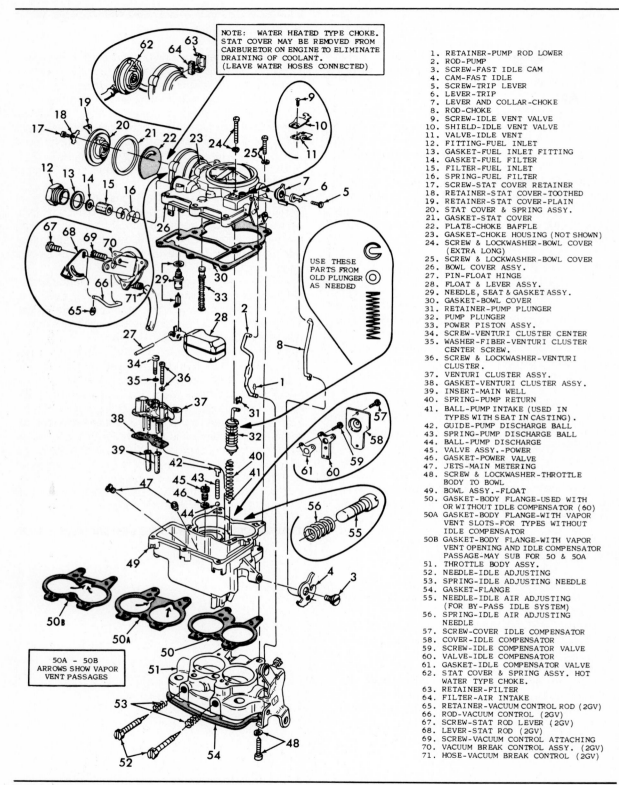

Figure 14-4. Typical Rochester Model 2G carburetor. (Borg-Warner)

Rochester Carburetors

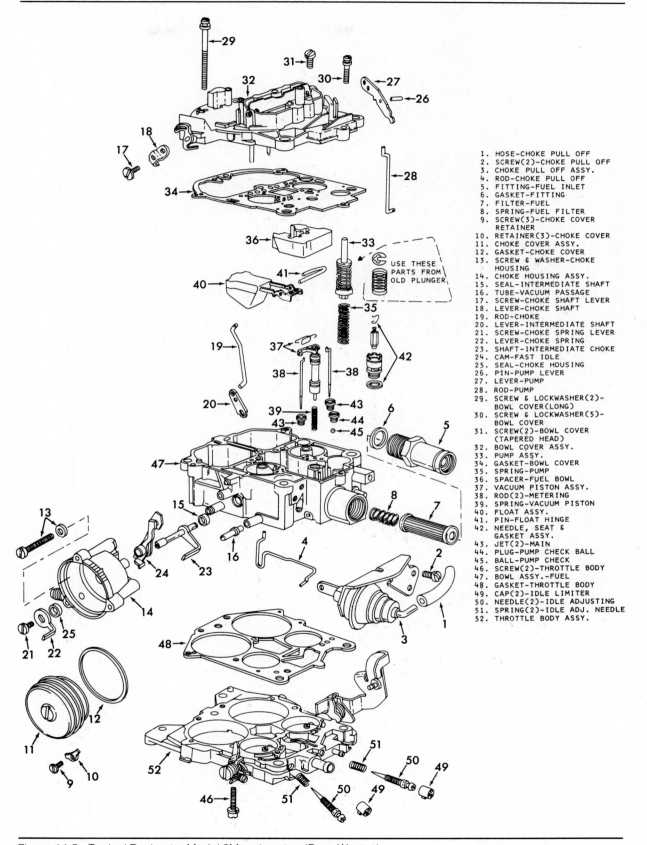

Figure 14-5. Typical Rochester Model 2M carburetor. (Borg-Warner)

Figure 14-6. Typical Rochester Model M2MC200 carburetor. (Rochester)

ROCHESTER 4GC FOUR-BARREL

The Model 4GC, figure 14-7, has an airhorn, a float bowl, and a throttle body. Each primary barrel is aided by a secondary which starts to operate at about half-throttle. The primary and secondary sides use independent fuel bowls and twin floats, as well as separate idle and main metering circuits. The secondary throttles are mechanically linked to the primary throttle valves. The primary side also has an accelerator pump, power enrichment and choke circuits, and an integral choke operated by a thermostatic coil spring. Small venturis are used in the primary side for fuel economy; large venturis in the secondary side provide more air flow. Some versions have a hot-idle compensator.

ROCHESTER QUADRAJET FOUR-BARREL (MODELS 4M, 4MC, 4MV, M4MC, M4ME, AND M4MEA)

The 4M series is a 2-stage, spread-bore design, figure 14-8, with large secondaries and small primaries. A small central fuel bowl serves both primary and secondary sides. The central bowl also reduces fuel evaporation during engine shutdown periods. The 3-piece design uses an aluminum throttle body. The primary side contains four basic fuel metering circuits: idle, main, accelerator pump, and power enrichment. The secondary side uses only a main metering system. Both primary and secondary sides use tapered metering rods with an air valve in the secondary side for metering control.

The wide use of this carburetor has resulted in a large number of different versions. Various models may use an integral choke, a remote choke, a manifold heated hot air choke, an electric-assist heated choke, a hot-idle compensator, altitude compensation, adjustable part throttle, pullover enrichment, vacuum break diaphragms, dashpots, and other devices.

Rochester Carburetors

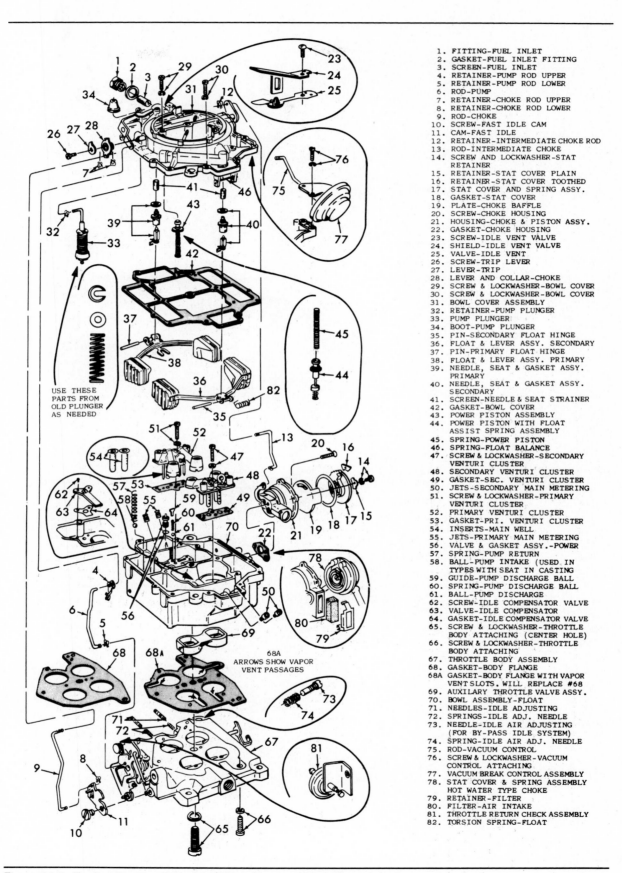

Figure 14-7. Typical Rochester Model 4GC carburetor. (Borg-Warner)

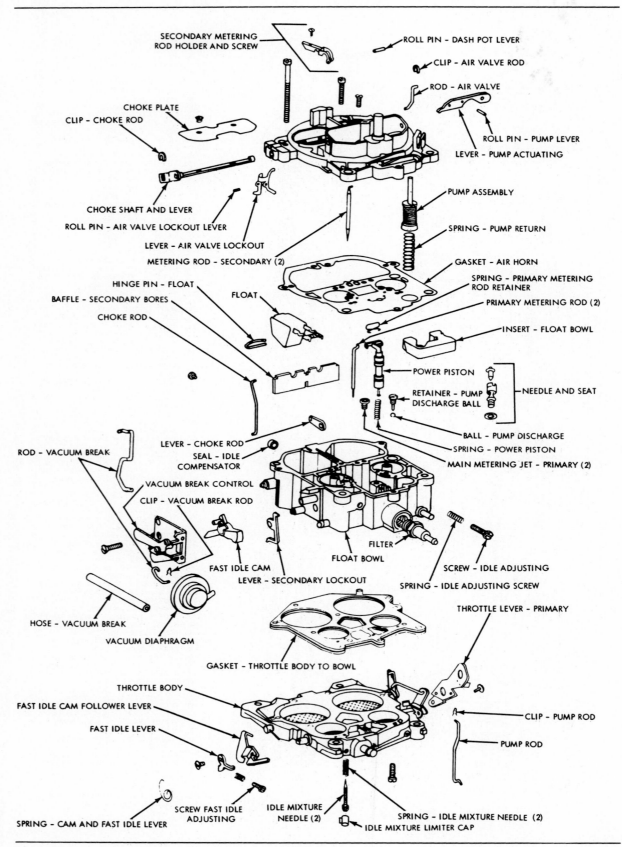

Figure 14-8. Typical Rochester Model 4MV carburetor. (Rochester)

PART FOUR

Emission Control Systems and Devices

Chapter Fifteen
Positive Crankcase Ventilation

Chapter Sixteen
Air Injection

Chapter Seventeen
Spark Timing Control

Chapter Eighteen
Exhaust Gas Recirculation

Chapter Nineteen
Catalytic Converters

Chapter 15
Positive Crankcase Ventilation

The problem of crankcase ventilation has existed since the beginning of the automobile. No piston ring, new or old, can provide a perfect seal between the piston and the cylinder wall. When an engine is running, the pressure of combustion forces the piston downward. This same pressure also forces gases and unburned fuel from the combustion chamber past the piston rings and into the crankcase. These gases are called crankcase vapors, or blowby.

Under perfect conditions, combustion of an engine's air-fuel mixture would completely consume all of the air and the fuel. It would leave only harmless byproducts, such as water vapor and carbon dioxide. However, combustion is seldom perfect and is usually incomplete. The byproducts of incomplete combustion include carbon monoxide (CO), hydrocarbons (HC), and oxides of nitrogen (NO_x).

These combustion byproducts, particularly unburned hydrocarbons, form blowby, figure 15-1. The crankcase must be ventilated to remove these vapors and gases. However, the crankcase on modern engines cannot be ventilated directly to the atmosphere, or the hydrocarbon vapors would add to air pollution. Positive crankcase ventilation (PCV) systems have been developed to ventilate the crankcase and to recirculate the vapors to the engine's induction system.

In this chapter, you will learn:
• The reasons for crankcase ventilation
• The differences between the open and the closed ventilation systems
• The parts of the modern PCV system
• How the PCV system ventilates the crankcase without polluting the air.

DRAFT TUBE VENTILATION

Blowby has three undesirable features:
1. It destroys the lubricating qualities of engine oil.
2. It causes sludge and varnish to form.
3. It helps cause formation of corrosive acids, which can damage engine parts.

After trying various ways of ventilating the engine crankcase, car makers at first settled on the road draft tube, figure 15-2. This is simply a tube connected to the engine crankcase that allows vapors to pass into the air. Fresh air to ventilate the crankcase enters through a vented oil filler cap. This air passes into the crankcase where it mixes with the vapors. Once the car is moving, a vacuum is created by the airflow past the road draft tube. This vacuum draws the crankcase vapors out into the atmosphere.

The road draft tube has three major shortcomings:
1. It works best only when there is a pressure

Positive Crankcase Ventilation

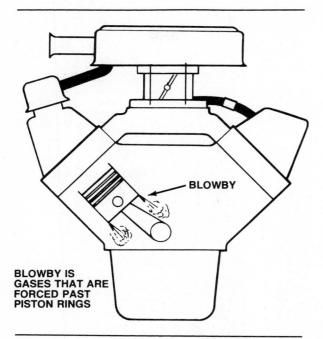

Figure 15-1. Piston rings do not provide a perfect seal. Combustion gases blow by the rings into the crankcase.

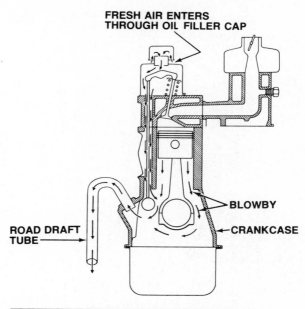

Figure 15-2. The road draft tube ventilates the crankcase to the atmosphere when the car is moving.

difference between the oil filler cap and the draft tube. This pressure depends on car movement. When the car is moving slower than about 25 mph, there is not enough vacuum to remove the vapors from the crankcase.
2. It passes the crankcase vapors directly into the atmosphere where they add to air pollution.
3. At higher vehicle speeds, too much crankcase ventilation will increase engine oil consumption because oil droplets are drawn out through the road draft tube.

POSITIVE CRANKCASE VENTILATION (PCV) SYSTEMS

The drawbacks of the road draft tube were eliminated when the controlled crankcase ventilation system was introduced. Controlled or positive crankcase ventilation relies on intake manifold vacuum to draw the vapors from the crankcase and into the intake manifold. This results in a positive movement of air through the crankcase whenever the engine is running. The vapors are then returned to the combustion chambers, where they are burned. These systems may be classified as open or closed PCV systems, depending upon their design.

Open PCV Systems

Positive crankcase ventilation, as we know it today, received its first major use on 1961 Cali-

Blowby: The leakage of combustion gases and unburned fuel past an engine's piston rings.

■ It Wasn't Always As Simple

The early PCV systems caused a good deal of grief and engine troubles for automakers. Many garages, even franchised dealers, ignored the PCV systems on 1963-64 cars. They required a lot of care and cleaning, and they clogged quickly when ignored. Contaminants remained in the crankcase, and sludge and moisture formed. This clogged oil lines and prevented adequate engine lubrication. The result was disaster for the engine, and major overhauls on engines still under warranty were often required.

The situation reached a crisis point for one major manufacturer, who stopped using PCV on its cars from the spring of 1964 until early in 1965. Auto engineers were frustrated by the problems PCV systems were creating. The systems had been designed to be simple and required only a minimum amount of service. But mechanics in the field completely ignored the emission control device, and engines began to fail.

These problems resulted in a crash project by the automakers. While engineers worked overtime developing a "better" PCV system, manufacturers started a program to educate dealers, servicemen, and car owners. The so-called "self-cleaning" PCV valve was developed and began appearing on mid-1965 models. This second-generation PCV system is practically the same one in use today.

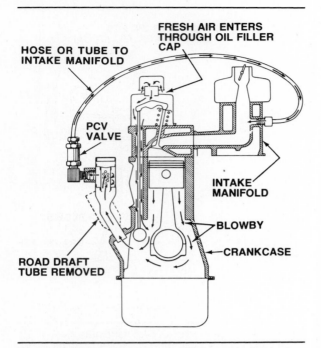

Figure 15-3. In a Type 1 open PCV system, fresh air enters through the oil filler cap. Crankcase vapors are returned to the intake manifold through a PCV valve and hose or tube.

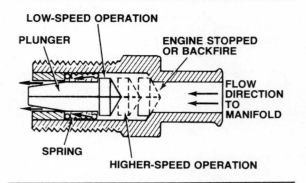

Figure 15-4. The PCV valve plunger responds to spring force and manifold vacuum to regulate the airflow rate.

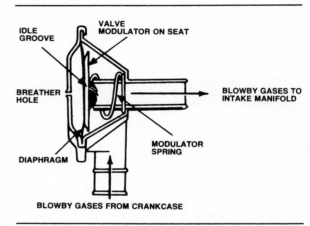

Figure 15-5. When the engine is at idle, the Type 2 PCV valve is closed by crankcase vacuum. Vapors flow through the idle groove at about 3 cubic feet per minute.

fornia cars. Open systems were installed on many 1963 models nationwide. Open crankcase ventilation systems can be divided into three types.

Type 1 system operation
A hose connects the crankcase with the intake manifold, figure 15-3. When the engine is running, fresh air is drawn into the crankcase through the vented oil filler cap. This air mixes with the crankcase vapors, travels to the intake manifold, and is drawn into the engine cylinders. Airflow from the crankcase is metered through a PCV valve, figure 15-4, that has a spring-operated plunger which controls the rate of airflow through the engine.

Type 1 open PCV systens were used as original equipment on many domestic passenger car engines through the mid-1960's.

Type 2 system operation
A Type 2 system is similar to a Type 1 system, but uses a special PCV valve and oil filler cap. The diaphragm-type PCV valve regulates the airflow according to crankcase vacuum. When a vacuum exists in the crankcase, the valve is closed, figure 15-5. When the crankcase is under pressure, figure 15-6, the valve opens. The oil filler cap has an orifice large enough to allow the right amount of air to enter but small enough to maintain crankcase vacuum. The orifice must remain open at all times for the system to work correctly.

Some imported cars, particularly British models in the mid-1960's, had Type 2 PCV systems as original equipment. Called a Smith's PCV valve, figure 15-7, it worked like the one we just described and could be opened for cleaning.

Type 3 system operation
In a Type 3 system, the crankcase is connected to the air cleaner by a hose, figure 15-8. Vapors are drawn from the crankcase by a slight suction in the air cleaner. No control valve is used. Fresh air enters through the oil filler cap, mixes with the crankcase vapors, and is drawn to the air cleaner. From the air cleaner, the vapors are then drawn through the carburetor and into the cylinders.

Positive Crankcase Ventilation

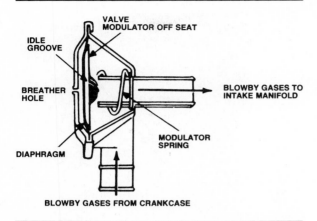

Figure 15-6. When the engine is at cruising speed, the Type 2 PCV valve is opened by crankcase pressure. The flow rate depends on the amount of blowby created by the engine.

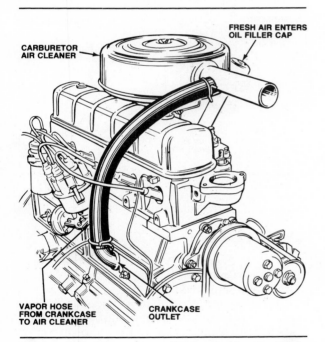

Figure 15-8. A Type 3 open PCV system has no PCV valve.

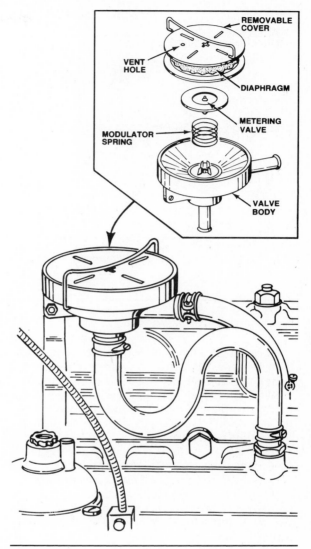

Figure 15-7. The Type 2 PCV valve used on some imported cars could be disassembled for cleaning.

■ Check PCV Valve For Fuel Dilution

Winter driving, with its quick trips and cold starts, can cause fuel dilution of the engine oil. This will not only thin out the oil, but the gasoline vapors fed to the intake manifold through the PCV system will produce a rich idle mixture. If you are working on a car with a rich idle mixture and you suspect fuel dilution of the oil, you can check it quickly.

Remove the PCV valve from the valve cover and let it draw in fresh air with the engine running. If the idle mixture becomes normal or leaner than normal, the engine oil probably has too much fuel dilution. Drive the car at highway speeds for up to half an hour to purge fuel vapors from the crankcase. Then readjust the idle mixture. As an alternative, drain the crankcase and refill with fresh oil.

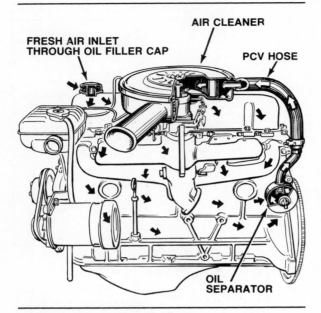

Figure 15-9. Ford Type 3 PCV systems had oil separators at the crankcase outlet.

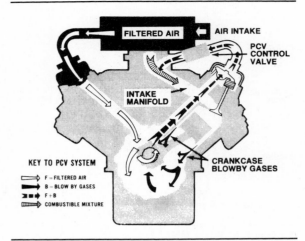

Figure 15-10. Type 4, sealed PCV system operation. (Chevrolet)

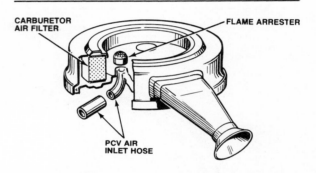

Figure 15-11. When PCV inlet air is drawn from the clean side of the air cleaner, a flame arrester is required in case of a backfire.

Because blowby reaches the carburetor *before* the air-fuel vaporization, this system tends to richen the air-fuel mixture. Carburetors used with a Type 3 system must be adjusted to make up for this richer mixture. Some imported cars use a "sealed" Type 3 system in which the oil filler cap is not vented.

Type 3 PCV systems were used as original equipment on some Ford and American Motors cars in the early 1960's. The systems on Ford 6-cylinder engines had oil separators at the crankcase outlets to minimize the amount of oil drawn through the PCV hose to the air cleaner, figure 15-9.

Limited efficiency of the open PCV system
Open PCV systems only partly control crankcase emissions. Since manifold vacuum decreases considerably under heavy load or acceleration, crankcase pressures build up. This forces some of the vapors into the atmosphere through the vented oil filler cap. Crankcase vapors also pass through the vented cap into the air if the system becomes clogged.

Open PCV systems do provide some benefits:
1. They promote longer engine life by removing most harmful vapors from the crankcase.
2. They reduce the amount of crankcase vapors which pollute the air.

Closed PCV Systems

Closed, or Type 4, PCV systems were required on all new California passenger cars in 1964. They were standard nationwide by 1968, and are still used today on all new domestically built cars. In a closed PCV system, figure 15-10, the oil filler cap is not vented to the atmosphere. Air for the crankcase is drawn through a hose from the air cleaner to one of the valve covers or to a crankcase inlet below the intake manifold. The dipstick is also sealed to prevent air from leaking into the crankcase.

Crankcase ventilation air may come from either the clean side (inside) or the dirty side (outside) of the carburetor air filter. When air is from the clean side using the air cleaner as a PCV filter, a flame arrester, figure 15-11, is used. This wire screen is in the PCV air intake line, either at the air cleaner or at the valve

Positive Crankcase Ventilation

189

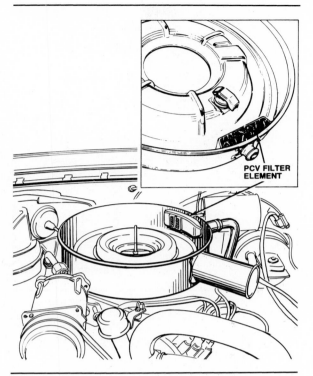

Figure 15-12. This PCV air filter is located in the air cleaner housing.

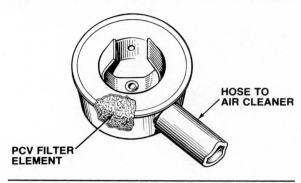

Figure 15-13. This PCV inlet air filter is located in the oil filler cap.

flame arrester is not required in this case, since the air cleaner filter does that.

Type 4 system operation
Under normal conditions, fresh air from the air cleaner passes through the inlet hose to the crankcase, figure 15-14. The fresh air mixes with the crankcase vapors, and passes through a PCV valve before being drawn into the intake manifold. Vapors that back up under certain conditions cannot escape from the closed system. If manifold vacuum drops, or if the system becomes clogged, extra crankcase vapors will reverse their direction. In the closed crankcase

■ **PCV System Service**

When a PCV system becomes restricted or clogged, the cause is usually an engine problem or the lack of proper maintenance. For example, scored cylinder walls or badly worn rings and pistons will allow too much blowby to pass. Start-and-stop driving requires more frequent maintenance and causes PCV problems more quickly than highway driving, as will any condition allowing raw fuel to reach the crankcase. Using the wrong grade of oil, or not changing the crankcase oil at periodic intervals will also cause the ventilation system to clog.

When a PCV system begins to clog, the engine tends to stall, idle roughly, or overheat. As ventilation becomes more restricted, burned plugs or valves, bearing failure, or scuffed pistons can result. Also look for an oil-soaked distributor or points, or leaking out around valve covers or other gaskets. Do not overlook the PCV system while troubleshooting. A partly or completely clogged PCV valve, or one of the incorrect capacity, may well be the cause of poor engine performance.

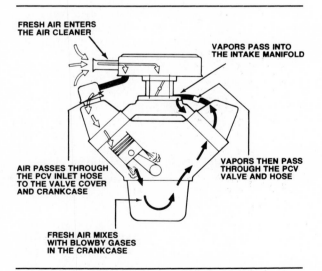

Figure 15-14. Closed PCV system operation under normal conditions.

cover, to prevent a crankcase explosion if the engine backfires.

When air is drawn from the dirty side of the air cleaner, a separate PCV air filter is used. This can be located in the air cleaner, figure 15-12, in the oil filler cap, figure 15-13, or in the inlet air hose where it connects to the valve cover. A

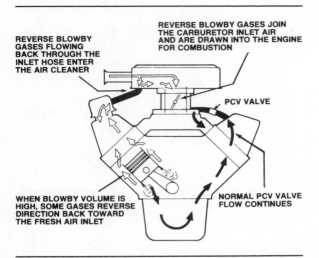

Figure 15-15. Closed PCV system operation under heavy load.

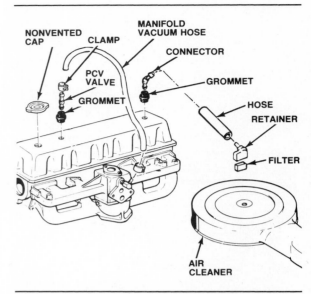

Figure 15-16. All late-model closed PCV systems share these common parts. (AMC)

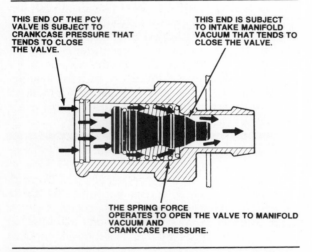

Figure 15-17. Spring force, crankcase pressure, and manifold vacuum work together to regulate PCV valve flow rate.

ventilation system, these excess vapors flow back to the air cleaner, figure 15-15, instead of passing out of the engine and into the atmosphere. Once in the air cleaner, they mix with incoming air and pass through the carburetor to be burned in the combustion chamber. This makes the closed system almost 100 percent effective in controlling crankcase emissions.

Closed PCV system efficiency

Closed crankcase ventilation provides three benefits:
1. It promotes longer engine life by removing harmful vapors from the crankcase.
2. It eliminates crankcase vapors that pollute the air.
3. It increases fuel economy by recirculating all unburned blowby back to the intake manifold.

ORIGINAL EQUIPMENT CLOSED PCV SYSTEMS

All new cars sold in the United States since 1968 have a Type 4, closed PCV system, figure 15-16. The design of closed PCV systems is essentially the same, regardless of the manufacturer. All use a PCV valve, an air inlet filter, and connecting hoses. The location of these parts may vary from one engine model to another, but all operate in the same manner.

Air Inlet Filter

PCV air inlet filters are usually installed in a retainer inside the air cleaner, figure 15-12, or in the oil filler cap, figure 15-13. They are made of wire gauze or polyurethane foam and are usually replaced at the same time as the air cleaner filter element.

Connecting Hoses

Two connecting hoses, figure 15-16, complete the closed PCV system. Fresh air travels from the air cleaner to the engine through the air inlet hose. Crankcase vapors travel from the engine to the intake manifold through the manifold vacuum hose. PCV hoses are made of special materials that resist oil vapors. Heater hose should *not* be used as a substitute.

PCV Valve

This one-way valve has a spring-operated plunger, figure 15-17, to control valve flow rate. Flow rate is set for each engine, and a valve for a different engine should not be substituted. This

Positive Crankcase Ventilation

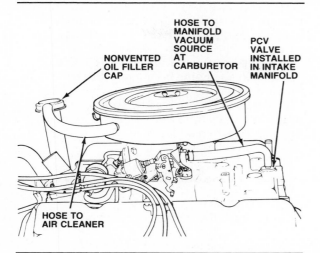

Figure 15-18. V-8 PCV system installation. (AMC)

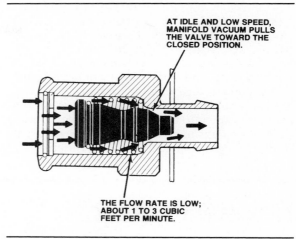

Figure 15-19. PCV valve airflow during cruising and light-load operation.

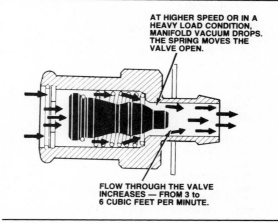

Figure 15-20. PCV valve airflow during acceleration and heavy-load operation.

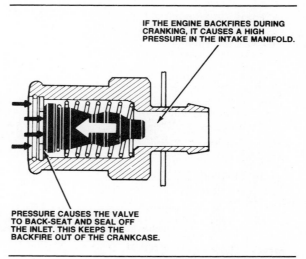

Figure 15-21. PCV valve operation in case of backfire.

setting is determined by the size of the plunger and the holes inside the valve. PCV valves are usually either in the valve cover, figure 15-16, or in the intake manifold, figure 15-18.

PCV valve operation

When manifold vacuum is high, the PCV valve restricts the airflow, figure 15-19, to keep a balanced air-fuel ratio. It also prevents high intake manifold vacuum from pulling oil out of the crankcase and into the intake manifold.

The PCV valve regulates the airflow through the crankcase under all driving conditions and speeds. Under high speed or heavy loads, the valve opens and allows maximum airflow, figure 15-20. If the engine backfires, the valve will close instantly, figure 15-21, to prevent a crankcase explosion.

RETROFIT SYSTEMS

Used cars sold in California must have a PCV system approved by the state. Those 1955-67 cars originally sold outside California with open PCV systems must be converted to a closed-type system, when sold as used cars within the state.

SUMMARY

Pressure in the engine cylinders forces combustion gases past the pistons. These gases, called blowby, settle in the engine crankcase, where they contaminate the lubricating oil and create harmful acids. Ventilation is necessary to remove the vapors from the crankcase. The draft tube system was used until the 1960's but it doesn't work well at low speeds. It also allows the vapors to pollute the air.

The recirculation of crankcase vapors to the intake manifold is called positive crankcase ventilation. Four types of positive crankcase ventilation, or PCV systems, have been used. Types 1, 2, and 3 are called open systems because vapors are forced into the atmosphere whenever manifold vacuum is low, or if the system is clogged. Type 4 is called a closed system because vapors cannot escape into the air under any conditions.

PCV systems use a fixed or variable metering device to regulate the airflow. The variable metering device, or PCV valve, is standard on modern PCV systems. It contains a spring-operated plunger and reacts to manifold vacuum.

It is important that the correct PCV valve be installed in an engine because of the lack of idle mixture adjustment on current carburetors. Differences in PCV valve idle airflow can greatly influence engine idle smoothness and vehicle driveability when incorrect valves are used.

Review Questions

Choose the single most correct answer. Compare your answers to the correct answers on page 248.

1. In a PCV system, crankcase vapors are recycled to the:
   a. Exhaust system
   b. Road draft tube
   c. Oil-filler breather cap
   d. Intake manifold

2. Which of these effects of blowby is not harmful to the engine:
   a. It destroys lubricating qualities of the engine oil
   b. It causes air pollution
   c. It causes sludge and varnish to form
   d. It causes formation of corrosive acids

3. Road draft tubes:
   a. Draw fresh air from the carburetor
   b. Work off intake manifold vacuum
   c. Are efficient at low speeds
   d. None of the above

4. Controlled crankcase ventilation was introduced in California in:
   a. 1959
   b. 1960
   c. 1961
   d. 1962

5. The illustration shows a:
   a. Type 1 PCV system
   b. Type 2 PCV system
   c. Type 3 PCV system
   d. Type 4 PCV system

6. Type 2 PCV systems have:
   a. A closed oil filler cap
   b. An open oil filler cap
   c. An oil filler cap with a preset orifice
   d. None of the above

7. In Type 3 PCV systems:
   a. The crankcase is connected to the intake manifold
   b. The PCV valve is a plunger type
   c. The PCV valve is a diaphragm type
   d. There is no PCV valve

8. The Type 3 PCV system:
   a. Is vented to the intake manifold
   b. Tends to make the air fuel mixture leaner
   c. Has no effect on fuel mixture
   d. Tends to make the fuel mixture richer

9. Closed (Type 4) PCV systems became standard nationwide in which year:
   a. 1964
   b. 1966
   c. 1967
   d. 1968

10. A separate flame arrester is used in a closed PCV system when inlet air is drawn from the:
    a. Clean side of the carburetor filter
    b. Dirty side of the carburetor filter
    c. The intake manifold
    d. The oil filler cap on a valve cover

11. The advantages of a closed PCV system are:
    a. It promotes longer engine life
    b. It eliminates crankcase vapors almost entirely
    c. It increases fuel economy
    d. All of the above

12. Which is not part of a Type 4 PCV system:
    a. A PCV valve
    b. A vented oil filler cap
    c. An air inlet filter
    d. A manifold vacuum hose

13. PCV system hoses are made from:
    a. Heater hose material
    b. Low-temperature-resistant nylon
    c. Oil resistant rubber material
    d. Fuel-resistant neoprene

14. The PCV valve operates in which of the following ways:
    a. Restricts airflow when intake manifold vacuum is high
    b. Increases airflow when intake manifold vacuum is low
    c. Acts as a check valve in case of carburetor backfire
    d. All of the above

Chapter 16
Air Injection

During the early days of emission control, **air injection** was an easy way to meet the required standards. Air injection, also known as the air pump system, figure 16-1, was one of the first add-on devices to help oxidize HC and CO exhaust emissions. By 1966, Chrysler was the only domestic automaker not using an air pump system on at least some cars. Early air injection systems contained many hoses and tubes placed across the engine. The external connecting lines made it difficult to work on the engine, and hose failure due to engine heat was common.

Chrysler chose an engine modification instead of air injection and proved that emission standards could be met without the add-on pump system. The other automakers profited from Chrysler's experience, and their cars of the late 1960's relied more on engine modifications and less on air injection. The use of air injection decreased on most engines until 1972. Then, to meet the stricter emission standards that year, even Chrysler was forced to install an air injection system on some engines.

As emission standards became even more strict during the 1970's, air injection became more popular. When the catalytic converter was introduced in 1975, it was hoped that the air pump system could be abandoned for good. But, the extra air provided by the injection system was found necessary to increase the catalytic action on many engines. With these changes, the air injection system now seems a permanent part of emission control for many engines.

In this chapter you will learn:
* The principles of air injection
* The parts used in a typical air injection system
* The changes that have been made to create second-generation systems for use with catalytic converters.

Manufacturers use the following names for their air injection systems:
* American Motors — Air Guard
* Chrysler — Air Injection System
* Ford — Thermactor Air Injection System
* General Motors — Air Injector Reactor (AIR).

Regardless of the name, the systems are simple, and all systems are basically the same. A belt-driven air pump supplies fresh air to the injector nozzles in the exhaust manifold or cylinder head. The air mixes with the hot exhaust leaving the engine. This helps the oxidation, or burning, reaction necessary to reduce HC and CO emissions.

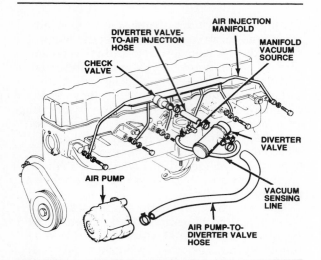

Figure 16-1. Typical late-model air injection system. (AMC)

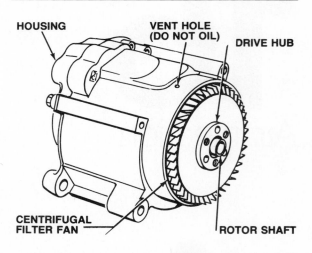

Figure 16-2. All air pumps used on domestic vehicles are made by the Saginaw Division of General Motors.

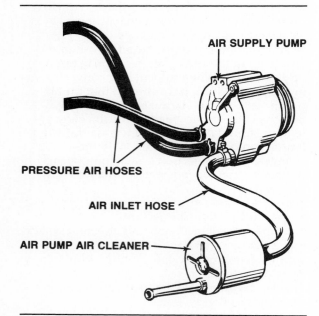

Figure 16-3. Older 3-vane pumps had separate inlet air cleaners.

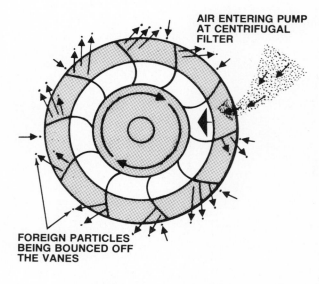

Figure 16-4. Dust and dirt are removed from the inlet air by centrifugal force.

BASIC SYSTEM DESIGN AND OPERATION

The basic air injection system contains the following parts:
1. Air supply pump with filter
2. Air manifolds and nozzles
3. Antibackfire valve
4. Check valve
5. Connecting hoses.

Air Supply Pump

The air pump, figure 16-2, is mounted at the front of the engine and driven by a belt from the crankshaft pulley. The pump pulls fresh air in through an external filter and pumps it under slight pressure to each exhaust port through connecting hoses. Adding this extra air to the hot HC and CO emissions in the exhaust manifold causes oxidation to take place, which helps change the HC and CO into H_2O and CO_2.

The Saginaw Division of General Motors makes the air pumps used on all domestic cars.

Air Injection

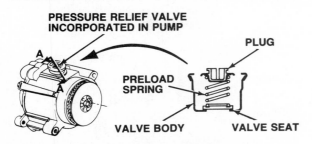

Figure 16-5. Some air pumps have built-in pressure relief valves. (AC-Delco)

Early Saginaw pumps (1966-67) were a 3-vane design and could be rebuilt if necessary. A Saginaw 2-vane design replaced the 3-vane pump on 1968 and later models. The 2-vane pumps cannot be rebuilt but must be replaced if they fail. Also, most older 3-vane pumps still in use are usually replaced with new 2-vane pumps if they fail.

The main difference between the two pumps is the way they filter the intake air. The 3-vane pump drew its fresh air supply through a separate air filter, figure 16-3, or from the clean side of the air cleaner. The 2-vane pump uses an impeller-type, centrifugal air filter fan, figure 16-2, mounted on the air pump rotor shaft. This is not a true filter, but cleans the air entering the pump by centrifugal force, figure 16-4. The relatively heavy dust particles in the air are forced in the opposite direction to the inlet airflow. The lighter air is then drawn into the pump by the impeller-type fan.

To keep pump pressure from becoming too high, many pumps use a pressure relief valve, figure 16-5, which opens at high engine speed. Some late-model pumps use a replaceable plastic plug to control the pressure setting of the relief valve. This pressure setting can be changed by putting in a plug with a different pressure setting. Pumps without a pressure relief valve use a diverter valve that has a relief valve.

Air Manifolds and Nozzles

In early air injection systems, air is delivered to the engine's exhaust system in one of two ways:
1. An external air manifold, figure 16-6, distributes the air through injection tubes to the exhaust port near each exhaust valve. This method is used mainly on smaller engines.
2. An internal air manifold, figure 16-7, distributes the air to the exhaust port near each exhaust valve through passages in the cylinder head or the exhaust manifold. This method is used mainly with larger engines.

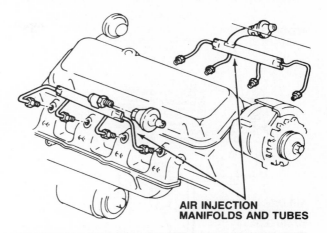

Figure 16-6. External air manifolds are used with many air injection systems. (Chevrolet)

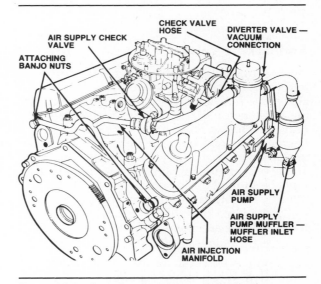

Figure 16-7. Some engines have air distribution passages built into the cylinder heads. (Ford)

The fresh air from the pump passes through the air injector tubes or the manifold to nozzles in the exhaust ports. These are made of stainless steel to resist the high exhaust temperatures.

Air Injection: A way of reducing exhaust emissions by injecting air into each of the exhaust ports of an engine. It mixes with the hot exhaust and oxidizes the HC and CO to form H_2O and CO_2.

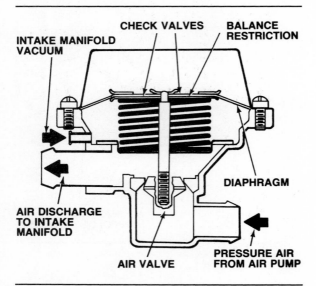

Figure 16-8. The gulp valve was used on early air injection systems.

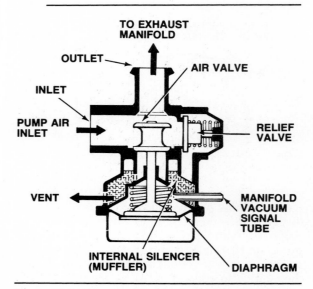

Figure 16-9. The diverter valve is used on most late-model air injection systems. This one also has a pressure relief valve.

Antibackfire Valve

During engine deceleration, high intake manifold vacuum richens the air-fuel mixture. If air is allowed to flow into the exhaust manifold during deceleration, it will combine with excess unburned fuel in the exhaust. The result is engine **backfire** — a rapid combustion of the unburned gases that can blow a muffler apart. To prevent engine backfire, the air pump flow must be shut off during deceleration. This is done by the antibackfire, or backfire suppressor, valve. Two kinds of valves have been used: the gulp valve and the diverter valve.

Gulp valve

Early air injection systems used a **gulp valve**, figure 16-8. When intake manifold vacuum is applied to the valve diaphragm, it causes the air valve to move. This redirects the pump air to the intake manifold to lean out the enriched air-fuel mixture during deceleration.

The gulp valve is connected to the intake manifold by two hoses. The large hose is the air discharge hose. The small hose is the sensing hose that sends manifold vacuum to the gulp valve to operate the diaphragm. There is a balance restriction, or bleed hole, inside the valve that equalizes pressure on both sides of the diaphragm after a few seconds. Even if manifold vacuum is high, the gulp valve only stays open for a few seconds until pressure equalizes.

Any sudden change in vacuum will operate the gulp valve. This is one of the undesirable features that led to its replacement. For example, the gulp valve will open when the engine starts. This can cause hard starting and a rough idle. Another problem with the gulp valve is that when the throttle is closed at high speeds when manifold vacuum is low, the valve may not open for a few seconds. The gulp valve has largely been replaced by the diverter valve on late-model air injection systems.

Diverter valve

The **diverter valve**, figure 16-9, is also called the dump valve or the bypass valve. Like the gulp valve, the diverter valve uses a diaphragm, operated by manifold vacuum, to redirect the airflow from the air pump. However, the pump air passes through the diverter valve continuously on its way to the air injection manifold. During deceleration, manifold vacuum operates the valve diaphragm to divert, or dump, the air directly to the atmosphere, not to the intake manifold. Some diverter valves vent the air to the engine air cleaner for muffling. Others vent it through a muffler and filter built into the valve. Because diverter valves do not affect the air-fuel mixture in the intake manifold, they are more troublefree than the gulp valves.

Some diverter valves also have a pressure relief valve to keep the pump from building up too much pressure in the system. This kind of diverter valve is used with an air pump that does not have a built-in relief valve.

Air Injection

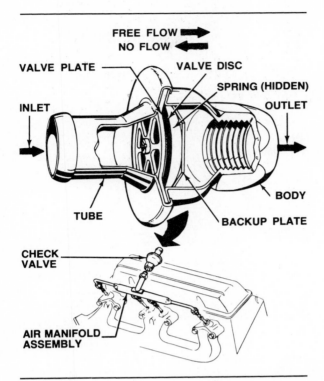

Figure 16-10. The check valve protects the system against reverse flow of exhaust gases. (AC-Delco)

Check Valve

A 1-way check valve, figure 16-10, protects the air pump from harmful reverse flow of exhaust gases from the engine. The check valve is between the air manifold and the diverter valve or the gulp valve. If exhaust pressure is higher than air injection pressure, or if the air pump fails, the check valve spring closes the valve to prevent the reverse flow of exhaust. A single check valve usually is installed on inline engines, and two valves (one per cylinder bank) usually are installed on V-type engines. However, some V-type engines have only one check valve, figure 16-7.

SECOND-GENERATION AIR INJECTION SYSTEMS

Air injection systems used with catalytic converters do the same thing as the basic air injection system just described. They help to oxidize the HC and CO in the exhaust gases by adding fresh air into the exhaust system. These newer, second-generation systems use many of the same parts as the non-catalytic systems. However, more controls are needed for effective use with the catalytic converter. Specific automakers' systems are explained in the following paragraphs.

Chrysler Catalyst Air Injection System

Some 1977 California engines use an air switching valve, figure 16-11, to switch air pump output from the exhaust ports to the right exhaust manifold. The valve is operated by an engine coolant vacuum switch. When the coolant warms up enough, the vacuum switch routes vacuum to the air switching valve. This causes the valve to switch the fresh air from the exhaust ports, "downstream" to the exhaust manifold. A bleed hole in the valve, figure 16-12, allows a small continuous airflow to the exhaust ports.

By switching the fresh airflow after engine warmup, the system helps oxidize HC and CO in the catalytic converter without interfering with the NO_x control by the EGR system.

Ford Thermactor Air Injection System

Ford air injection systems used with catalytic converters have become quite complicated since 1975. We could not possibly cover all the varia-

---

Backfire: The accidental combustion of gases in an engine's intake or exhaust system.

Gulp Valve: A valve used in an air injection system to prevent backfire. During deceleration it redirects air from the air pump to the intake manifold where the air leans out the rich air-fuel mixture.

Diverter Valve: Also called a dump valve. A valve used in an air injection system to prevent backfire. During deceleration it "dumps" the air from the air pump into the atmosphere.

■ Don't Oil The Air Pump

No air injection system is completely quiet. Usually pump noise increases in pitch as engine rpm increases. If the drive belt is removed and the pump shaft turned by hand, it will squeek or chirp. Many who work on their own cars and even some mechanics are not aware that air injection pumps are permanently lubricated, and require no periodic maintenance.

Suppose you pinpoint the air pump as the source of the noise. It would seem that a few squirts of oil would silence it. See those three small holes in the housing? While these are actually vents, it is easy to mistake them from oiling points. Don't oil them. More than a few pumps have failed because someone assumed that taking "good" care of the pump would make it last longer!

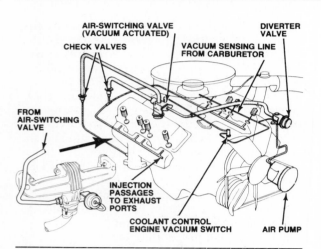

Figure 16-11. Chrysler introduced this air switching system in 1977. (Chrysler)

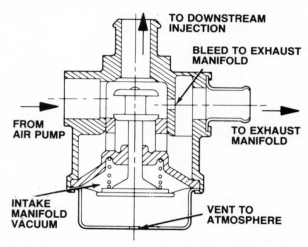

Figure 16-12. Cutaway of Chrysler's air switching valve. (Chrysler)

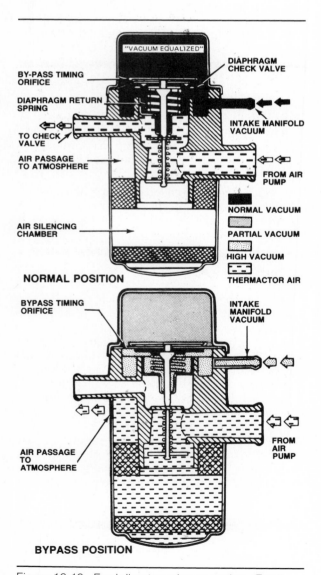

Figure 16-13. Ford diverter valve operation. (Ford)

tions here, but the major parts are described in the following paragraphs.

A large-capacity air pump without a relief valve is used on all 1975 and later Ford air injection systems. The system relief valve is built into the diverter valve. Ford uses three different kinds of diverter valves on late-model systems. (Ford calls its diverter valves "timed air bypass valves".)

One Ford valve, figure 16-13, works like the usual diverter valve we just described. Ford introduced another diverter valve in 1975 that uses manifold vacuum to hold the valve open and deliver air to the exhaust, figure 16-14. During cruising, manifold vacuum is applied to the valve diaphragm. During deceleration, long periods of idle, and cold engine operation, vacuum is cut off from the valve. This dumps the air pump output to the atmosphere. Various vacuum differential valves, vacuum delay valves, and bleed valves are used to control the vacuum to the diverter valve.

The vacuum differential valve, figure 16-15, is in the vacuum line to the diverter valve. Sudden changes in vacuum cause the differential valve to dump the vacuum that goes to the diverter valve. This causes the diverter valve to cut off the air injection and dump the air pump output to the atmosphere. This particular diverter valve, figure 16-14, can also be operated with a vacuum delay valve, an air temperature vacuum switch, and a ported vacuum signal taken above the throttle plates in the carburetor.

The third Ford diverter valve, figure 16-16, has a vacuum differential valve built into it. The top of the valve also has a vacuum vent. When the vent is closed, the valve acts as a normal diverter valve. When the vacuum vent is open,

Air Injection

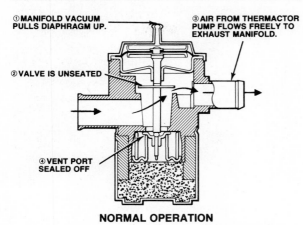

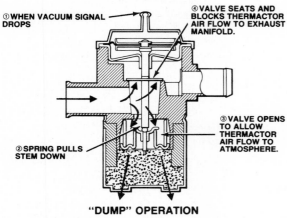

Figure 16-14. This Ford diverter valve, introduced in 1975, uses vacuum to hold the air passage *open* to the exhaust manifold. (Ford)

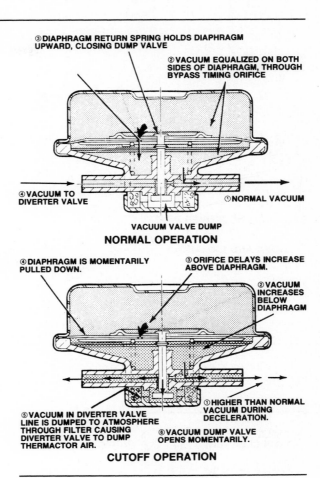

Figure 16-15. This vacuum differential valve dumps the vacuum from the diverter valve whenever vacuum increases suddenly. (Ford)

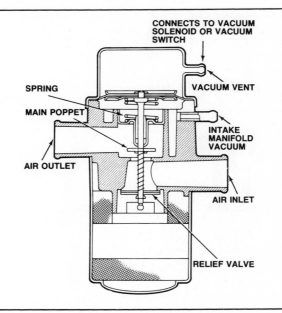

Figure 16-16. This Ford diverter valve has a vacuum vent that allows air to be dumped during cold engine operation. (Ford)

any manifold vacuum above 4 inches of Mercury causes the diverter valve to dump the air pump output. An air temperature electric switch and an electric solenoid open the vent line at cold temperatures.

General Motors Air Injection Reactor System

General Motors air injection system has been modified on some converter-equipped engines. The newer system is called the converter air injection reactor (CAIR) or converter air injection (CAI) system. The AIR designation for the older system may also be called manifold air injection (MAI).

The air pump, the diverter valve, and the check valve are the same in both converter and non-converter systems. In the CAI or CAIR system, air is injected into the exhaust pipe, downstream from the exhaust manifold, figure 16-17. This downstream injection delivers the

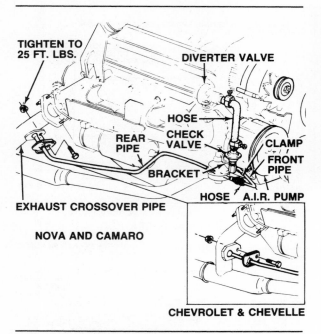

Figure 16-17. General Motors' converter air injection delivers the air to the exhaust pipe downstream from the exhaust manifold. (Chevrolet)

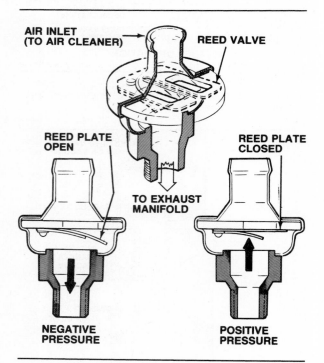

Figure 16-18. A pulse air injection valve opens when negative pressure occurs in the exhaust manifold. (Ford)

Figure 16-19. Pulse air injection was pioneered on Chevrolet's Cosworth Vega in 1975.

air directly to the catalytic converter to aid HC and CO oxidation. Keeping the air out of the exhaust manifold also keeps the combustion chamber temperature lower, which reduces NO_x emissions.

On some late-model GM cars (mainly those without catalytic converters), the air is still injected into the exhaust manifold. On these engines, this is the hottest place in the exhaust system to oxidize HC and CO.

PULSE AIR INJECTION

Since 1975, the use of pulse air injection has increased on GM, Chrysler, and Ford products. Pulse air injection systems do not use an air pump. Instead, they use a pulse air valve, figure 16-18, which is similar to the check valve in an air pump system. It is a spring-loaded, diaphragm or reed valve that is connected to the exhaust system.

Each time an exhaust valve closes, there is a period when pressure in the manifold drops below atmospheric pressure. During these low pressure (vacuum) pulses, the air injection valve opens to admit fresh air to the exhaust. When exhaust pressure rises above atmospheric pressure, the valve acts as a check valve and closes.

The pulse air injection valve works best at low engine speed when extra air is needed most by the catalytic converter. The vacuum pulses occur too fast at high speed for the valve to follow, and the internal spring simply keeps the valve closed.

Pulse air injection was first used on Chevrolet's Cosworth Vega in 1975 and 1976, figure 16-19. The system was added to other GM 4-cylinder engines in 1977, figure 16-20. Chrysler introduced a similar system in 1977, called an air aspirator system. The pulse injection (air aspirator) valve draws its fresh air from the clean side of the engine air filter, figure 16-21. In 1977, Ford installed a pulse air injection valve

Air Injection

201

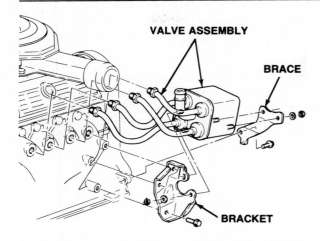

Figure 16-20. Pulse air injection was installed on the Chevette and other GM small cars in 1977. (Chevrolet)

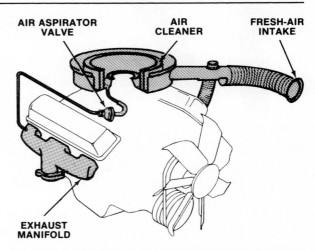

Figure 16-21. Chrysler's air aspirator system.

on the 300-cid 6-cylinder engines in some vans. This system was added to the 6-cylinder engines in Fairmonts and Zephyrs in 1978.

SUMMARY

Air injection is one of the oldest methods used to control HC and CO exhaust emissions. The injected air mixes with hot exhaust gas as it leaves the combustion chambers to further oxidize HC and CO emissions. All air injection systems used on domestic cars operate in essentially the same way, regardless of manufacturer. With slight modifications, the air injection system also is used on engines equipped with catalytic converters to promote converter action.

The pulse air injection system relies upon exhaust pulsations instead of an air pump to draw fresh air into the exhaust.

Review Questions

Choose the single most correct answer. Compare your answers to the correct answers on page 248.

1. The American Motors air injection system is called:
   a. Air Injection System
   b. Air guard
   c. Thermactor Air Injection System
   d. Air Injection Reactor

2. The main reason for an air injection system is to:
   a. Oxidize HC and CO exhaust emissions
   b. Reduce NO_x exhaust emissions
   c. Eliminate crankcase emissions
   d. Eliminate evaporative HC emissions

3. Oxidation of HC and CO emissions produces:
   a. HCO and CO_2
   b. H_2CO_3 and H_2O
   c. H_2O and CO_2
   d. All of the above

4. Which of the following is true:
   a. All air pumps can be rebuilt
   b. Two-vane pumps can be rebuilt
   c. No air pumps can be rebuilt
   d. Three-vane pumps can be rebuilt

5. Two-vane pumps have:
   a. An impeller-type fan for filter
   b. An integral wire mesh filter
   c. A hose to the clean side of the air cleaner
   d. A separate air filter

6. Air injection nozzles are made of:
   a. Copper
   b. Stainless steel
   c. Aluminium
   d. Vanadium

7. The two types of air injection backfire suppressor valves are:
   a. The check valve and the gulp valve
   b. The gulp valve and the diverter valve
   c. The diverter valve and the relief valve
   d. The diverter valve and the check valve

8. The Chrysler air switching valve is controlled by:
   a. A vacuum solenoid
   b. A reed valve
   c. A bypass timing orifice
   d. A coolant vacuum switch

9. This illustration shows:
   a. A relief valve
   b. A solenoid operated vacuum valve
   c. A gulp valve
   d. A diverter valve

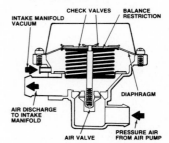

10. New GM air injection systems are called:
    a. MAI
    b. CAIR
    c. Timed Bypass System
    d. Vacuum Differential System

11. Pulsed air injection works best at:
    a. Low speeds
    b. High speeds
    c. Idle
    d. Deceleration

Chapter 17
Spark Timing Control Systems

Proper timing of the ignition spark can help car engines reduce exhaust emissions and meet U.S. Federal HC and NO_x standards. General Motors began using auxiliary **spark timing control** in some 1970 cars, but the systems achieved widespread use in 1971 to 1974 cars. Each automaker developed slightly different spark timing controls, according to engine requirements and emission standards for each model year, but the systems and devices all operate on the same principles.

This chapter explains these principles of spark timing control systems, their effect on engine emission levels, and how the specific manufacturers' systems function.

SPARK TIMING AND COMBUSTION

We already learned that a spark from the ignition system ignites the compressed air-fuel mixture in the combustion chamber. The burning flame front spreads through the combustion chamber as a continuous and progressive action, not an explosion. Approximately 3 milliseconds (0.003 second) elapse from the instant the air-fuel mixture ignites until its combustion is complete.

The ignition spark must occur early enough so that the combustion pressure reaches its maximum just after top dead center, when the piston is beginning its downward power stroke. Combustion should be completed by about 10° atdc. If the spark occurs too soon before top dead center, the rising piston will have to push against combustion pressure. If the spark occurs too late, the force on the piston will be reduced. In either case, power is lost. In extreme cases, the engine could be damaged. Ignition must start at the proper instant for maximum power and efficiency.

As engine speed increases, piston speed increases. If the air-fuel ratio remains relatively constant, the fuel burning time will remain constant. However, at greater engine speeds, the piston will travel farther during this burning time. Ignition timing must be changed to make sure that the highest combustion pressure occurs at the proper piston position.

For example, consider an engine, figure 17-1, that requires 0.003 second for the fuel charge to burn and that achieves maximum power if the burning is completed at 10° atdc.

• At an idle speed of 625 rpm, position A, the crankshaft rotates about 11 degrees in 0.003 second. Therefore, timing must be set at 1° btdc to allow enough burning time.
• At 1,000 rpm, position B, the crankshaft rotates 18 degrees in 0.003 second. Ignition should begin at 8° btdc.
• At 2,000 rpm, position C, the crankshaft ro-

Spark Timing Control Systems

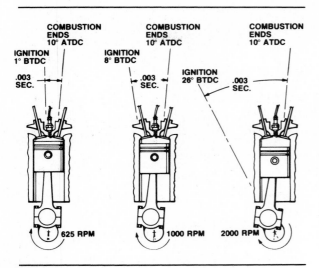

Figure 17-1. An example of ignition timing.

tates 36 degrees in 0.003 second. Spark timing must be advanced to 26° btdc.

Change in timing is called spark advance, or ignition advance, and is typically the job of the distributor's mechanical and vacuum advance units.

SPARK TIMING AND EMISSION CONTROL

Spark timing affects combustion temperature. Firing the spark at precisely the right instant, as in the example above, will create the maximum possible amount of heat and pressure and the maximum possible engine power. Unfortunately, this great heat also creates a large amount of NO_x exhaust emissions.

An efficient combustion process starts with an advanced spark. High-temperature combustion occurs at an early point during the process. However, NO_x emission levels are increased by higher combustion temperatures. Later in the combustion process, the exhaust gases are cooler. They do not heat up the exhaust manifold as much. But, this cooler exhaust temperature allows a great amount of HC emissions to be exhausted.

Spark timing controls keep ignition timing retarded during idle and low-speed operation, when the air-fuel mixture is rich. Retarded timing reduces the peak combustion temperature because ignition occurs when cylinder pressure is lower. This helps to reduce NO_x formation. At the same time, the greatest combustion temperatures occur during the end of combustion. This results in higher exhaust temperatures, which reduce the amount of HC in the exhaust.

SPARK TIMING EMISSION CONTROL SYSTEMS

All of the systems we are about to describe work with the distributor vacuum advance. Early controls concentrated on advancing or retarding the ignition timing under certain operating conditions. Later systems simply delay vacuum advance at low and intermediate speeds, and allow it during high-speed cruising. Electronically controlled ignition timing is discussed at the end of this chapter.

When catalytic converters were introduced in 1975, many spark timing control systems were no longer needed. Converters reduce HC and carbon monoxide (CO) emissions, and exhaust gas recirculation (EGR) provides more effective control of NO_x emissions.

Early Distributor Controls

Early emission-control equipment advanced or retarded ignition timing under certain engine operating conditions, usually during starting, deceleration, and idle.

The deceleration vacuum advance valve, figure 17-2, was used in the middle to late 1960's on Chrysler, Ford, AMC, and Pontiac products with manual transmissions. In a car with a manual transmission, the air-fuel mixture becomes extremely rich upon deceleration or gear shifting.

This valve momentarily switches the vacuum operating the vacuum advance chamber from a low vacuum source at the carburetor to a high manifold vacuum source during deceleration, then back to the low carburetor vacuum source. This prevents over-retard during deceleration or gear shifting which could cause high CO emissions in some engines.

The deceleration vacuum advance valve was discontinued in the early 1970's because it was most effective against CO emissions, but not against HC and NO_x emissions. As emission limits for HC and NO_x became stricter, manufacturers developed other devices which proved effective against all three major pollutants.

The distributor retard solenoid was used in 1970 and 1971 on some Chrysler products with V-8 engines and automatic transmissions. This electric solenoid is attached to, and controls the action of, the distributor vacuum advance unit, figure 17-3.

Spark Timing Control: A way of controlling exhaust emissions by controlling ignition timing. Vacuum advance is delayed or shut off at low and medium speeds, reducing NO_x and HC emissions.

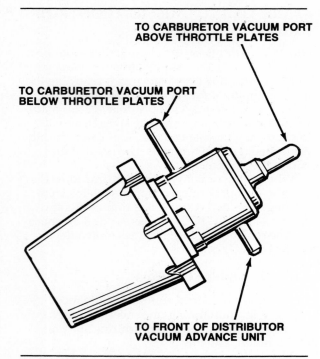

Figure 17-2. A deceleration vacuum advance valve.

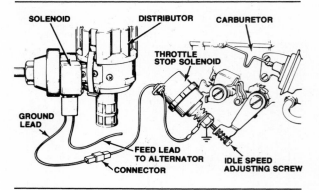

Figure 17-3. A distributor retard solenoid.

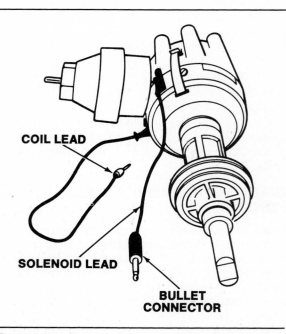

Figure 17-4. The installation of a distributor advance solenoid.

The solenoid is energized by contacts mounted on a carburetor throttle stop solenoid. When the throttle is closed, the carburetor solenoid is touched by the idle adjusting screw to complete the ground circuit. The contacts in the carburetor solenoid carry current to the distributor solenoid windings. Since the distributor solenoid's armature is connected to the vacuum diaphragm, movement of the distributor solenoid shifts the breaker plate in the retard direction.

When the engine speed increases, the idle adjusting screw breaks contact with the carburetor solenoid. Current flow to the distributor solenoid is stopped, and normal vacuum advance is allowed.

The distributors on some 1972 and 1973 Chrysler V-8 engines have a spark timing advance solenoid that aids better starting by advancing the spark by 7½ degrees. The solenoid is mounted in the distributor vacuum unit, figure 17-4. It is activated by power from the starter relay at the same terminal that sends power to the starter solenoid. The advance solenoid is activated only while the engine is being cranked.

The starting advance solenoid is not an emission control device by itself, but it allows lower basic timing settings which help in emission control while providing an advanced timing setting for quicker starting.

Principles of Recent Systems

The most common types of control systems used in the early to mid-1970's are:
- Vacuum delay valves
- Speed- and transmission-controlled timing.

Vacuum delay valves

The vacuum delay valve "filters" the carburetor vacuum, making it take longer to get to the distributor advance mechanism. Generally, vacuum must be in the system for 15 to 30 seconds before it is allowed to affect the advance mechanism.

One method of vacuum delay is used in Ford's spark delay valve (SDV) system, figure 17-5. Here, vacuum must work its way through a **sintered**, or sponge-like, metal disc to reach the distributor. Many GM engines also use this type of spark delay valve.

Spark Timing Control Systems

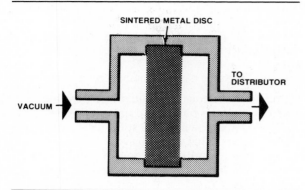

Figure 17-5. A cross-sectional view of the type of vacuum delay valve that uses sintered metal to slow the application of vacuum.

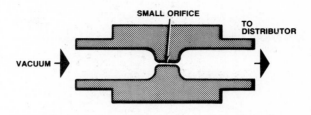

Figure 17-6. A cross-sectional view of the type of vacuum delay valve that uses a small orifice to delay the application of vacuum.

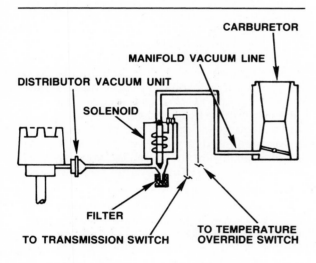

Figure 17-7. A simplified transmission-controlled spark system.

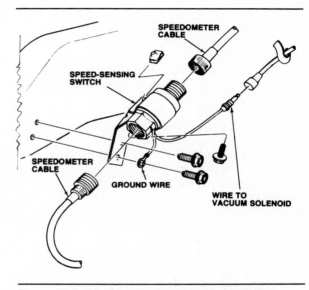

Figure 17-8. Most speed-controlled spark systems use a speed-sensing switch such as this. (Cadillac)

Another method of vacuum delay is used in Chrysler's orifice spark advance control (OSAC) system, figure 17-6. A restriction is placed in the vacuum line to delay vacuum buildup.

Manufacturers often use vacuum delay valves along with other emission control systems. All valves operate on one of the two principles described above.

Speed- and transmission-controlled timing
These systems prevent any distributor vacuum advance when the car is in a low gear or is traveling slowly. A solenoid controls the application of carburetor vacuum to the advance mechanism, figure 17-7. Current flow through the solenoid winding is controlled by a switch that reacts to the car's operating conditions.

A control switch used with a manual transmission can sense shift lever position. A control switch used with an automatic transmission will usually work from hydraulic fluid pressure. Both systems prevent any vacuum advance when the car is in a low or intermediate gear.

A speed-sensing switch may be connected to the vehicle speedometer cable, figure 17-8. The switch signals an electronic control module when the vehicle speed is below a certain level. The module triggers a solenoid that controls engine vacuum at the distributor.

Both vacuum-delay systems and speed- and transmission-controlled systems usually have an engine temperature bypass. This allows normal vacuum advance at high and low engine temperatures. Before March, 1973, some systems had an **ambient temperature** override switch. Most of these switches were dis-

---

Sintered: Welded together without using heat to form a porous material, such as the metal disc used in some vacuum delay valves.

Ambient Temperature: The temperature of the air surrounding a particular device or location.

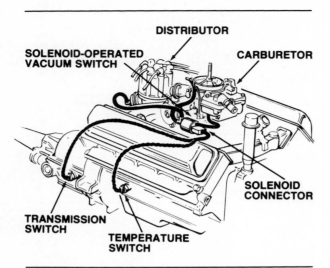

Figure 17-9. A typical General Motors transmission-controlled spark system.

continued at the direction of the Environmental Protection Agency (EPA). Later temperature override systems sense coolant temperature or under-hood temperature.

The systems using these principles are known by many different trade names, and are all somewhat different. The following paragraphs describe the major systems used by domestic manufacturers.

General Motors

Transmission-controlled spark (TCS)

The TCS system was introduced by GM on its 1970 automobiles. A solenoid-operated vacuum switch near the carburetor, figure 17-9, prevents vacuum advance in low and intermediate gears. Vacuum is blocked, and the vacuum advance unit is vented to the atmosphere.

In high gear, the solenoid closes the vent and opens the vacuum port. This sends vacuum to the distributor advance unit. The solenoid is controlled by the transmission switch, which senses shift lever position on manual transmissions and fluid pressure on automatics.

Transmission control switch and solenoid operation on GM systems varies for different models and years. In some systems, the transmission switch is closed in low and intermediate gears to activate the solenoid and deny vacuum advance. In other systems, the switch is open and the solenoid deactivated to deny vacuum advance. In either case, switch and solenoid positions change from their low-gear positions to allow vacuum advance when the transmission is shifted into high gear.

A coolant temperature switch provides manifold vacuum to the distributor at low coolant temperatures. Some GM systems also have a hot coolant override switch for vacuum advance in all gears at high coolant temperatures. A time delay relay opens the solenoid vacuum switch during starting and for the first few seconds after the engine is started. This provides vacuum advance for easier starting and warmup.

Combination emission control (CEC) valve

The CEC valve is part of the transmission-controlled spark system on some 1971 through 1973 GM cars. This valve provides vacuum spark advance control and deceleration throttle position control in high gear. The CEC valve is mounted on the side of the carburetor, figure 17-10, and is a simple solenoid with a vacuum valve at one end and a throttle check rod extending from the other.

When the transmission is in low and intermediate gears, the solenoid is deenergized and vacuum to the distributor advance unit is shut off. The vacuum is vented to the atmosphere through a filter on the CEC valve.

When the transmission is in high gear, the TCS switch energizes the CEC solenoid to allow vacuum to the distributor advance unit. At the same time, the CEC throttle rod extends to hold the throttle open by a predetermined amount during high-gear deceleration. When the transmission is shifted to a lower gear, the CEC valve is deenergized and the throttle rod retracts, allowing a normal idle and cutting off vacuum.

Distributor vacuum control switch

The distributor vacuum control switch, figure 17-11, is used on some Oldsmobiles along with a normally closed TCS switch. The vacuum control switch does the dual jobs of a TCS solenoid and a thermostatic vacuum switch. When the engine is at normal operating temperatures, it permits vacuum spark advance only when the transmission is in high gear. When engine coolant temperature increases, the switch sends full manifold vacuum to the distributor, regardless of gear.

When the transmission is in low or intermediate gears, the TCS switch is closed to energize the solenoid. The solenoid seals off the carburetor port and vents the distributor vacuum unit to the atmosphere. During high-gear operation, the deenergized solenoid seals off the vent port and applies carburetor vacuum to the distributor.

In case of engine overheating, due to prolonged idling with a retarded spark, the thermostatic portion of the switch seals off the vent port and applies manifold vacuum to the distributor.

Some Buick engines use a combination thermostatic vacuum switch and TCS solenoid that works the same as the distributor vacuum control switch we just described above.

Spark Timing Control Systems

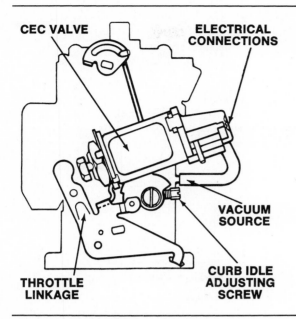

Figure 17-10. The General Motors combination emission control valve.

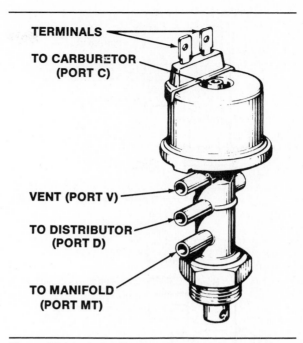

Figure 17-11. A distributor vacuum control switch.

Speed control switch (SCS) system
The SCS system, used on 1972 Cadillacs and Pontiacs, controls distributor vacuum advance by the speed of the car rather than by the transmission gear it is in.

A speed sensor switch in the speedometer cable or in the transmission speedometer drive gear controls a normally deenergized vacuum solenoid. At low speeds, the vacuum solenoid is energized because the speed switch is closed. The distributor vacuum unit is vented to the atmosphere, denying vacuum spark advance. At higher speeds, the speed switch opens and the solenoid is deenergized. Normal vacuum advance is allowed. If the engine begins to overheat, a thermostatic vacuum switch applies manifold vacuum to the distributor regardless of vehicle speed.

American Motors

Before 1971, AMC products used dual-diaphragm vacuum units and deceleration valves to control exhaust emissions. Some models have a thermostatic vacuum valve, figure 17-12, that allows normal vacuum advance at low engine temperatures. As engine temperature increases, the valve connects manifold vacuum to the distributor advance unit.

After 1971, many AMC products use a speed- or transmission-controlled spark timing system similar to General Motors' TCS. A control switch reacts to the car's operating conditions and controls current flow to a solenoid vacuum valve, figure 17-13.

■ Are Spark Timing Controls Confusing?

As you study this chapter, you may be confused by the various spark control systems. If so, here is a summary which may help you understand this form of emission control.

The greatest amount of exhaust emissions are produced at idle and during low-speed operation. *ALL* spark control systems are designed to reduce these emissions by preventing vacuum advance at idle, and when the car is in the lower gears. This means that a spark control system should work *ONLY* when the engine is at normal operating temperature, and the car is operating at low speeds or in the low gears. When the engine is at normal operating temperature and the car is at cruising speed or in high gear, vacuum advance takes place just as it would without a spark control system.

ALL spark control systems have some form of temperature control to prevent the system from operating if the engine is too cold or too hot.

Here are three troubleshooting tips:
1. If vacuum advance *is* present when the engine is at idle or in the low gears, then a speed sensor, transmission switch, control module, or solenoid is defective.
2. If vacuum advance is *not* present when the engine is in high gear, then a speed sensor, transmission switch, control module, or solenoid is defective.
3. If vacuum advance is *not* present when the engine is cold or overheated, then a temperature sensor or switch is defective.

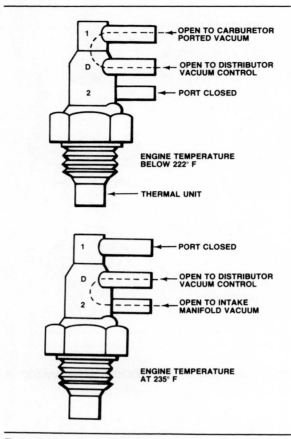

Figure 17-12. American Motors' thermostatic vacuum valve. (AMC)

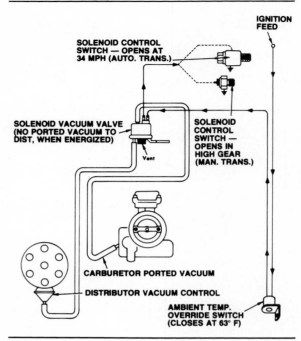

Figure 17-13. The AMC transmission-controlled spark system. (AMC)

Transmission control switches on AMC systems are closed in low and intermediate gears to energize the solenoid and block vacuum advance. In high gear, the switch opens, the solenoid deenergizes, and vacuum advance is applied.

Before March 1973, cars with automatic transmissions used a speed-sensing control switch. On later cars, fluid pressure operates the automatic transmission control switch. On cars with manual transmissions, the position of the gear lever activates the control switch.

Some AMC cars have an ambient temperature override switch that allows normal vacuum advance at low temperatures. Later cars may have a coolant temperature override switch that applies manifold vacuum to the distributor when coolant temperature is low, and carburetor vacuum when coolant temperature is high. When this coolant switch is used along with a transmission control switch, vacuum from the carburetor is controlled but vacuum from the manifold is *always* applied when coolant temperature is low.

Chrysler

NO$_x$ spark control

Chrysler cars for 1971 and 1972 sold in California have a transmission-controlled spark system similar to those of GM and AMC. The Chrysler NO$_x$ control system for cars with manual transmissions has a normally open transmission switch and a deenergized solenoid vacuum valve. When the transmission is in low and intermediate gears, the switch is closed and the solenoid is energized to cut off vacuum advance. When the transmission is in high gear, the switch opens and the solenoid is deenergized to provide vacuum advance.

An ambient temperature switch was used on 1971 cars to allow vacuum advance at all times when the temperature is low. This switch was discontinued in 1972.

The NO$_x$ control system on cars with manual transmissions uses an electric control unit, figure 17-14, to operate the solenoid vacuum valve. The control unit is a reversing relay that receives signals from grounding switches: a temperature switch, a speed switch, and a vacuum switch (the vacuum switch was removed in 1972). All switches must be open to activate the control unit, energize the solenoid, and cut off vacuum advance.

Orifice spark advance control (OSAC) system

The OSAC valve is used on 1973 and later models. It is in the vacuum line between the

Spark Timing Control Systems

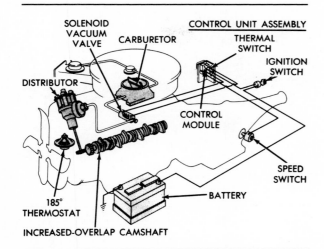

Figure 17-14. Chrysler's NO$_x$ control system.

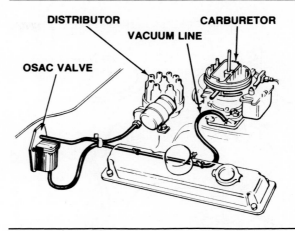

Figure 17-15. The Chrysler orifice spark advance control system, in its original location on the firewall. Later versions placed the OSAC valve in the air cleaner.

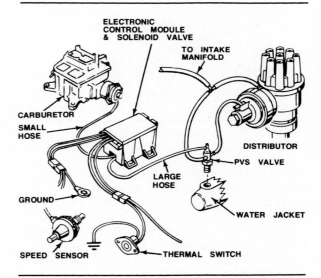

Figure 17-16. Ford's electronic distributor modulator system.

carburetor and the distributor, figure 17-15. It contains a small orifice that delays the rate of vacuum buildup in the distributor vacuum unit during acceleration from idle to part throttle. When the throttle is closed, vacuum is cut off to the OSAC valve. A check valve within the OSAC valve opens to relieve vacuum in the line to the distributor.

The first OSAC valves used in 1973 had a temperature control device that bypassed OSAC operation and allowed immediate vacuum advance at low ambient temperature. The temperature control device was eliminated in March of 1973, and the OSAC valve was moved from its original location on the firewall to the air cleaner cover.

Some OSAC installations include a thermal ignition control valve, which is another type of coolant-sensing valve. It is installed in the engine cooling system, and the vacuum lines from the OSAC valve to the distributor run through it. A manifold vacuum line also runs to the thermal valve. When coolant temperature increases, the thermal valve applies full manifold vacuum to the distributor.

When the catalytic converter was introduced in 1975 cars, the OSAC valve was eliminated from many models. However, it continues on some Chrysler cars.

Ford

Electronic distributor modulator (EDM) system
The EDM system was used on some 1970 and 1971 Ford vehicles. The system allows vacuum spark advance only during high-gear operation.

The system consists of a speed sensor, an ambient temperature switch, a coolant temperature switch, an electronic control module, and a 3-way solenoid valve, figure 17-16.

The speed sensor is connected between two sections of the speedometer cable. A rotating magnet, driven by the cable, turns inside a stationary winding. Above a selected speed, the voltage created by the magnet is sent to the electronic control module.

The ambient temperature switch is mounted near either front door hinge and senses outside air temperature. It will override the speed sensor signal when air temperature is low, to permit the distributor vacuum advance unit to operate normally.

The coolant temperature switch is also called the distributor vacuum control valve or the ported vacuum switch. It overrides the module to apply intake manifold vacuum to the distributor when coolant temperature is high.

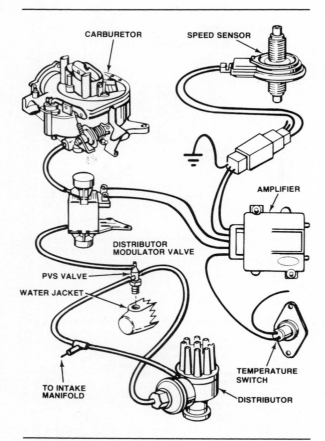

Figure 17-17. The Ford electronic spark control system.

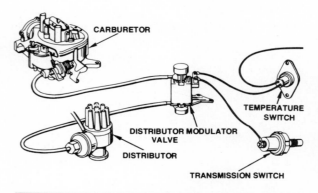

Figure 17-18. The Ford transmission-regulated spark system.

The electronic control module contains the 3-way solenoid valve and is mounted in the passenger compartment under the dash. The module receives signals from the speed sensor and ambient temperature switch. It determines whether or not the solenoid will be energized.

If the ambient temperature is high enough, the solenoid will be energized at speeds above 23 mph. Ported vacuum from the carburetor is then applied to the distributor vacuum unit. On deceleration, the reduced voltage signal from the speed sensor deactivates the solenoid. Carburetor vacuum is then shut off from the advance unit.

Electronic spark control (ESC) system
The ESC system, used on some 1972 Ford products, is similar to the EDM system just described. The main difference is that the ESC module is separate from the solenoid valve, figure 17-17, while the earlier EDM module contained the solenoid valve.

Transmission-regulated spark (TRS) system
The TRS system was introduced by Ford in 1972 and, like the EDM and ESC systems, denies vacuum spark advance at low speed. The system consists of a 2-way distributor solenoid valve, a transmission switch, and an ambient temperature switch, figure 17-18. In some applications, a spark delay valve is also used.

The distributor solenoid valve is installed between the carburetor and the distributor. In low and intermediate gears, the transmission switch is closed to energize the solenoid and cut off the vacuum to the distributor.

When the transmission is shifted into high gear, the transmission switch opens, the solenoid is deenergized, and vacuum is applied to the distributor.

At low temperatures, the ambient temperature switch contacts open to deenergize the solenoid and allow normal vacuum advance in all gears. As temperature increases, the contacts close, and the solenoid is controlled by the transmission switch.

On some versions, the vacuum lines are routed through a ported vacuum switch like that used in the EDM and ESC systems. When coolant temperatures increase, manifold vacuum is applied to the distributor by the ported vacuum switch.

When exhaust gas recirculation was introduced in 1973, the TRS system on some cars was modified to become the TRS+1 system. The "+1" part of the system controls EGR valve operation.

The TRS and TRS+1 systems were discontinued in 1974.

Spark delay valve (SDV) system
This system, figure 17-19, uses a spark delay valve to restrict the vacuum applied to the carburetor during acceleration. The valve is installed between the carburetor and the distributor, and vacuum must work its way through a sintered metal disc to reach the advance diaphragm. When carburetor vacuum decreases, a check valve in the SDV opens to relieve vacuum from the advance unit.

On engines built before March 1973, a delay

Spark Timing Control Systems

valve bypass is used to allow full advance at low ambient temperature. An ambient temperature switch controls a solenoid vacuum valve. At low temperature, the switch closes to energize the solenoid and apply vacuum to the distributor, bypassing the SDV.

Temperature-activated vacuum (TAV) system
The TAV system is used on some Ford products to match the vacuum spark advance to engine requirements by switching between two vacuum sources, figure 17-20. The 3-way vacuum solenoid valve is connected to the distributor, the carburetor spark port, and the carburetor EGR port.

An ambient temperature switch controls the 3-way valve. At high temperatures, the switch is closed, the solenoid is energized, and EGR vacuum controls the spark advance. At lower temperatures, the switch is open, the solenoid is deenergized, and carburetor spark port vacuum controls the vacuum advance.

A variation of the TAV system is the cold-temperature-activated vacuum (CTAV) system. In this system, the ambient temperature switch is located in the air cleaner. The temperature required to energize the solenoid is greater than in the TAV system, and a latching relay is used to allow only one cycle of the temperature switch each time the ignition is turned on.

ELECTRONICALLY CONTROLLED TIMING

Extreme accuracy from the ignition system is needed to meet emission standards and fuel mileage requirements. Centrifugal and vacuum advance devices often cannot react fast enough to changes in engine operating conditions. Manufacturers are developing computer-controlled ignition systems to gain this necessary accuracy.

In these systems, an electronic control module receives signals from various sensors. These signals may include information on coolant temperature, atmospheric pressure and temperature, throttle position and rate of change of position, and crankshaft position. Integrated circuits in the control module are programmed to interpret this information and calculate the proper ignition timing for each individual spark.

The systems now used work with the manufacturers' standard solid-state ignition systems. Some changes are made to the standard ignition, because it no longer has to control spark timing, but many parts remain the same.

Two types of computer-controlled ignition systems are used by domestic manufacturers. One type depends on distributor shaft rotation to send a crankshaft position signal to the control module. The other type receives crankshaft

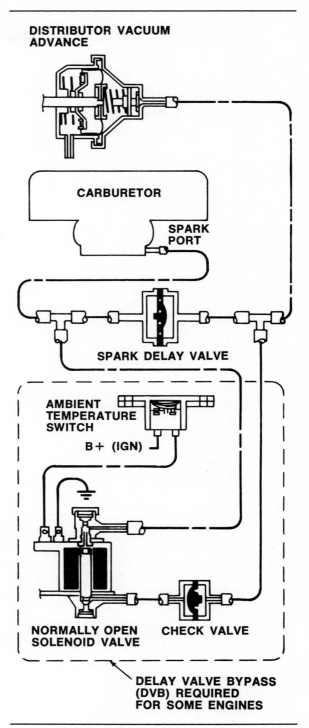

Figure 17-19. Ford's spark-delay valve system.

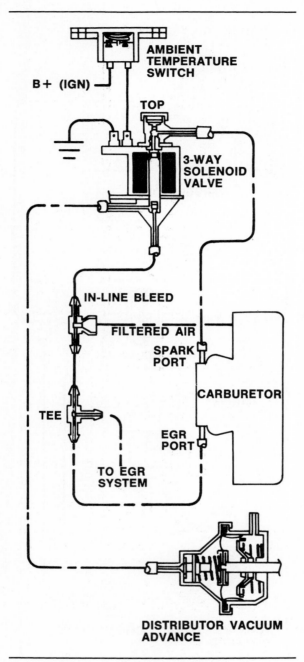

Figure 17-20. Ford's temperature-activated vacuum system.

position information from a sensor mounted near the crankshaft, figure 17-21. The sensor reacts to the rotation of a special disc attached to the crankshaft itself.

Signals taken directly from the crankshaft are more accurate than those taken from the distributor shaft. The gears or chain driving the camshaft and the gears driving the distributor shaft have tolerances, or looseness. While small in actual measurements, these tolerances can combine to cause a significant difference between crankshaft position and ignition timing.

These high-technology systems are still being developed, studied, and evaluated. It may be a number of years before they are accepted for industry-wide use, if ever. Chrysler, GM, and Ford are using such systems on a limited-production basis. They are called:
- Chrysler Electronic Lean-Burn (ELB)
- GM Microprocessed Sensing and Automatic Regulation (MISAR)
- Ford Electronic Engine Control (EEC).

The electronic timing-regulation function of all the systems is similar, but the electronics are fundamentally different. The Lean-Burn system uses an analog computer, while MISAR and EEC use digital microprocessors.

The practical difference between **analog** and **digital** electronics in this kind of application is that a digital computer can instantly alter timing, for example, from 1 to 65 degrees. An analog computer must do far more calculations to make such an adjustment. Since an electronic spark advance adjustment takes only a few milliseconds, this fact is not really significant to the driver or serviceman. However, a digital system is more flexible and, at this time, more economical to build than an analog system.

Chrysler

Chrysler's ELB has been used on some V-8 engines since 1976. Its use has steadily expanded since then.

The spark control computer is mounted on the air cleaner, figure 17-22. Early models have two printed circuit boards: the ignition schedule module and the ignition control module. Computers for 1977 318-cid engines and all 1978 and later systems have only one circuit board that does both jobs.

The ignition schedule module receives signals from various engine sensors and computes them to determine the exact spark timing required. It then directs the ignition module to advance or retard the timing accordingly.

Six or seven sensors, figure 17-23, feed information to the computer:
- The Start pickup coil in the distributor provides a fixed amount of advance during cranking.
- The Run pickup coil supplies a basic timing signal and allows the computer to determine engine speed. The system for 1977 318-cid engines and all 1978 and later systems use only one pickup coil to provide all timing signals to the computer.
- The coolant temperature sensor on the water pump housing signals the computer when coolant temperature is low.
- The air temperature sensor is located inside

Spark Timing Control Systems

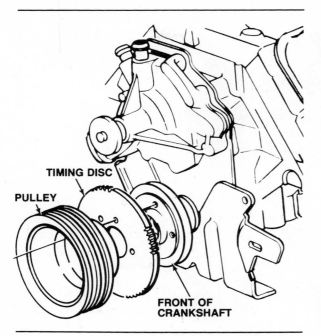

Figure 17-21. Crankshaft position signals can be taken directly from the crankshaft.

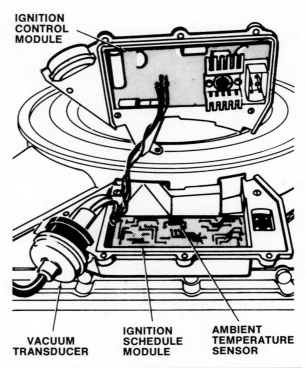

Figure 17-22. The Electronic Lean-Burn computer contains two separate modules.

the computer. It is a **thermistor** that provides a varying amount of resistance with changing air temperature. As temperature increases, resistance decreases.
• The carburetor switch sensor tells the computer if the engine is at idle or off idle.

The remaining two sensors are **transducers**, devices that change mechanical movement to an electrical signal. The transducer has a coil and a movable metal core. A small amount of voltage applied to the coil will vary in strength as the core moves within the coil. This varying voltage signal is interpreted by the computer.
• The throttle position transducer has a core connected to the throttle lever. Core movement tells the computer the position and rate of change of position of the throttle plates.
• The vacuum transducer contains a diaphragm that is exposed to engine vacuum. Movement of the diaphragm moves the core and signals the computer of changes in engine vacuum.

General Motors

The MISAR system, figure 17-24, was introduced on 1977 Oldsmobile Toronados. The microprocessor is contained in a control module that is mounted under the instrument panel in the passenger compartment. The module monitors signals from these engine sensors:
• Coolant temperature
• Manifold vacuum
• Atmospheric pressure
• Crankshaft speed and position.

The coolant sensor is not a simple on-off switch, but rather a thermistor, supplying varying amounts of resistance with changes in temperature. As temperature increases, resistance decreases.

The vacuum sensor is located in the control module. The sensor is a solid-state unit and is connected to the intake manifold by a vacuum line. A second line connected to the module is open in the engine compartment, to provide an atmospheric pressure signal.

Analog: Describing a computer that uses similar (analogous) electrical signals to make its calculations.

Digital: Describing a computer that makes calculations with quantities represented electronically as digits.

Thermistor: A semiconductor whose resistance decreases as heat is applied, opposite to the reaction of a normal conductor.

Transducer: Device that changes one form of energy into another form of energy. In an ignition system, they may sense a mechanical movement and change it to an electrical signal.

214   **Chapter Seventeen**

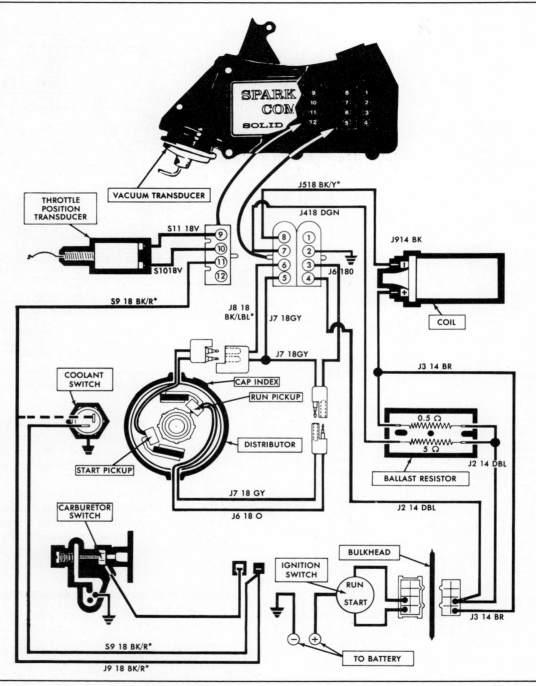

Figure 17-23. A circuit diagram of the Electronic Lean-Burn system. (Chrysler)

On 1977 systems, crankshaft speed and position signals are provided by a rotating disc and a stationary sensor on the front of the engine, figure 17-25. In 1978, the crankshaft speed and position sensor was moved into the distributor.

Ford

In 1978, Ford introduced its electronic engine control (EEC) system on the Lincoln Versailles, figure 17-26. The system controls both spark timing and EGR valve operation. A digital microprocessor installed in the passenger compartment, figure 17-27, receives signals from various sensors.

Spark Timing Control Systems

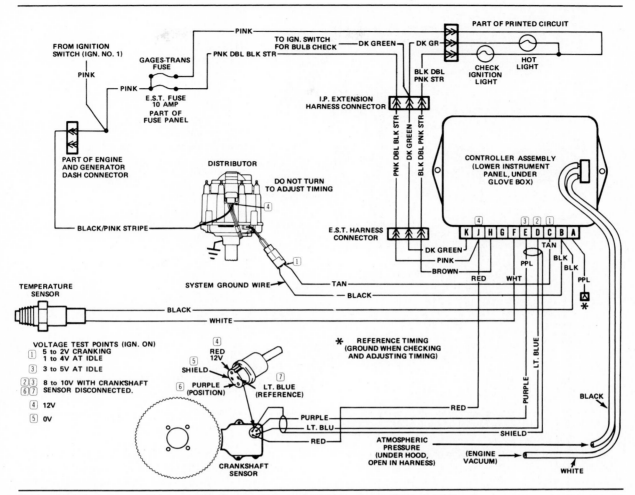

Figure 17-24. A circuit diagram of the General Motors MISAR system. (Oldsmobile)

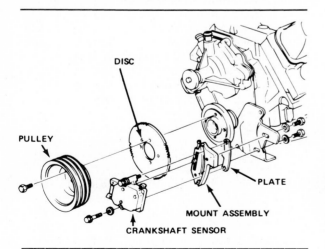

Figure 17-25. The MISAR system takes crankshaft position signals from a special disc on the front of the crankshaft. (Oldsmobile)

Figure 17-26. The control module of Ford's Electronic Engine Control system.

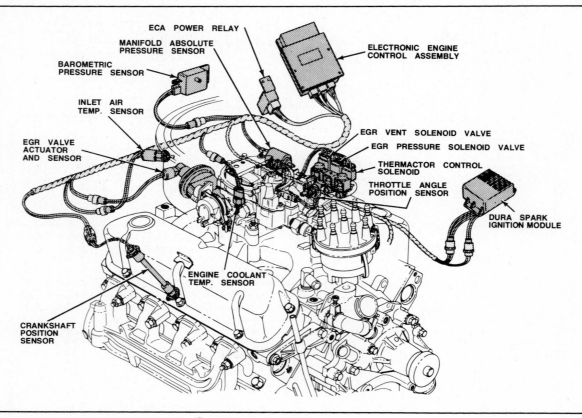

Figure 17-27. Ford's Electronic Engine Control system. (Ford)

An electromagnetic pickup at the flywheel end of the crankshaft signals crankshaft position and speed. It also replaces the pickup coil and trigger wheel of the distributor. Other information is fed to the computer by a throttle position sensor, a coolant temperature sensor, a barometric pressure sensor, an inlet temperature sensor, a manifold pressure sensor, and an EGR valve position sensor.

The module determines the optimum spark timing and EGR valve operating mode. A spark timing signal is sent to the Dura-Spark solid-state ignition control module. This module is similar to that used with the standard Ford Dura-Spark ignition.

SUMMARY

Exhaust emissions are greatest during idle and part-throttle conditions. The best way to reduce HC emissions during these periods is to retard the ignition timing. Retarded spark timing creates higher temperatures at the end of the combustion process and allows more time for the mixture to vaporize. This maintains adequate burning of the air-fuel mixture and lowers the HC and CO emissions. Retarded spark timing also reduces peak combustion pressures and combustion chamber temperatures. This reduced NO_x formation.

Many different spark timing emission control systems have been used. Regardless of the variation of their design and operation, these systems all regulate distributor vacuum advance. Vacuum advance is generally allowed only at cold startup, during high gear operation, and if the engine overheats.

Auxiliary spark timing control systems are not used on all engines. Development of exhaust gas recirculation (EGR) systems, catalytic converters, and the expanded use of air injection systems have gradually reduced the use of TCS and SCS systems on late-model engines.

Electronic spark timing systems regulate all ignition timing for the best combination of emission control, fuel economy, and performance.

Spark Timing Control Systems

Review Questions

Choose the single most correct answer. Compare your answers to the correct answers on page 248.

1. Retarded spark timing:
   a. Occurs during idle and low-speed operation
   b. Reduces peak combustion temperature
   c. Reduces NO_x emissions
   d. All of the above

2. The deceleration vacuum advance valve was most effective in limiting which type of emission:
   a. CO
   b. NO_x
   c. HC
   d. All of the above

3. The distributor retard solenoid was used by:
   a. GM in 1973
   b. AMC in 1971
   c. Chrysler in 1971-72
   d. Ford in 1977

4. A sintered metal disc is used with:
   a. The CEC system
   b. The SDV system
   c. Speed-controlled spark systems
   d. Transmission controlled spark systems

5. Speed- and transmission-controlled timing prevent distributor vacuum advance when a car is:
   a. In high gear
   b. Traveling at high speed
   c. In low gear
   d. None of the above

6. Vacuum control solenoids respond to:
   a. Transmission fluid hydraulic pressure
   b. Shift lever position
   c. Vehicle speed
   d. Any of the above

7. The GM TCS system is controlled by:
   a. A transmission switch
   b. A throttle-vacuum switch
   c. A speedometer switch
   d. A diverter valve

8. The CEC valve:
   a. Holds the throttle open in low gear
   b. Provides vacuum advance during idle
   c. Is deenergized in high gear
   d. Holds the throttle open during high-gear deceleration

9. The distributor vacuum control switch applies full vacuum:
   a. At all times
   b. At increased engine coolant temperature regardless of transmission gear position
   c. Only at idle
   d. Only in low gear

10. In the SCS system at low speeds:
    a. The solenoid is deenergized
    b. The speed switch is open
    c. The speed switch is closed
    d. The solenoid is energized and the speed switch is open

11. The illustration shows:
    a. A thermostatic vacuum valve
    b. A distributor vacuum control switch
    c. A CEC valve
    d. A temperature override switch

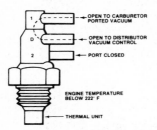

12. AMC cars with automatic transmissions used a speed-sensing control switch:
    a. Before 1971
    b. After 1975
    c. Before 1973
    d. After 1976

13. Which is *not* part of the Chrysler NO_x control system:
    a. A speed switch
    b. A vacuum switch
    c. A temperature switch
    d. A throttle switch

14. The OSAC valve on late-model Chryslers is located:
    a. Between the engine and the carburetor
    b. Between the engine and the exhaust manifold
    c. Between the carburetor and the distributor
    d. In the cooling system

15. The EDM system has:
    a. A speed sensor
    b. An electronic control module
    c. A 3-way solenoid valve
    d. All of the above

16. In the EDM system, the ported vacuum switch is:
    a. The ambient temperature switch
    b. The 3-way solenoid valve
    c. The coolant temperature switch
    d. All of the above

17. The TRS system was introduced by:
    a. Chrysler
    b. Ford
    c. GM
    d. AMC

18. In the TRS system, the ambient temperature switch deenergizes the solenoid at:
    a. Low temperatures
    b. High temperatures
    c. High temperatures and low speeds
    d. High temperatures and high speeds

19. In the TAV system, the spark advance is controlled at high temperatures by:
    a. Manifold vacuum
    b. The electronic module
    c. EGR vacuum
    d. A latching relay

20. Chrysler's electronically controlled timing system is called:
    a. Microprocessed Sensing and Automatic Regulation
    b. Electronic Engine Control
    c. Electronic Lean-Burn
    d. Cold Temperature Activated Vacuum

21. In the ELB system, which of the following is a transducer:
    a. The Start pickup sensor
    b. The throttle position sensor
    c. The air temperature sensor
    d. The Run pickup sensor

22. The MISAR system was introduced in the:
    a. AMC Pacer
    b. Lincoln Versailles
    c. The Oldsmobile Toronado
    d. The Cadillac Seville

Chapter 18
Exhaust Gas Recirculation

Exhaust gas recirculation (EGR) is an emission control system designed to reduce the amount of oxides of nitrogen (NO_x) produced during combustion. When NO_x is present in the atmosphere and acted upon by sunlight, it combines with hydrocarbons to form photochemical smog, the prime air pollutant.

In this chapter, you will learn:
• The principles of EGR systems
• How a basic EGR system works to reduce NO_x formation
• EGR system variations used by domestic auto makers.

NO_x FORMATION

Under normal circumstances, nitrogen and oxygen do not combine unless temperatures exceed 2,500° F. When ignition timing is correct, maximum heat and pressure are created in an engine's combustion chambers. Whenever combustion chamber temperatures are higher than 2,500° F, nitrogen and oxygen combine rapidly to form large amounts of NO_x.

Since peak combustion chamber temperature is controlled by ignition timing, the spark timing control systems discussed in Chapter 17 were the first attempts to meet NO_x control requirements. By retarding the spark timing slightly, less pressure and heat are produced. This holds combustion chamber temperatures below the level at which NO_x forms rapidly. Spark timing control systems were used to control NO_x formation until 1972, when Federal test procedures to determine NO_x levels were changed. For this reason, **exhaust gas recirculation** systems were devised as a better way to control NO_x.

How EGR Works

Although small amounts of NO_x are formed at temperatures below 2,500° F, these quantities can be easily controlled. But once combustion chamber temperatures reach 2,500° F or higher, more NO_x is formed.

There are two ways in which peak combustion chamber temperatures can be held down to prevent NO_x formation. One is to retard spark timing slightly, as was done during 1971-72. The other, and more efficient way, is to dilute the incoming air-fuel mixture with a small amount of an **inert gas** to lower the combustion chamber temperature. Because exhaust gases are relatively inert, they are used to dilute the air-fuel mixture.

This is done by routing small quantities of exhaust gas (6 to 10 percent) from the engine's exhaust ports to the intake manifold, figure 18-1. This exhaust gas dilutes the incoming air-

Exhaust Gas Recirculation

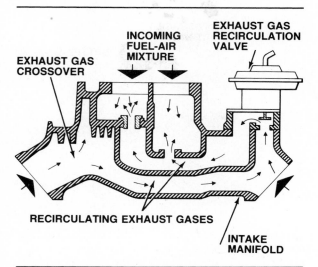

Figure 18-1. Basic exhaust gas recirculation methods.

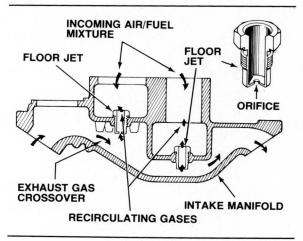

Figure 18-2. Chrysler's floor jet EGR system.

fuel mixture in the cylinder. Since exhaust gas contains no oxygen, the resulting air-fuel-exhaust gas mixture is not as powerful when ignited, and so it does not create as much heat as an undiluted air-fuel mixture would produce.

How EGR Affects Combustion

Since it does not require much exhaust gas to cool down peak combustion temperatures, recirculation must be held to very low quantities. Even when the EGR valve used to reroute the exhaust gas is wide open, the orifice through which the gas passes is very small.

Because the amount of NO_x produced at low engine speeds is very small, exhaust recirculation is not needed or desirable at idle. It is also undesirable during high-speed driving at wide-open throttle, if efficient operation and good driveability are to be maintained. Maximum recirculation is necessary only during cruising and acceleration at speeds between 30 to 70 mph, when NO_x formation is greatest. Engine temperature is also a determining factor in recirculation. When engine temperature is low, NO_x formation is also low, and recirculation is eliminated to produce fast warmup and better driveability.

SYSTEM COMPONENTS AND OPERATING PRINCIPLES

The devices used to recirculate exhaust gas are discussed below. Since system designs and controls differ from one manufacturer to another, we will discuss each of the carmakers later in this chapter.

EGR Floor Jets

The floor jet system, figure 18-2, is used on some 1972-73 Chrysler-built engines. On V-8 engines, stainless steel jets are threaded into the floor of the intake manifold under the carburetor. With 6-cylinder engines, the jet is in the intake manifold hot spot beneath the carburetor. The jets provide an opening between the exhaust passage and the intake manifold. In this way, manifold vacuum controls how much exhaust is drawn into the intake system through the preset jet orifice. Floor jets are the simplest of all EGR system designs. However, they also are unsatisfactory because they allow exhaust gas to enter the intake manifold at all times. This causes rough engine operation at idle and during cold engine warmup.

EGR Valves

Introduced on 1972 Buicks, the EGR valve is a spring-loaded, vacuum-operated, poppet-type valve, figure 18-3. Modulating valves and tapered-stem valves are also used, but their operation is about the same as the poppet type. This valve meters the exhaust gas entering the intake system. The EGR valve is mounted on the intake manifold, figure 18-4, or on a plate under the carburetor, figure 18-5. The valve may be connected to the intake and exhaust systems by internal passages in the intake mani-

Exhaust Gas Recirculation (EGR): A way of reducing NO_x emissions by directing unburned exhaust back through an engine's intake.

Inert Gas: A gas that will not undergo chemical reaction.

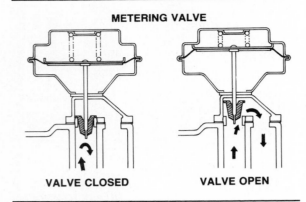

Figure 18-3. Typical EGR valve.

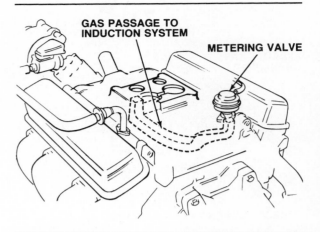

Figure 18-4. EGR metering valve mounted on intake manifold.

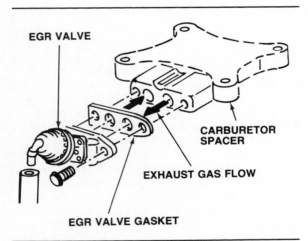

Figure 18-5. This EGR valve is mounted on a spacer that goes between the carburetor and the intake manifold.

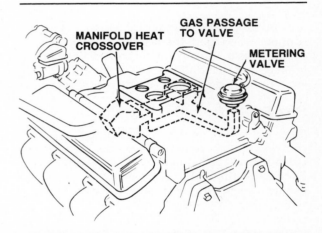

Figure 18-6. Exhaust gas passages to the EGR valve.

fold, figures 18-4 and 18-6, or in some cases by external steel tubing. The EGR valve is held by the spring in the closed position. The valve is opened by ported or venturi vacuum from the carburetor, depending upon the system design.

Ported Vacuum Systems

EGR systems controlled by ported vacuum use a slot-type port in the carburetor throttle body above the throttle plate, figure 18-7. This port is connected to the EGR valve by a vacuum line. With the throttle plate closed, no vacuum is transmitted. As the throttle plate opens, the port is exposed to increasing manifold vacuum. The amount of exhaust gas flow depends on manifold vacuum, throttle position, and exhaust gas backpressure. Since ported vacuum cannot be more than intake manifold vacuum, recirculation at wide-open throttle is prevented. The valve opens at a point greater than the relatively weak manifold vacuum produced during wide-open throttle operation.

Venturi Vacuum Systems

EGR systems controlled by venturi vacuum use a vacuum port at the throat of the carburetor venturi to provide a control vacuum, figure 18-8. Since this control vacuum is very weak, a vacuum amplifier boosts it enough to operate the EGR valve. The amount of exhaust gas flow depends mainly on engine intake airflow. It is also affected by intake vacuum and exhaust gas backpressure.

Recirculation at wide-open throttle is prevented by a relief valve, or dump diaphragm, that compares venturi and manifold vacuum. When the diaphragm senses that the throttle is wide open, the stored vacuum in the amplifier is vented to the atmosphere. This limits the vacuum output reaching the EGR valve to that provided by manifold vacuum. Since the EGR valve opens at a point greater than the relatively weak manifold vacuum, the valve remains closed at wide-open throttle.

Exhaust Gas Recirculation

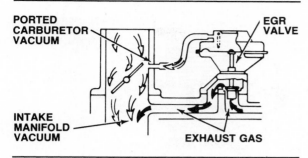

Figure 18-7. This EGR valve is operated by ported vacuum.

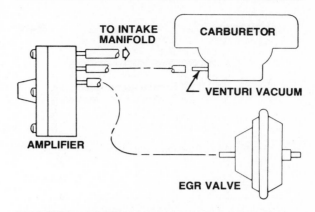

Figure 18-8. EGR systems controlled by venturi vacuum require a vacuum amplifier. (Ford)

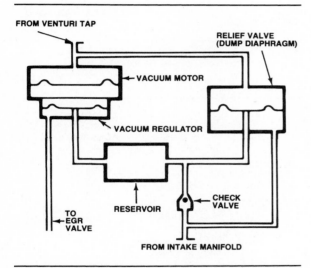

Figure 18-9. A vacuum amplifier is controlled by venturi vacuum and uses manifold vacuum to operate the EGR valve. (Ford)

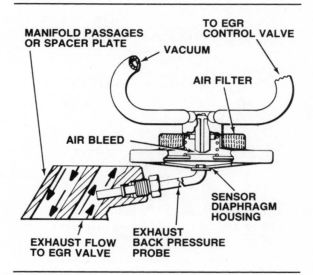

Figure 18-10. Typical EGR backpressure transducer.

Vacuum amplifiers

The vacuum amplifier, figure 18-9, converts the weak venturi control vacuum into one that is strong enough to operate the EGR valve. It does this by storing manifold vacuum in a reservoir inside the amplifier unit. This guarantees enough vacuum, regardless of variations in manifold vacuum. Whenever the venturi vacuum is equal to, or greater than, manifold vacuum, a relief valve (dump diaphragm) vents the reservoir, cancelling the output EGR signal.

If the vacuum amplifier works perfectly, it would produce an accurate, repeatable, and almost precise proportion between venturi airflow and EGR flow. But, vacuum amplifiers are not as commonly used on late-model EGR systems because storage reservoirs tend to leak in actual use.

Backpressure Transducer

This diaphragm-operated sensor unit, figure 18-10, has a tube that extends into an exhaust gas passage. When high exhaust backpressure is sensed through the tube, the diaphragm closes an air bleed hole in the EGR vacuum line. This provides maximum EGR during acceleration, when backpressure is high. As backpressure drops, a spring moves the sensor diaphragm to reopen the vacuum line bleed. Since this decreases the vacuum at the EGR valve, the amount of exhaust gas recirculated is also reduced.

Modulating Devices

Engineers have developed various methods of modulating, or adjusting, EGR valve operation in relation to engine operating conditions. High and low ambient temperature vacuum modulators weaken the vacuum signal to the EGR valve. A coolant temperature override switch or valve, figure 18-11, may be used to eliminate EGR vacuum below certain engine operating

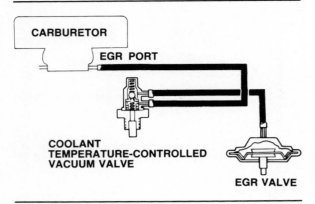

Figure 18-11. The coolant-temperature-controlled vacuum valve cuts off vacuum to the EGR valve when the engine is cold. (Ford)

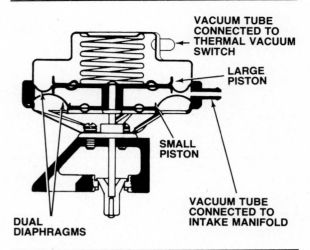

Figure 18-12. The dual-diaphragm EGR valve modulates the exhaust gas flow.

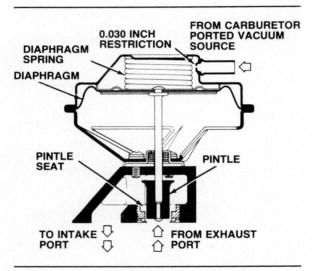

Figure 18-13. This American Motors EGR valve is typical of most single-diaphragm EGR valves. (AMC)

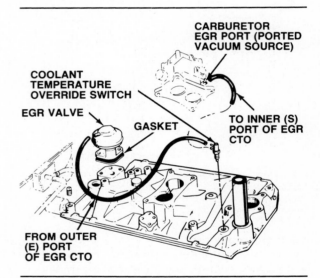

Figure 18-14. Basic AMC V-8 EGR installation. (AMC)

temperatures. Some venturi-vacuum-controlled systems have a time delay solenoid to shut off vacuum to the EGR valve for about half a minute after the ignition is turned on.

A vacuum-bias valve is used to bleed off part of the EGR vacuum signal under high-manifold-vacuum conditions to eliminate **high-speed surge**. A dual-diaphragm EGR valve, figure 18-12, is sometimes used. This uses manifold vacuum to help the valve spring offset the carburetor vacuum under certain cruising conditions.

AMERICAN MOTORS EGR SYSTEMS

The basic AMC EGR system, as introduced on 1973 models, uses a vacuum-operated EGR valve. When the carburetor throttle is opened beyond the idle position, ported vacuum is applied to the normally closed EGR valve diaphragm, figure 18-13. Moving upward against coil spring pressure, the diaphragm opens the **pintle valve**. This valve permits exhaust gas to be drawn into the engine intake from the manifold crossover exhaust passage on V-8 engines, figure 18-14, or from below the carburetor heat riser on 6-cylinder engines, figure 18-15. The EGR valve is closed during idle and deceleration to prevent a rough idle. California cars use an exhaust backpressure sensor, figure 18-10.

American Motors also uses the following override and modulator devices to control EGR system operation.

Exhaust Gas Recirculation

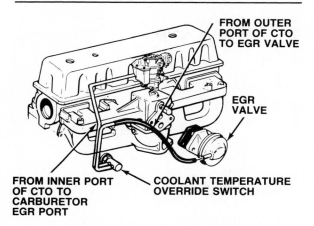

Figure 18-15. Basic AMC 6-cylinder EGR installation. (AMC)

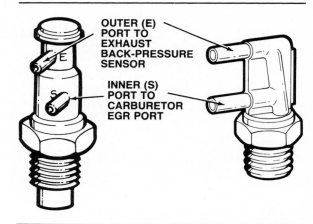

Figure 18-16. American Motors' 2-port EGR coolant temperature override (CTO) switch. (AMC)

Coolant Temperature Override Switch

This EGR coolant temperature override (CTO) prevents normal EGR valve operation when engine coolant temperature is below 115° F. The 3-port switch originally used has been replaced by a 2-port switch, figure 18-16, on 1974 and later models. At coolant temperatures above 115° F (160° F on some models), vacuum is permitted to reach the EGR valve.

Low- and High-Temperature Vacuum Signal Modulators

Used only on 1973 models, these modulators are connected to the EGR vacuum control line. When air temperature is under 60° F, the low-temperature modulator weakens the vacuum to the EGR valve. This decreases the amount of exhaust gas flow. At temperatures above 115° F, the high-temperature modulator restores vacuum to the EGR valve and increases exhaust gas flow.

Exhaust Backpressure Sensor

Fitted only on California cars during 1973-75, use of the backpressure sensor, figure 18-10, was extended to all 1976 and later models. The sensor permits EGR flow only when the engine is at normal operating temperature and exhaust backpressure is high. If backpressure is not high enough, and the EGR-CTO switch does not open, the EGR vacuum signal is vented to the atmosphere.

Restrictor Plate

A stainless steel restrictor plate, figure 18-17, is used on some 1974-75 California models, and on all 1976-77 cars. Located between the intake

High-Speed Surge: A sudden increase in engine speed caused by high manifold vacuum pulling in an excess air-fuel mixture.

Pintle Valve: A valve shaped much like a hinge pin. In an EGR valve, the pintle is attached to a normally closed diaphragm. When ported vacuum is applied, the pintle rises from its seat and allows exhaust gas to be drawn into the engine's intake system.

■ **Engine Modifications Can't Do The Whole Job**

Since the automobile was discovered to be the biggest source of air pollution, many changes in engine design have been made to "clean it up." These engine modifications, such as EGR systems, have reduced exhaust emissions, but it is impossible to eliminate the major cause of HC emissions simply by changing the engine design. This is because the major cause of HC emissions is the effect of the "quench area" on combustion.

The quench area is the inner surface of the combustion chamber. When the ignition flame front passes through the combustion chamber, it burns the fuel charge as it goes until the quench area is reached. This is a thin layer between .002" and .010" thick at the edge of the combustion chamber. When the flame front reaches the quench area, it is snuffed out because the quench area is so close to the cylinder head water jacket that the temperature there is too low for combustion to continue. Consequently, hydrocarbons within the quench area do not burn. They are ejected from the cylinder on every exhaust stroke along with the exhaust gases formed by combustion and enter the atmosphere as pollutants.

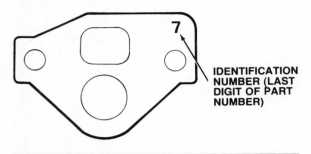

Figure 18-17. American Motors' EGR restrictor plate. (AMC)

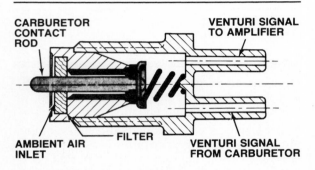

Figure 18-19. Wide-open-throttle EGR dump valve. (Chrysler)

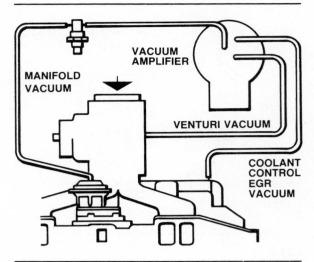

Figure 18-18. Chrysler's external vacuum reservoir.

manifold and spacer, the plate is adjusted for each engine and exhaust system. It limits EGR flow rate and improves driveability.

CHRYSLER EGR SYSTEMS

Chrysler used the floor jet system, figure 18-2, on some 1972-73 models. Stainless steel jets with a preset orifice are installed in the floor of the intake manifold beneath the carburetor. Engine vacuum draws a small amount of exhaust gas into the intake to dilute the air-fuel mixture. This system was supplemented in 1973 with both the ported and venturi vacuum systems. The floor jet system was discontinued after 1973.

Venturi Vacuum System

The venturi vacuum system was changed on some 1975 and later engines to include an external vacuum reservoir mounted on a bracket and attached to the vacuum amplifier, figure 18-18. After the internal vacuum reservoir has been vented, or dumped, the external reservoir supplies manifold vacuum for exhaust recirculation until the internal reservoir can be refilled.

Wide-Open-Throttle Dump Valve

A wide-open throttle dump valve, figure 18-19, was added to some engines in March, 1976. These engines use a delay-dump amplifier, which can hold the EGR valve open too long at wide-open throttle. Mounted on the carburetor, the wide-open-throttle EGR dump valve overrides the delay dump amplifier at wide-open or near wide-open throttle by mechanically bleeding the vacuum to the amplifier. This closes the EGR valve immediately.

Coolant-Control EGR Valve

A coolant-control EGR valve (CCEGR) added to the 1974 system blocks vacuum flow to the EGR valve whenever coolant temperature is less than that specified for the particular engine. Some 1974 and later engines use an EGR delay timer. This electric timer, mounted on the firewall, operates an engine-mounted vacuum solenoid to prevent exhaust gas recirculation for about half a minute after the ignition is turned on.

EGR Maintenance Reminder System

An EGR maintenance reminder system, figure 18-20, is on 1975 engines only. This mileage counting device signals the driver every 15,000 miles to have the EGR system inspected. An instrument panel reminder light comes on at 15,000 mile intervals, and remains lighted until the switch attached to the speedometer cable is reset manually.

Exhaust Gas Recirculation

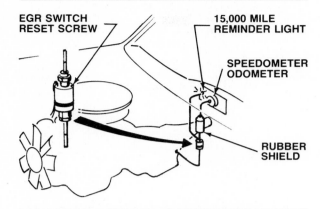

Figure 18-20. Chrysler's EGR maintenance reminder system.

FORD EGR SYSTEM

The basic Ford EGR system, figure 18-21, was introduced on all Ford-built passenger cars in 1973, except those using the 1,600-cc and 2,000-cc engines. The basic system has three main parts: an EGR valve, a temperature-controlled vacuum valve, and a carburetor spacer.

EGR Valve

Three types of EGR valves are used: poppet, modulating, and tapered stem. The poppet valve contains a spring-loaded diaphragm, a valve and valve stem, and a flow restrictor. Ported carburetor vacuum opens the valve at a specific vacuum level and allows exhaust gas to enter the valve. Exhaust gas flow to the combustion chambers is controlled by the flow restrictor in the valve body inlet port.

The modulating valve, figure 18-22, has an extra disc valve on the stem below the main valve, and it operates much like a poppet valve. However, at a specific vacuum level, the lower disc valve restricts exhaust gas flow to the valve chamber. This modulating action improves the driveability of some engines.

The tapered-stem valve operates in the same way. Exhaust gas flow is modulated by the tapered stem as it gradually unseats to permit an increasing gas flow.

The three valve types are not interchangeable between engine types or model years.

Ford uses various ways to deliver exhaust gas to the EGR valve. On many 4- and 6-cylinder engines, exhaust gas is drawn from the exhaust manifold through a stainless steel tube. Most V-8 engines use a passage in the exhaust crossover of the intake manifold. This routes the gas through the carburetor spacer to the EGR valve.

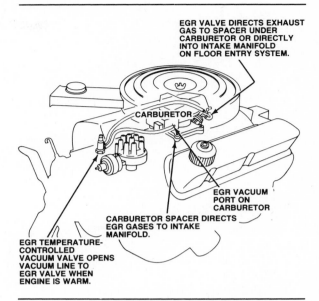

Figure 18-21. Basic Ford EGR system. (Ford)

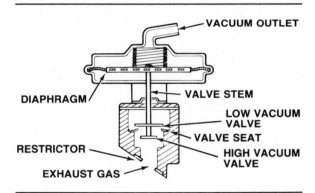

Figure 18-22. Ford modulating EGR valve. (Ford)

The EGR valve then meters the gas back through a separate passage in the spacer to the carburetor primary venturis, where it is mixed with the air-fuel mixture.

Some 1974 engines use a floor-entry system, figure 18-23. The intake manifold has two passages cast in the floor under the intake runners. One connects the exhaust crossover to the EGR valve. The other connects the valve with two holes in the manifold floor directly under the carburetor primary venturis. This allows exhaust gas to mix with the air-fuel mixture before entering the combustion chambers. This is a valve-controlled system and is not the same as Chrysler's floor jet EGR system, which does not use a valve.

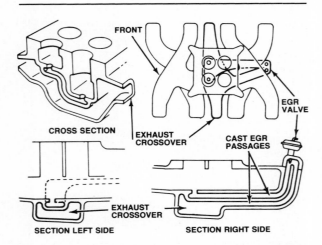

Figure 18-23. Ford 1974 floor entry EGR system. (Ford)

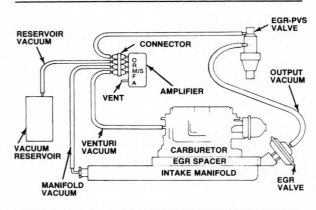

Figure 18-25. Ford venturi vacuum control subsystem with single-connector amplifier. (Ford)

Carburetor Spacer

Ford's carburetor spacer plate contains separate inlet and outlet passages to the EGR valve. The valve is mounted at one end of the plate. The spacer is installed between the carburetor and the intake manifold with top and bottom gaskets to prevent leaks. Spacer plates are not interchangeable between engine types.

System Modifications

Basic EGR system operation is modified by the following subsystems.

High-speed EGR modulator subsystem
Some Ford V-8 engines have this subsystem, figure 18-24, to improve driveability at speeds above 64 mph. This subsystem shuts off carburetor vacuum to the EGR valve to prevent recirculation. A speed sensor driven by the speedometer cable signals an electronic module. When road speed exceeds 64 mph, the module closes the open solenoid valve. This, in turn, closes the EGR vacuum port and vents the vacuum outside the valve. As road speed drops below 64 mph, the module deenergizes the solenoid valve. This closes the vent and opens the EGR vacuum line.

Venturi vacuum control
Some 1974 6-cylinder engines use venturi vacuum rather than ported vacuum. A vacuum amplifier, figure 18-25, is connected between the EGR valve and intake manifold vacuum. Once airflow through the venturi is sufficient, venturi vacuum opens the amplifier. This allows manifold vacuum to open the EGR valve.

The EGR valve remains closed at idle, since there is no venturi vacuum produced at idle. Whenever venturi vacuum is equal to or greater than manifold vacuum, such as under wide-open throttle, the vacuum amplifier dumps the output vacuum to the EGR valve, causing the

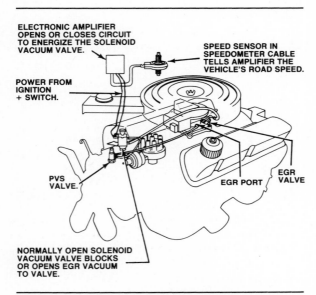

Figure 18-24. Ford high-speed EGR modulator subsystem. (Ford)

Temperature-Controlled Vacuum Valve

A temperature-controlled vacuum valve in the engine cooling system cuts off vacuum to the EGR valve when the engine is cold, figure 18-21. Ford calls this valve a ported vacuum switch (PVS). It works the same as American Motors' coolant temperature override (CTO) switch and Chrysler's coolant-controlled EGR valve (CCEGR). The valve may be installed in a heater hose or in the intake manifold water jacket. Different temperature settings are used for different Ford engines. The valves are color coded for identification.

Exhaust Gas Recirculation

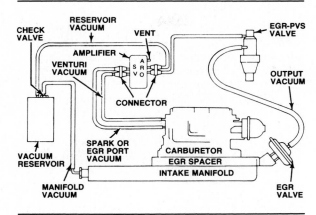

Figure 18-26. Ford venturi vacuum control subsystem with dual-connector amplifier. (Ford)

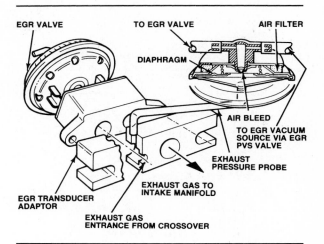

Figure 18-27. EGR backpressure transducer (sensor). (Ford)

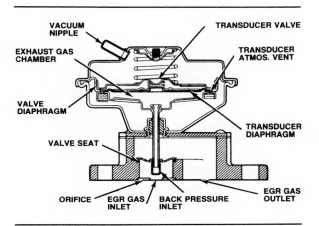

Figure 18-28. This Ford EGR valve has a built-in exhaust backpressure transducer. (Ford)

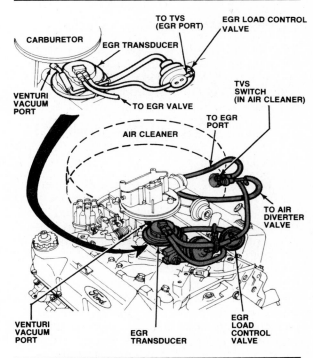

Figure 18-29. Ford air-temperature vacuum switch and EGR load control valve installation on a 400-cid V-8. (Ford)

valve to close. Some of the amplifiers contain an output bias. This is a bleed designed for quicker EGR valve opening.

A single-connector amplifier with all ports on one side is used with 1974-75 systems, figure 18-25. A dual-connector amplifier with ports on both sides is used on 1975 and later systems, figure 18-26.

Backpressure transducer
Late 1975 and most later engines use a backpressure sensor, or transducer, figure 18-27. This is connected between the EGR valve and the intake manifold. It modulates exhaust gas flow by varying the vacuum to the EGR valve, according to exhaust backpressure.

On some 1977 models, the backpressure transducer is within the EGR valve, figure 18-28. This combination unit has an internal exhaust gas chamber and a transducer diaphragm to sense exhaust backpressure through a hollow valve stem. This combination EGR valve-transducer constantly meters exhaust flow according to exhaust backpressure.

Temperature control
Some 1976 and later V-8 engines use an air-temperature vacuum switch (TVS) mounted in the air cleaner, figure 18-29, instead of the coolant PVS. The TVS also controls the vacuum flow to the diverter valve in the thermactor air injection system. Since the switch is closed at temperatures below about 60° F, exhaust gas does not recirculate on a cold engine.

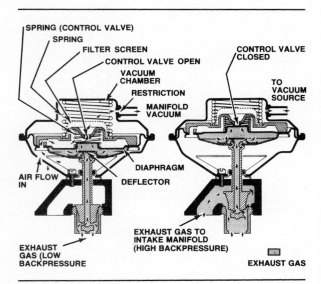

Figure 18-30. General Motors' EGR valve with built-in backpressure transducer. (Buick)

GENERAL MOTORS EGR SYSTEMS

Exhaust gas recirculation was introduced by Buick on 1972 cars with manual transmissions and on all 1972 cars sold in California. All 1973 and later GM engines use an EGR system to reduce NO_x. Although similar to the EGR systems used by other automakers, GM systems vary slightly between the company divisions.

Buick Division

The 1972 engines use ported vacuum to control the EGR valve. Exhaust gas is drawn from the manifold crossover exhaust channels and into the engine intake. Since ported vacuum is used as a control source, the valve is closed during idle and deceleration.

A temperature-control valve, in the carburetor-to-EGR valve vacuum line on 1973 models, prevents the valve from opening until underhood temperatures reach 60° F. The EPA ordered this removed, and a temperature vacuum switch (TVS) sensitive to engine coolant temperature replaced it on engines built after March 15, 1973. A dual-diaphragm EGR valve, figure 18-12, is used on some 1974 California engines.

An exhaust backpressure transducer is used on some 1976 models. The transducer is combined with the EGR valve, figure 18-30, for 1977 cars. EGR systems using the backpressure transducer, or the combined transducer and valve, regulate the timed vacuum to the EGR valve according to exhaust backpressure level.

Cadillac Division

The 1973 Cadillac EGR system is essentially the same as that used by Buick. The inline temperature-sensitive switch was replaced after March 15, 1973, by a different temperature vacuum switch, which Cadillac calls a thermal vacuum valve (TVV). A backpressure transducer is on 1974 California cars, and on all 1975 models that have the air pump (AIR) system.

The backpressure transducer is used on all 1976 EFI-equipped cars and all California cars. For 1977, three valves are used:
1. The combined transducer-EGR valve, figure 18-30, on California cars
2. A standard EGR valve operated by ported vacuum on all non-California cars
3. A standard EGR valve and a separate backpressure transducer on all EFI cars. The EFI-EGR valve has an orifice gasket for flow control.

Chevrolet Division

The basic 1973 EGR system was changed to add a TVS to the thermostat housing on 1974 models. The TVS improves cold engine starting by preventing recirculation until coolant temperature reaches a specific level. A dual-diaphragm EGR valve, figure 18-12, is on some engines. The combined transducer-EGR valve, figure 18-30, is on 1977 California and high-altitude cars.

Oldsmobile Division

The 1973-74 Oldsmobile EGR system is basically the same as that used by Cadillac, including the use of a backpressure transducer on California cars. A variety of temperature vacuum switches (EGR-TVS) and temperature control valves (EGR-TCV) are used on 1975 and later engines. Some are interconnected with the early fuel evaporation (EFE) system to switch vacuum from the EFE valve to the EGR valve at a specific temperature. An EGR check valve (EGR-CV) is used on some California engines with the backpressure transducer. The EGR-CV holds the highest ported vacuum reached. This keeps the EGR valve open during hard acceleration, when the backpressure transducer bleed is closed.

On some 1977 cars, the separate backpressure transducer has been replaced by the transducer-EGR valve, figure 18-30. A vacuum delay valve (EGR-VDV) is also used in the carburetor-to-EGR valve line on some V-8 engines to delay the vacuum from bleeding down too fast at the EGR valve. Metering the vacuum through a 0.005-inch orifice prevents its full loss for four seconds.

Exhaust Gas Recirculation

Pontiac Division

The basic 1973 Pontiac EGR system has three variations. Two of these are electrically interconnected with spark advance: the transmission-controlled spark (TCS) system on V-8 engines, and the combined emission control (CEC) system on 6-cylinder engines with manual transmissions. Those 6-cylinder engines with automatic transmissions use the normal ported vacuum EGR system. The TCS-EGR and CEC-EGR systems operate according to two basic rules:

1. When the TCS system is operating, there is no EGR.
2. When the EGR system is operating, there is no distributor vacuum advance.

TCS and CEC system descriptions and diagrams are in Chapter 17.

TCS and EGR systems operate separately on 1973½ models, with a temperature vacuum valve (TVV) providing temperature control. A dual-diaphragm EGR valve, figure 18-12, is on some 1974 California engines, as well as a vacuum bias valve (VBV) to reduce surging. An exhaust backpressure transducer is used to modulate recirculation according to exhaust backpressure.

The system becomes far more complex on 1975 engines, with a variety of TVV, TCV, VDV, and backpressure transducers used. Virtually every engine type has a unique system in terms of devices and operational temperatures. On 1976 engines, the TVV is replaced by a heat-sensitive snap disc valve. This is housed in a stamped steel shield and attached to the intake manifold on most engines. Since the snap disc valve senses engine radiant heat, it shuts off vacuum to the EGR valve on cold engine startup for improved driveability. For 1977, the primary change is the use of the combined transducer-EGR valve, figure 18-30, on some engines.

SUMMARY

Combustion chamber temperatures that exceed 2,500° F cause nitrogen and oxygen to combine rapidly, forming large quantities of NO_x, a prime air pollutant. To reduce NO_x as much as possible, automakers use exhaust gas recirculation (EGR) systems to meter small quantities of exhaust gas into the incoming air-fuel mixture. This dilutes the fuel charge and results in lower combustion chamber temperatures.

Each automaker uses some variation of the basic system, to achieve exhaust gas recirculation. Various temperature sensors, flow valves, and other control devices in the EGR system modulate the flow of exhaust gas to maintain driveability while reducing NO_x emissions.

Review Questions
Choose the single most correct answer.
Compare your answers to the correct answers on page 248.

1. Photochemical smog is a result of:
   a. Sunlight + NO_x + HC
   b. Sunlight + NO_x + CO_2
   c. Sunlight + CO + HC
   d. Sunlight + NO_x + CO

2. NO_x forms in an engine under:
   a. High pressure and low temperature
   b. Low pressure and low temperature
   c. High temperature and high pressure
   d. All of the above

3. Which of the following is a relatively inert gas:
   a. NO_x
   b. HC
   c. Exhaust gases
   d. Air

4. EGR is most desirable at:
   a. Speeds of 30-70 mph
   b. Idle speeds
   c. High speeds
   d. Low engine temperature

5. Which of the following is *not* part of an exhaust gas recirculation system:
   a. Chrysler's floor jets
   b. Buick's EGR valve
   c. Ported vacuum systems
   d. Slow-idle solenoid

6. Which is *not* true of EGR valves:
   a. They operate on venturi vacuum systems
   b. They operate on ported vacuum
   c. They may be mounted on the intake manifold
   d. They operate at wide-open throttle

7. Which is *not* part of the AMC EGR system:
   a. Low- and high-temperature vacuum signal modulators
   b. Floor jets
   c. Restrictor plate
   d. CTO switch

8. A Chrysler EGR delay timer prevents exhaust gas recirculation:
   a. During wide-open throttle
   b. At idle
   c. For 30 seconds after ignition
   d. At medium speeds

9. The Ford temperature-controlled vacuum valve is called:
   a. CTO
   b. CCEGR
   c. PVS
   d. OSAC

10. The high-speed EGR modulator subsystem shuts off the EGR valve above:
    a. 24 mph
    b. 34 mph
    c. 64 mph
    d. 84 mph

11. The Ford TVS, in addition to controlling EGR vacuum, controls:
    a. The temperature control override
    b. The ported vacuum switch
    c. The diverter valve vacuum
    d. All of the above

12. The combined transducer and EGR valve was introduced in which Buick model year:
    a. 1975
    b. 1976
    c. 1977
    d. 1978

Chapter 19
Catalytic Converters

To meet the strict exhaust emission limits of the late 1970's, automakers have turned to the catalytic converter. The catalytic converter is installed in the exhaust system between the exhaust manifold and the muffler. The converter looks from the outside something like a small muffler. It has no moving parts. It simply forms a chamber in the exhaust system through which the exhaust passes. Inside the converter, the exhaust flows through the catalyst material, which turns the exhaust pollutants into harmless byproducts of combustion.

In this chapter, you will learn:
1. How the catalytic converter is built
2. How it works
3. How to make sure that the converter does its job efficiently.

REDUCING EMISSIONS

One way of lowering hydrocarbon (HC) and carbon monoxide (CO) is to increase the combustion temperature, which causes more complete burning. But, as combustion temperatures rise, so does the formation of the third major pollutant, oxides of nitrogen NO_x. A second method of turning exhaust gas into non-polluting materials is the use of the **catalytic converter**, figure 19-1. By sending the exhaust gas through a catalyst in the presence of oxygen, the HC and CO compounds unite with the oxygen, resulting in two harmless byproducts of the catalytic reaction: water vapor (H_2O) and carbon dioxide (CO_2).

The Oxidation Reaction

A **catalyst** is a substance that starts or increases a chemical reaction, while remaining unchanged by the reaction. Since it only encourages rather than takes part in the reaction, the catalyst is never used up.

To change HC and CO into harmless materials, the catalytic elements platinum and palladium start an **oxidation** or burning reaction in the catalytic converter. Oxidation is the addition of oxygen to a material. If there is not already enough oxygen in the exhaust, an air pump supplies extra air. The oxidation mixes the HC and CO with oxygen to form H_2O (water) and CO_2 (carbon dioxide).

Considerable heat is generated by the oxidation process. The heat of the catalyst will range from 900° to 1,600° F, and the exhaust gas at the outlet end of the converter will be 50° to 200° F higher than the inlet end. By 1978, automakers were able to reduce the outside, or "skin," temperature of converters by as much as 300° F. The use of smaller engines and changes in engine timing were responsible for this reduction in temperatures. In spite of the intense heat,

Catalytic Converters

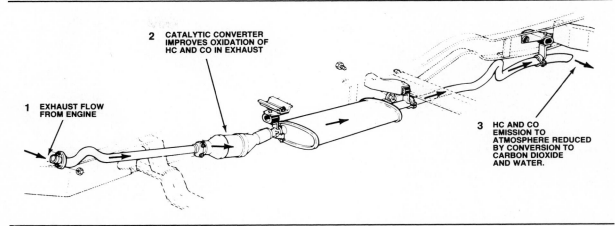

Figure 19-1. Typical catalytic converter installation. (Ford)

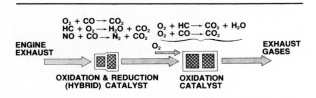

Figure 19-2. Catalytic oxidation and reduction reactions. (Chrysler)

however, oxidation does not generate the flame and radiant heat associated with a simple burning reaction.

The Reduction Reaction

The oxidation catalytic reaction we just mentioned has no effect on oxides of nitrogen. NO_x control requires a separate reaction, called reduction, rather than oxidation. **Reduction** is the opposite of oxidation. Reduction is the chemical removal of oxygen from a material. The reduction reaction changes NO_x to harmless nitrogen (N_2) and CO_2 by chemically promoting the switch of oxygen from the NO_x to the CO compound. The elements rhodium and platinum are used as reduction catalysts.

Because the oxidation and the reduction reactions oppose each other, it is hard for both to occur at the same time and in the same place. An oxidation and a reduction catalyst can be combined in the same converter. However, a second oxidation catalyst is also required for complete emission control. An oxidation-reduction catalyst is often called a 2-stage, a 3-way (because it works on all three major pollutants), or a hybrid, catalyst. A hybrid catalyst and a second oxidation catalyst can be installed in opposite ends of the same converter housing or in two separate converters, figure 19-2. The first hybrid converters were used by Volvo and Saab on a limited number of cars sold in California in 1977.

The hybrid converter works best to reduce NO_x when the CO level in the exhaust is between 0.8 and 1.5 percent. As the CO level increases or decreases from that percentage, the hybrid converter's efficiency decreases, figure 19-3. Exhaust gas first passes through the hybrid converter, where the NO_x is changed to N_2 and CO_2. It then goes through the oxidation converter, where the HC and CO are changed to H_2O and CO_2.

Oxidation Converter Design

Catalytic converters are made of two stamped metal pieces welded together to form a round or oval shell. Each contains the catalytic element, figure 19-4. The outer housing is made of aluminized or stainless steel because it must be able to withstand the high temperatures of oxidation.

The catalytic element generally used is platinum, or a mixture of platinum and palladium. These two **noble metals** best meet the requirements of an effective catalyst: durability, operat-

---

Catalytic Converter: A device installed in an exhaust system that converts pollutants to harmless byproducts through a catalytic chemical reaction.

Catalyst: A substance that causes a chemical reaction, without being changed by the reaction.

Oxidation: The combining of an element with oxygen in a chemical process that often produces extreme heat as a byproduct.

Reduction: A chemical process in which oxygen is taken away from a compound.

Noble Metals: Metals, such as platinum and palladium, that resist oxidation.

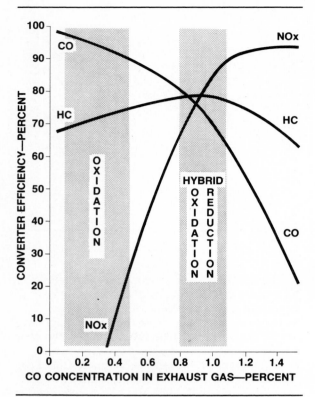

Figure 19-3. Catalyst operating characteristics. (Chrysler)

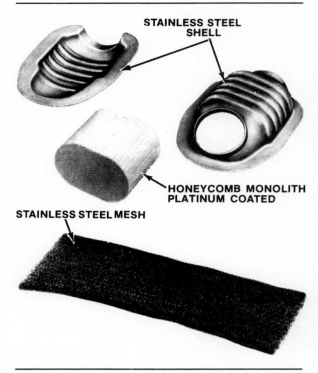

Figure 19-4. The parts of a typical catalytic converter. (Chrysler)

ing temperature, and chemical activity. The catalyst is deposited on an aluminum oxide **substrate** through which the exhaust gas flows. This substrate must provide a catalyst support which can withstand high temperatures.

Two forms of substrate material are used:
1. Tiny pellets of aluminum oxide
2. A honeycomb **monolith** of aluminum oxide. It can be either a laminated (sandwiched) design or an **extruded** design.

The catalyst element is deposited on the surface of the substrate material. Both kinds of substrate material — pellets or monolith — provide several thousand square yards of catalyst surface area over which the exhaust gases flow.

In converters using a pellet substrate, the gases flow over the top and down through the substrate layers. With the monolithic substrate material, a diffuser inside the converter shell allows a uniform flow of exhaust gases over the entire area of the substrate. If a diffuser were not used, the gases would tend to flow only through a central portion of the substrate.

Monolithic substrate material can break easily when subjected to shock or severe jolts. To prevent damage to the core, it is placed inside a stainless steel mesh, figure 19-4, which acts as a cushion. This also protects the core from thermal shock caused by temperature extremes, and keeps it properly positioned during final assembly of the converter shell.

Reduction Converter Design

A typical hybrid converter uses two substrates of different sizes, and coated with platinum and rhodium. These substrates are surrounded by the protective stainless steel mesh and are encased in a stainless steel shell. Other types of reduction converters are likely to be made available in the coming years. But, given the designs now available, the main differences between the oxidation and the reduction converters remains the type of catalyst used and the positioning of the reduction converter in the exhaust system ahead of the oxidation converter.

CONVERTER OPERATING PRECAUTIONS

There are several ways in which a catalytic converter can be damaged or destroyed. Using the wrong fuel, bad air-fuel mixtures, excessive combustion heat — all of these can damage a converter. We will look at these in detail.

Catalytic Converters

Fuel Requirements

Cars with catalytic converters must use unleaded fuel. The lead and phosphorous additives used in leaded gasoline will greatly reduce the catalyst's efficiency. When leaded fuel is used, the lead plates the catalyst and forms a coating that prevents the exhaust gases from reaching the catalyst. Returning to unleaded gasoline will allow the converter to regain some, but not all, of its efficiency. How much it regains depends on how long leaded fuel is used. However, there is no known method at present of testing a converter's efficiency. Continued use of leaded fuel will eventually cause the catalyst to lose its ability to promote oxidation of HC and CO.

To prevent the use of leaded fuel in cars with catalytic converters, the unleaded gasoline pump nozzles in service stations and the fuel tank filler necks on cars are smaller in diameter than those using leaded fuel. In addition, cars which require unleaded fuel also have labels reading "Unleaded Gasoline Only" next to the gas tank filler and on the instrument panel in the driver's compartment. A United States Federal law makes it illegal to sell leaded fuel for use in converter-equipped cars.

Engine Condition

Excessively high temperatures can reduce converter life and destroy the core. Converters operate best at internal temperatures up to 1,500° F. At higher temperatures, the catalyst will start to break up or melt. Although temperature-protection systems are used on some vehicles, they work only under certain conditions. Proper engine maintenance is needed to prevent overheating the converter.

When an engine is in poor condition, or needs a tune-up, the exhaust gas fed to the converter will contain too much raw fuel. This will cause the converter to become a catalytic furnace. If more than two spark plugs misfire at the same time, raw fuel is pumped into the catalytic converter. This misfiring over a long period of time causes internal converter heat to rise rapidly.

Excessive temperatures may also be caused by improper use of an engine in good operating condition. Long idling periods are one of the worst running conditions, since more heat is developed when an engine runs at idle for long periods, than when driving at normal highway speeds. Idle periods of more than 10 minutes should be avoided. It is much better to shut off the engine and restart it when required.

Other Precautions

To avoid excessive catalyst temperatures and the possibility of fuel vapors reaching the converter, follow these rules:
1. Do not attempt to start the engine on compression by pushing the vehicle. Use jumper cables instead.
2. Do not crank an engine for more than 60 seconds when it is flooded or firing intermittently.
3. Do not turn off the ignition switch when the car is in motion.
4. Fix problems such as engine dieseling, heavy surging, repeated stalling or backfiring, and other such indications of below-normal performance immediately.
5. Avoid disconnecting spark plugs to test the ignition whenever possible. If an oscilloscope is not available, do not run the engine for more than 30 seconds with the plug wire off or shorted.

Substrate: The layer, or honeycomb, of aluminum oxide upon which the catalyst (platinum or palladium) in a catalytic converter is deposited.

Monolith: A large block. In a catalytic converter, the monolith is made like a honeycomb to provide several thousand square yards of catalyst surface area.

Extruded: Shaped by forcing through a die.

■ Catalytic Converter Odors

Although catalytic converters control HC and CO emissions, they also produce other undesirable emissions in small quantities. For example, gasoline has a little bit of sulfur in it. This reacts with the water vapor inside a converter to produce hydrogen sulfide. This toxic byproduct has the distinct odor of rotten eggs. The smell is usually noticed while the engine is warming up, or during deceleration.

When the odor is very strong at normal operating temperatures, it may mean that the engine is out-of-tune and is running too rich. In many such cases, the carburetor idle mixture screws are not set right, although the problem also may be caused by a high fuel level or some other carburetor problem. Since the odor does not necessarily mean an incorrect mixture adjustment, the carburetor should not be adjusted simply to get rid of the odor. Changing brands of gasoline may help control the odor in some cases, since the amount of sulfur present in gasoline varies from one brand to another.

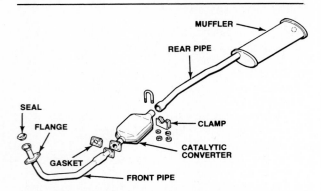

Figure 19-5. This catalytic converter is installed with a bolted flange at one end and a clamp at the other. (AMC)

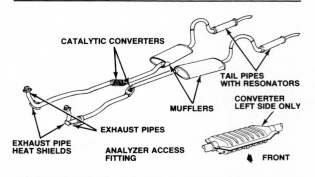

Figure 19-6. This dual exhaust system has two catalytic converters. (Chrysler)

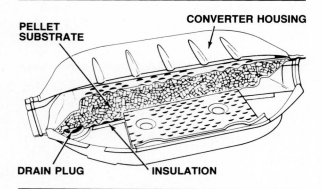

Figure 19-7. The AC pellet-type catalytic converter is used by American Motors and General Motors.

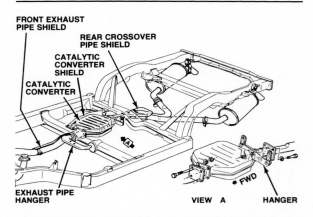

Figure 19-8. General Motors catalytic converter installation with a heat shield. (Chevrolet)

OXIDATION CONVERTER INSTALLATIONS

Federal regulations require that a catalytic converter last a minimum of 50,000 miles. No service should be required during this period. All oxidation converters are installed between the engine and the muffler. Reduction converters must be installed between the engine and the oxidation converter. Most converters are located under the passenger compartment. They are connected to the exhaust system in one of three ways:
1. A bolt-together flange, figure 19-5
2. A slip fit with a clamp, figure 19-5
3. Welded to the exhaust pipes.
Most single exhaust systems have only one converter. Dual exhaust systems may have a converter on one side only or a separate converter on each side, figure 19-6. The catalytic converters used by domestic automakers are discussed below.

American Motors and General Motors Systems

AMC and GM cars use pellet converters made by the AC Division of General Motors, figure 19-7. Since the converter housing is insulated, it remains at about the same temperature as the outside of the muffler. Thus, only a few cars, such as the Corvette, figure 19-8, require a heat shield. The pellet catalyst can be removed from the converter housing through a drain plug with special equipment. This permits replacement with a fresh catalyst charge after 50,000 miles or if the catalyst material is damaged.

Chrysler Systems

Chrysler products use the monolithic converter, figure 19-9. In case of damage or after 50,000 miles, the entire unit must be replaced. Since this type of converter transmits quite a lot of heat through the housing, heat shields, figure 19-10, are used to protect the passenger com-

Catalytic Converters

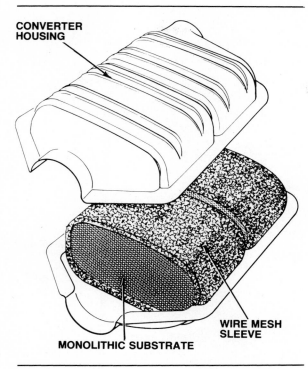

Figure 19-9. Chrysler catalytic converter with a monolithic substrate.

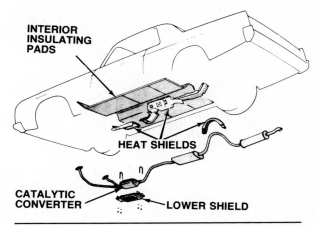

Figure 19-10. Chrysler catalytic converter heat shield installation. (Chrysler)

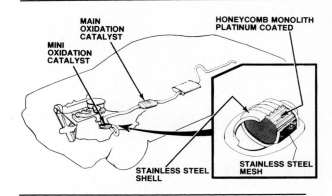

Figure 19-11. Chrysler mini-converter installation.

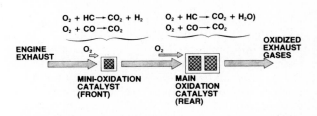

Figure 19-12. Two-stage oxidation reaction with Chrysler's mini-converter. (Chrysler)

partment, the automatic transmission and its cooler lines, the torsion bars, the rear shock absorbers, and other chassis components.

Size, shape, and placement of these shields vary according to the car model. The underbody floor pan above the converter is heavily shielded, and more heat insulation is placed under the floor mats. Those models with both a converter and an air pump system use even more shielding, since the air pump causes a converter to operate at a higher temperature than one without the pump. Air-pump-equipped models, as well as models sold in California, also use a protective grillwork under the converter. This guard prevents direct housing contact with burnable materials. It also helps cool the converter housing by radiating heat to the airflow beneath the car.

On some 1977 and later models, Chrysler uses a small oxidation-type converter welded into the engine exhaust pipe 6 to 12 inches from the exhaust manifold, figure 19-11. This mini-catalytic converter is called an underhood unit and is used along with the regular, or main underfloor, converter. The mini-converter is of the same design as the regular converter and begins the oxidation of HC and CO before they reach the main underfloor converter. This results in more complete oxidation of the exhaust gas, figure 19-12.

Some 1975 models use a catalyst protection system (CPS), figure 19-13. This prevents fuel from reaching the converter during closed-throttle deceleration after high-speed cruising. The CPS electronic speed switch determines engine rpm. When engine speed reaches approximately 2,000 rpm, the switch energizes a throttle positioner solenoid to extend its plunger.

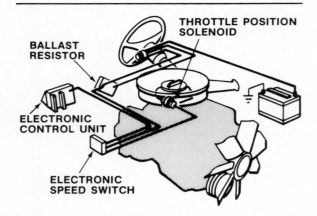

Figure 19-13. Chrysler's catalyst protection system.

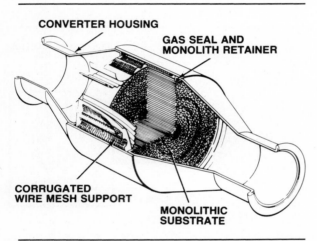

Figure 19-14. Ford's catalytic converter with a monolithic substrate.

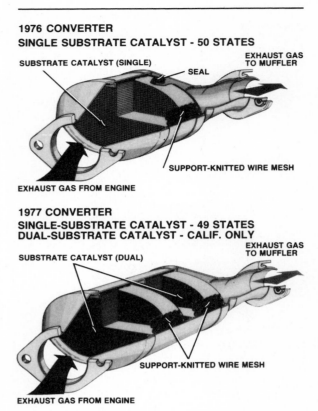

Figure 19-15. Ford's single- and dual-substrate catalytic converters. (Ford)

If the accelerator pedal is released when engine speed is above 2,000 rpm, the solenoid plunger holds the throttle valve open at the fast-idle position. This provides intake air to balance the air-fuel mixture and speed up combustion to protect the converter from overheating. Once engine speed drops below 2,000 rpm, the speed switch deenergizes the solenoid, and the throttle returns to a normal slow-idle position as the solenoid plunger retracts. The CPS system was discontinued as a running change on 1975 models.

Ford Systems

Ford products use a monolithic converter, figure 19-14. Some 1977 and later cars have converters with two monolithic substrates, figure 19-15. The entire unit must be replaced after 50,000 miles or in case of damage. Converters are located underneath the seating area floor pan on some models, and under the toeboard area on other models. Since this type of converter transmits a high level of heat through the housing, various formed and perforated exhaust shields are welded or clamped in place, according to the particular model.

The Ford catalytic converter system needs a secondary air source to keep enough oxygen in the exhaust for complete oxidation of the HC and CO. This is provided by the Thermactor air injection system, figure 19-16, which we discussed in Chapter 16. We are interested in the Thermactor system in this chapter because of the role it plays to protect the catalyst from overheating.

A 2-port solenoid vacuum valve, figure 19-17, which is normally closed, replaces a normally open valve of the same type used before 1975 with other Ford emission systems. This new valve does the same thing as the older one: it opens and closes the vacuum circuit according to an electrical signal. The solenoid vacuum valve controls the air-bypass operation of the diverter valve by controlling vacuum to a vacuum differential valve (VDV).

The solenoid vacuum valve may be energized in several different ways, depending on which car model it's on. Some vehicles use a tempera-

Catalytic Converters

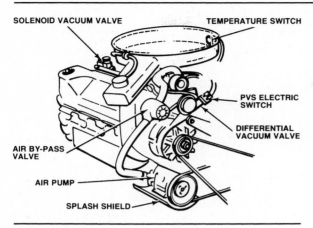

Figure 19-16. Ford's air injection system supplies extra air to the catalytic converter.

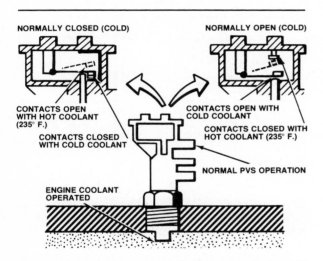

Figure 19-18. This electric ported vacuum switch is used in some Ford catalytic converter air injection systems.

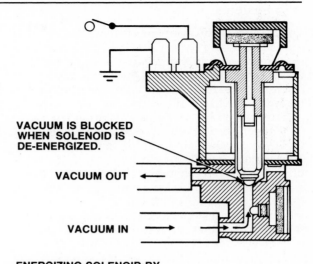

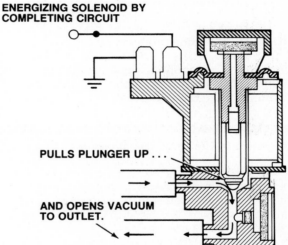

Figure 19-17. This solenoid valve controls vacuum to the air injection diverter valve on some Fords with catalytic converters.

ture switch in the air cleaner. Those with catalyst protection use a floor-mounted switch, or either a normally open or a normally closed electric ported vacuum switch (PVS), figure 19-18. The switch works with the solenoid vacuum valve for bypass operation of the diverter valve if coolant temperature reaches or exceeds 235° F.

Whether a floor-mounted switch or a PVS is used, the result is the same. The switch or PVS activates the solenoid vacuum valve. This causes the vacuum differential valve to temporarily dump vacuum to the diverter valve during deceleration. It also prevents backfire when the exhaust mixture is too rich. In each case, the VDV prevents the converter from overheating and damaging the catalyst.

OXIDATION-REDUCTION CONVERTER INSTALLATIONS

The catalytic converters described in the previous section are all oxidation converters. They were first used on 1975 cars, and they are still being used. In 1977, Volvo and Saab introduced the first oxidation-reduction converters to control all three major exhaust emissions. In 1978, two domestic automakers — Ford and General Motors — began using hybrid, or 2-stage, converters to control HC, CO, and NO_x. These installations are described in the following paragraphs.

Ford Systems

The 2.3-litre, 4-cylinder engine used in California Pintos and Bobcats for 1978 has a closed-loop, feedback emission control system, figure 19-19. It has an oxidation-reduction catalytic converter, an electronically controlled carburetor, and an exhaust gas oxygen sensor.

The "closed-loop feedback" is provided by the oxygen sensor, which continuously monitors the unburned air-fuel ratio in the exhaust. This signal is sent to the electronic control module, which also receives signals from a throttle-angle switch, two vacuum switches, and the ignition control side of the ignition coil. The module sends a signal to the solenoid regulator that controls the vacuum supply to the carburetor power valve. This provides the precise control of the intake air-fuel ratio needed by the hybrid converter.

The front half of the converter has an oxidation-reduction catalyst that controls HC, CO, and NO_x. It uses platinum and rhodium as catalysts on a monolithic substrate. The rear half of the converter has only a platinum catalyst to further oxidize HC and CO.

■ Future Of The Three-Way Converter

The 3-way catalytic converter is the newest development in the fight against high emissions. While the ordinary platinum-palladium converter has no effect on NO_x emissions, the addition of rhodium to the converter makes it effective in reducing all three major pollutants — HC, CO, and NO_x. Although the 3-way converter arrived with some 1977-78 models, its future already is in considerable doubt.

Today's 3-way catalytic converters use five times more rhodium at a ratio of about 5:1. Yet there is about 20 times more rhodium than platinum in the earth. Obviously, there is not enough rhodium to supply the auto industry's massive needs for very long. This is especially true when you realize that current platinum productions will not be enough to meet the demand beyond 1981.

Automakers are confident that world supplies of platinum and rhodium (none of which are located in the U.S.) are far greater than those now being mined. However, they do say that the cost of a converter may rise to $600 or more in the next decade. Scientists are aware of the problem. They have already found that a 3-way converter can be made with the same platinum-to-rhodium ratio found in nature. What they have not yet determined is whether such a converter can lower emissions enough to meet the legal limits.

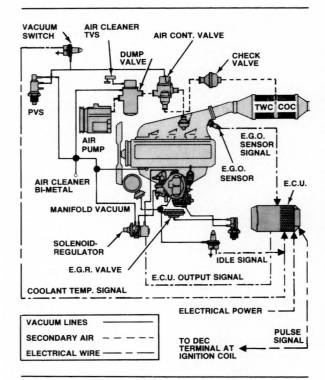

Figure 19-19. For 1978, Ford introduced this 3-way converter on some 4-cylinder engines. It works along with a closed-loop feedback system that uses an exhaust gas oxygen sensor and an electronically controlled carburetor. (Ford)

During engine warmup, the air injection pump supplies extra air to the exhaust manifold so that it passes through the converter. When the engine reaches operating temperature, a vacuum signal from the temperature vacuum switch redirects the air injection to the middle of the converter. The extra air is then supplied *only* to the oxidation catalyst in the rear half of the converter.

General Motors Systems

Some of GM's 1978 California small cars also have a closed-loop feedback emission control system. It works about the same as the Ford system. It is installed on all 1978 California Sunbirds with the 151-cid 4-cylinder engine, as well as some Skyhawks with the 231-cid V-6. This is the first production use of a closed-loop system on a V-type engine.

Catalytic Converters

SUMMARY

Catalytic converters in the exhaust system promote a chemical reaction to change polluting byproducts of combustion into harmless water and carbon dioxide. Two types of converters are used: oxidation and reduction. The oxidation-type removes HC and CO emissions from the exhaust gases; reduction converters remove NO_x.

Converters use a monolithic or pellet-type substrate coated with platinum, palladium, or rhodium to provide a surface for the oxidation or reduction reaction to take place. These catalysts can be damaged by too much heat, the use of leaded fuel, or too much unburned fuel reaching their surface. There is no way to determine the efficiency of a given converter, nor do they require any service during the 50,000 mile life-span required by U.S. Federal regulations.

Catalytic converters were first used in 1975. Some early converters have a temperature protection system to prevent converter damage from overheating. These were abandoned on most 1976 and later vehicles. All Chrysler and Ford converter systems use a series of heat shields to protect chassis components from heat damage. AMC and GM cars use the insulated pellet converter, whose housing does not transmit excessive heat, and does not require heat shields in most cases.

Hybrid oxidation-reduction catalytic converters were introduced on some 1977-78 cars to control all three major exhaust emissions: HC, CO, and NO_x.

Review Questions

Choose the single most correct answer. Compare your answers to the correct answers on page 248.

1. Catalytic converters:
   a. Increase the HC content in exhaust emissions
   b. Neither add nor remove the oxygen from exhaust emissions
   c. Improve oxidation of HC and NO
   d. Improve oxidation of HC and CO

2. The catalyst material in an oxidation catalytic converter is:
   a. Platinum or palladium
   b. Aluminum oxide
   c. Stainless steel
   d. Lead oxide

3. A catalyst:
   a. Slows a chemical reaction
   b. Heats a chemical reaction
   c. Increases, but is not consumed by, a chemical reaction
   d. Combines with the chemicals in the reaction

4. A reduction reaction:
   a. Adds oxygen to a compound
   b. Removes oxygen from a compound
   c. Reduces HC in exhaust gases
   d. Removes H_2O from exhaust gases

5. The outer shell of the oxidation catalyst is made of:
   a. Aluminum oxide pellets
   b. Platinum or palladium
   c. A honeycomb monolith
   d. Aluminized or stainless steel

6. A catalyst is deposited on:
   a. A stainless steel shell
   b. An aluminum oxide substrate
   c. A stainless steel mesh
   d. All of the above

7. Cars with catalytic converters are required to use:
   a. Premium leaded fuel
   b. Unleaded fuel
   c. Low-lead fuel
   d. Any of the above

8. Which of the following will damage the converter:
   a. Push starting the engine
   b. Engine misfiring
   c. Long idling periods
   d. All of the above

9. Which of the following is *not* true for cars with catalytic converters:
   a. Engines should not be cranked for more than 60 seconds
   b. Spark plugs should always be disconnected to test ignition
   c. Engine dieseling, surging, and stalling should be fixed immediately
   d. Ignition should not be turned off while car is moving

10. The reduction converter must be located:
    a. Between the engine and the oxidation converter
    b. Between the oxidation converter and the muffler
    c. After the muffler
    d. Any of the above

11. Chrysler introduced a miniconverter in which model year:
    a. 1975
    b. 1976
    c. 1977
    d. 1978

12. Ford introduced the dual substrate catalytic converter:
    a. In California in 1976
    b. In California in 1977
    c. Nationally in 1976
    d. Nationally in 1977

13. Hybrid converters were first used on domestic cars in:
    a. 1975
    b. 1976
    c. 1977
    d. 1978

14. In a Ford closed-loop feedback system the air injector:
    a. Supplies extra air to the exhaust manifold during normal operation
    b. Supplies extra air only to the oxidation converter during normal operation
    c. Cuts air to the manifold during warmup
    d. All of the above

15. The first production use of closed-loop systems on V-type engines were introduced by:
    a. Ford in 1978
    b. AMC in 1975
    c. GM in 1978 California Sunbirds
    d. All Chrysler 1978 models

NIASE Mechanic Certification Sample Test

This sample test is similar in format to the series of tests given by the National Institute for Automotive Service Excellence (NIASE). Each of these exams covers one of eight areas of automobile repair and service. The tests are given every fall and spring in about 250 cities throughout the United States.

For a mechanic to earn certification in a particular field, he or she must successfully complete one of these tests, and have at least two years of "hands-on" experience (or a combination of work experience and auto mechanics training). Successfully finishing all eight tests earns the mechanic the certification as a General Automobile Mechanic. More than 120,000 mechanics have taken these tests since the program was begun in 1972.

In the following sample test, some of the questions were provided by NIASE. *All* of the questions follow the form of the national exams. Learning to take this kind of test will help you if you plan to apply for certification later in your career.

For more information, write to:
National Institute for
  Automotive Service Excellence
1825 K Street N.W.
Washington, D.C. 20006

1. Pressure and volume tests of a mechanical fuel pump are both below specs.

   Mechanic A says that an air leak in the fuel line between the tank and pump could be the cause.

   Mechanic B says a plugged fuel tank pick-up filter could be the cause.

   Who could be right?

   a. A only
   b. B only
   c. Either A or B
   d. Neither A nor B

2. A compression test has been made on a 6-cylinder engine. Cylinders 3 and 4 have readings of 10 psi. The other cylinders all have readings between 130 and 135 psi.

   Mechanic A says this could be caused by a blown head gasket.

   Mechanic B says this could be caused by wrong valve timing.

   Who could be right?

   a. A only
   b. B only
   c. Either A or B
   d. Neither A nor B

3. The exhaust manifold heat riser valve is stuck in the open position.

   Mechanic A says this can cause poor gas mileage.

   Mechanic B says this can cause the intake manifold vacuum to be lower than normal.

   Who is right?

   a. A only
   b. B only
   c. Both A and B
   d. Neither A nor B

4. Mechanic A says the setup shown below is used to check carburetor float level.

   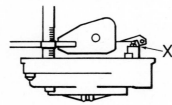

   Mechanic B says tang "X" should be bent to adjust carburetor float level.

   Who is right?

   a. A only
   b. B only
   c. Both A and B
   d. Neither A nor B

5. While the engine is running, a mechanic pulls the PCV valve out of the valve cover and puts his thumb over the valve opening. There are no changes in engine operation.

   Mechanic A says the PCV valve could be stuck in the open position.

   Mechanic B says the hose between the intake manifold (carburetor base) and the PCV valve could be plugged.

   Who is right?

   a. A only
   b. B only
   c. Either A or B
   d. Neither A nor B

6. Mechanic A says the automatic choke is opened by manifold vacuum.

   Mechanic B says the automatic choke is closed by spring force.

   Who is right?

   a. A only
   b. B only
   c. Both A and B
   d. Neither A nor B

7. All of these statements are true about the operation of the EGR (exhaust gas recirculation) emission control system *EXCEPT*:

NIASE Sample Test

a. The EGR valve allows exhaust gases to enter the intake manifold
b. The EGR valve is open at engine idle
c. The EGR system reduces NO$_x$ (oxides of nitrogen) emissions
d. The EGR system reduces combustion chamber temperatures

8. A car with an air pump emission control system backfires when decelerating. Which of these should the mechanic check?

   a. The operation of the exhaust manifold check valve
   b. The output pressure of the air pump
   c. The operation of the diverter or gulp valve
   d. The air manifolds for restrictions

9. Which of these is LEAST LIKELY to cause a car to hesitate (stumble) when the gas pedal is depressed quickly?

   a. Retarded ignition timing
   b. Low carburetor float level
   c. Leaking carburetor accelerator pump check valve
   d. Leaking carburetor power valve

10. All of these statements are true about the device shown below EXCEPT:

    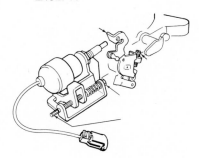

    a. It controls engine idle speed
    b. It can cause dieseling (after-run) when not adjusted correctly
    c. It opens the carburetor throttle plates when the ignition is on
    d. It controls engine fast idle during warmup

11. Which of these statements is true about adjusting valve lash on an engine with solid lifters?

    I. Too much valve lash can cause poor engine performance.
    II. Not enough valve lash can cause valve burning.

       a. I only
       b. II only
       c. Both I and II
       d. Neither I nor II

12. The valve shown below controls ignition timing:

    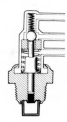

    a. Advance during deceleration
    b. Retard during high speed driving
    c. Advance during high temperature idle
    d. Retard during high temperature idle

13. The three *major* air pollutants emitted by motor vehicles are:

    a. Sulfates, particulates, carbon dioxide
    b. Sulfates, carbon monoxide, nitrous oxide
    c. Carbon monoxide, oxides of nitrogen, hydrocarbons
    d. Hydrocarbons, carbon dioxide, nitrous oxide

14. Completely closed PCV systems were required nationwide in the United States beginning in which model year:

    a. 1961
    b. 1964
    c. 1968
    d. 1971

15. Generally, a fuel pump should deliver a fuel volume of:

    a. One pint in one minute at 500 rpm
    b. One quart in one minute at 500 rpm
    c. One gallon in one minute at 500 rpm
    d. Twenty ounces in one minute at 500 rpm

16. When the fuel filter in the illustration below is installed, the arrow must point toward the:

    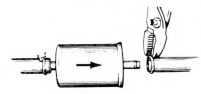

    a. Fuel pump
    b. Fuel tank
    c. EEC canister
    d. Carburetor

17. The recommended service for paper element air filters is to:

    a. Lightly oil the filter paper at the recommended intervals
    b. Blow them out with compressed air at regular intervals
    c. Replace them at regular intervals
    d. Rotate them in the air cleaner housing at regular intervals

18. When the engine is off, the hot air passage remains open and the cold air passage remains closed in a vacuum-operated air cleaner. This could be caused by:

    a. High ambient air temperature
    b. A broken vacuum motor spring
    c. Low ambient air temperature
    d. The heat control valve stuck closed

19. Which of the following is *not* a carburetor float adjustment:

    a. Float toe
    b. Float reach
    c. Float drop
    d. Float level

20. A choke adjustment specification that calls for the choke to be set on "Index" means:

    a. The choke rod or link must be bent to a specified length
    b. The vacuum break link must be bent until the U-bend is closed
    c. The index mark on the choke cover must be aligned with a similar mark on the housing
    d. The choke link must be placed in the index hole in the choke lever

21. Choke vacuum break adjustment on many carburetors is made by:

    a. Rotating the choke cover to align the index mark with a mark on the housing
    b. Bending the vacuum break link at the U-bend
    c. Opening or closing the vacuum break air bleed
    d. Repositioning the choke rod in the fast-idle link

22. A test lamp connected between the electric assist choke lead and ground on a Ford choke does not light when the engine is running. This could be caused by:

    a. A defective heating element
    b. A defective thermostatic switch
    c. A defective ground connection at the choke
    d. Current not being available at the choke lead connection

23. An ohmmeter connected between the intake manifold and the choke control switch lead on the Chrysler choke shown below indicated infinite resistance.

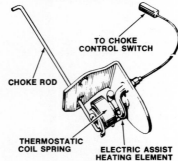

This could be caused by:
a. An open circuit in the choke heating element
b. A short circuit between the heating element and the manifold
c. A normal heating element in good operating condition
d. A defective control switch

24. Which of the following operations must be performed *before* carburetor idle speed and mixture adjustment:
a. Mechanical valve lifter adjustment
b. Manifold heat control valve service
c. Complete ignition system service
d. All of the above

25. On a car with a throttle stop solenoid, normal slow idle is usually adjusted with the solenoid:
a. Deenergized and retracted
b. Energized and extended
c. Removed from the carburetor
d. None of the above

26. Idle speed adjustment on a carburetor with an idle air bypass circuit is made by turning a large screw that:
a. Varies the throttle valve position
b. Regulates fuel flow through the idle circuit
c. Varies airflow through the bypass passage
d. Opens an air-fuel passage in the secondary idle circuit

27. Adjusting idle mixture by turning the mixture screws to obtain the smoothest idle and then turning them in leaner until the speed drops a specified amount is called the:
a. Lean best idle method
b. Lean drop method
c. One-quarter turn rich method
d. None of the above

28. The Ford artificial enrichment tests are performed with:
a. The air cleaner in place and the air injection system disconnected
b. The air cleaner in place and the air injection system connected
c. The air cleaner removed and the air injection system disconnected
d. The air cleaner removed and the air injection system connected

29. If the rpm gain measured during artificial enrichment tests is less than specified:
a. The air-fuel ratio is too lean
b. The engine is overheated
c. The air-fuel ratio is too rich
d. The air injection system is not working correctly

30. When an engine is run at about 1,800 rpm with no load and the vacuum hose is removed from the EGR valve, engine speed remains unchanged. This probably means:
a. The EGR valve is working correctly
b. Excessive vacuum is being applied to the EGR valve
c. The EGR valve is not opening
d. None of the above

31. An engine idles roughly and stalls on light acceleration. When the vacuum line is disconnected from the EGR valve, the problem disappears. This probably means:
a. The EGR valve is stuck closed
b. The EGR valve is getting a weak vacuum signal
c. The EGR valve diaphragm spring is broken
d. The EGR valve is incorrectly opening at idle

32. High HC exhaust emissions are often due to:
a. An overheated engine
b. A restricted air cleaner
c. Ignition system problems
d. An inoperative EGR system

33. Normal HC and high CO emissions at idle are probably *NOT* due to:
a. A lean misfire
b. A restricted air filter
c. A restricted PCV system
d. The choke stuck partially closed

34. Air injection pumps should be oiled:
a. At regular intervals recommended by the carmaker
b. Only if they become noisy
c. Only when a defective check valve is replaced
d. Never

35. During inspection of an air injection system, a brittle, burned hose is found on the upsteam side of a check valve. This may be due to:
a. The hose being positioned too close to the engine
b. The air pump developing excessive air pressure
c. The injection nozzles being plugged and allowing backpressure to build up in the system
d. The check valve allowing exhaust gas to leak back into the air injection system

36. Which of the following precautions should be observed when servicing a car with a catalytic converter:
a. Avoid prolonged cranking
b. Avoid running the engine for prolonged periods in a shop with poor air circulation
c. Avoid push starting cars with manual transmissions
d. All of the above

37. Which of the following operations should be performed before testing engine compression:
a. Be sure battery is fully charged and engine cranks freely
b. Be sure engine is at normal temperature and throttle is held open
c. Disconnect or group ignition system
d. All of the above

38. A cylinder power balance test is performed while the engine is running by:
a. Shorting the ignition of each cylinder and noting rpm drop
b. Shorting the ignition of each cylinder and noting rpm gain
c. Running the ignition for each cylinder in an open circuit condition and noting the available secondary voltage
d. Opening a vacuum port at the carburetor or intake manifold and noting rpm gain

39. During a cylinder power balance test of a 4-cylinder engine, rpm-drop readings of 110, 115, 75, and 120 were recorded for cylinders 1 through 4, respectively. The results of a compression test on cylinders 1 through 4 were 135 psi, 139 psi, 90 psi, and 142 psi. When oil was injected into number 3 cylinder, compression rose to 125 psi. The problem could be:
a. Burned exhaust valves
b. A blown head gasket leaking into the water jacket
c. Broken piston rings
d. Collapsed hydraulic valve lifters

Glossary of Technical Terms

Adsorption: A chemical action by which liquids or vapors are gathered on the surface of a material. In a vapor storage canister, fuel vapors are attached to the surface of charcoal granules.

Air-Fuel Ratio: The ratio of air to gasoline by weight in the air-fuel mixture drawn into an engine.

Air Injection: A way of reducing exhaust emissions by injecting air into each of the exhaust ports of an engine. It mixes with the hot exhaust and oxidizes the HC and CO to form H_2O and CO_2.

Ambient Temperature: The temperature of the air surrounding a particular device or location.

Analog: Describing a computer that uses similar (analogous) electrical signals to make its calculations.

Antiknock Value: The characteristic of gasoline that helps prevent detonation or "knocking."

Antioxidant Inhibitors: A gasoline additive that prevents the formation of gum.

Armature: The movable part in a relay. The revolving part in a generator or motor.

Atomization: Breaking down into small particles or a fine mist.

Atmospheric Pressure: The pressure on the earth's surface caused by the weight of air in the atmosphere. At sea level, this pressure is 14.7" psi at 32° F (0° C).

Backfire: The accidental combustion of gases in an engine's intake or exhaust system.

Backpressure: The resistance, caused by turbulence and friction, that is created as a gas or liquid is forced through a passage.

Baffle: A plate or obstruction that restricts the flow of air or liquids. The baffle in a fuel tank keeps the fuel from sloshing as the car moves.

Bimetal Temperature Sensor: A device made of two strips of metal welded together. When heated, one side will expand more than the other, causing it to bend.

Blowby: The leakage of combustion gases and unburned fuel past an engine's piston rings.

Boost: A measure of the amount of air pressurization, above atmospheric, that a supercharger can deliver.

Bore: The diameter of an engine cylinder.

Bottom Dead Center: The exact bottom of a piston's stroke. Abbreviated: bdc.

Camshaft Overlap: The brief period of time during which a camshaft lobe may be forcing one valve open before another lobe has allowed another valve to close.

Carbon Monoxide: An odorless, colorless, tasteless poisonous gas. A major pollutant given off by an internal combustion engine.

Carburetor Icing: A condition that is the result of the rapid vaporization of fuel entering a carburetor; the temperature drops enough to freeze the water particles in the airflow.

Catalyst: A substance that causes a chemical reaction, without being changed by the reaction.

Catalytic Converter: A device installed in an exhaust system that converts pollutants to harmless byproducts through a catalytic chemical reaction.

Catalytic Cracking: A process of refining oil so that the gasoline which is produced has a low sulfur content.

Catalytic Oxidation: The oxidation of exhaust emissions that occurs in a catalytic converter. This chemical reaction changes HC and CO to H_2O, CO_2 and other harmless byproducts.

Check Valve: A valve that permits flow in only one direction.

Clearance Volume: The volume of a combustion chamber when the piston is at top dead center.

Combination Valve: A valve used on the fuel tanks of some Ford cars that allows fuel vapors to escape to the vapor storage canister, relieves fuel tank pressure, and lets fresh air into the tank as fuel is withdrawn. Similar to a liquid-vapor separator valve.

Compression Ratio: The total volume of an engine cylinder divided by its clearance volume.

Detonation: Also called knocking, pinging. An unwanted explosion of an air-fuel mixture caused by high heat and compression.

Diaphragm: A thin, flexible wall that separates two spaces, such as the diaphragm in a mechanical fuel pump.

Dieseling: A condition in which extreme heat in an engine's combustion chamber continues to ignite excess fuel after the ignition switch has been turned off.

Digital: Describing a computer that makes calculations with quantities represented electronically as digits.

Displacement: A measurement of the volume of air displaced by a piston as it moves from bottom to top of its stroke. Engine displacement is the piston displacement multiplied by the number of pistons in an engine.

Diverter Valve: Also called a dump valve. A valve used in an air injection system to prevent backfire. During deceleration it "dumps" air from the air pump into the atmosphere.

Dynamometer: A device that measures the power of an engine or motor.

Eccentric: Off center. A shaft lobe which has a center different from that of the shaft.

Evaporative Emission Control (EEC): A way of controlling HC emissions by collecting fuel vapors from the fuel tank and carburetor fuel bowl vents and directing them through an engine's intake system.

Exhaust Gas Recirculation (EGR): A way of reducing NO_x emissions by directing unburned exhaust back through an engine's intake.

Extruded: Shaped by forcing through a die.

Firing Order: The order in which combustion occurs in the cylinders of an engine.

Float Valve: A valve that is controlled by a hollow ball floating in a liquid, such as in the fuel bowl of a carburetor.

Flooding: A condition caused by heat expanding the fuel in a fuel line. The fuel pushes the carburetor inlet needle valve open and fills up the bowl even when more fuel is not needed.

Four-Stroke Engine: The Otto cycle engine. An engine in which a piston must complete four strokes to make up one operating cycle. The strokes are: intake, compression, power, and exhaust.

Fuel Injector: A nozzle that meters, atomizes, and injects fuel at high pressure into a combustion chamber.

Gulp Valve: A valve used in an air injection system to prevent backfire. During deceleration it redirects air from the air pump to the intake manifold where the air leans out the rich air-fuel mixture.

Headers: Exhaust manifolds on high-performance engines that reduce backpressure by using larger passages with gentle curves.

High-Speed Surge: A sudden increase in engine speed caused by high manifold vacuum pulling in an excess air-fuel mixture.

Hydrocarbon: A chemical compound made up of hydrogen and carbon. A major pollutant given off by an internal combustion engine. Gasoline, itself, is a hydrocarbon compound.

Impeller: A rotor or rotor blade used to force a gas or liquid in a certain direction under pressure.

Inert Gas: A gas that will not undergo chemical reaction.

Integrated Circuits: Electronic circuits in which the parts have been miniaturized and assembled together to greatly reduce the number of parts in the circuit.

Liquid-Vapor Separator Valve: A valve in some EEC fuel systems that separates liquid fuel from fuel vapors.

Lobes: The rounded protrusions on a camshaft that force, and govern, the opening of the intake and exhaust valves.

Manifold Vacuum: Low pressure in an engine's intake manifold below the carburetor throttle valve.

Monolith: A large block. In a catalytic converter, the monolith is made like a honeycomb to provide several thousand square yards of catalyst surface area.

Noble Metals: Metals, such as platinum and palladium, that resist oxidation.

Normally Aspirated: An engine that uses normal vacuum to draw in its air-fuel mixture. Not supercharged.

Octane Rating: The measurement of the antiknock value of a gasoline.

Orifice: A small opening in a tube, pipe, or valve.

Oxidation: The combining of an element with oxygen in a chemical process that often produces extreme heat as a byproduct.

Oxides Of Nitrogen: Chemical compounds of nitrogen and oxygen given off by an internal combustion engine. They combine with hydrocarbons to produce smog.

Particulates: Liquid or solid particles such as lead and carbon that are given off by an internal combustion engine as pollution.

Percolation: The bubbling and expansion of a liquid. Similar to boiling.

Photochemical Smog: A combination of pollutants which, when acted upon by sunlight, forms chemical compounds that are harmful to human, animal, and plant life.

Pintle Valve: A valve shaped like a hinge pin. In an EGR valve, the pintle is attached to a normally closed diaphragm. When ported vacuum is applied, the pintle rises from its seat and allows exhaust gas to be drawn into the engine's intake system.

Plenum Chamber: The area of an intake manifold that the air-fuel mixture from the carburetor enters before it is distributed to the cylinders.

Poppet Valve: A valve that plugs and unplugs its opening by axial motion.

Ported Vacuum: Vacuum immediately above the throttle valve in a carburetor.

Positive Crankcase Ventilation: A method of controlling engine emissions by directing crankcase vapors (blowby) back through an engine's intake system. Abbreviated: PCV.

Potentiometer: A variable resistor.

Glossary of Technical Terms

Pressure Drop: A loss of pressure between two points.

Pressure Differential: A difference in pressure between two points.

Pressure-Vacuum Relief Valve: A valve that will correct a pressure differential between two points, such as the valve used in the filler cap of some fuel tanks.

Pulsating: To expand and contract rhythmically and regularly.

Purge Valve: A vacuum-operated valve used to draw fuel vapors from a vapor storage canister.

Reciprocating Engine: Also called piston engine. An engine in which the pistons move up and down or back and forth, as a result of combustion of an air-fuel mixture at one end of the piston cylinder.

Reduction: A chemical process in which oxygen is taken away from a compound.

Reed Valve: A one-way check valve. A reed, or flap, opens to admit a fluid or gas under pressure from one direction, while closing to deny movement from the opposite direction.

Runners: The short branches of an intake manifold that connect the manifold's plenum chamber to the engine's inlet ports.

Scavenging: A slight suction effect caused by a vacuum drop through a well designed header system. Scavenging helps pull exhaust gases out of an engine cylinder.

Siamesed: Joined together. A siamesed port on an intake manifold is a single port that supplies the air-fuel mixture to two cylinders.

Sintered: Welded together without using heat to form a porous material, such as the metal disc used in some vacuum delay valves.

Siphoning: The flowing of a liquid as a result of a pressure differential, without the aid of a mechanical pump.

Solenoid: An iron core with a wire coil surrounding it. The core moves when electrical current is applied to the coil. Used to convert electrical energy to mechanical energy.

Spark Timing Control: A way of controlling exhaust emissions by controlling ignition timing. Vacuum advance is delayed or shut off at low and medium speeds, reducing NO_x and HC emissions.

Stoichiometric Ratio: An ideal air-fuel mixture for combustion in which all oxygen and all fuel are completely burned.

Stratified Charge Engine: An engine that uses 2-stage combustion: the first combustion is a rich air-fuel mixture in a precombustion chamber, then combustion of a lean air-fuel mixture in the main combustion chamber.

Stroke: One complete top-to-bottom or bottom-to-top movement of an engine piston.

Substrate: The layer, or honeycomb, of aluminum oxide upon which the catalyst (platinum or palladium) in a catalytic converter is deposited.

Sulfur Oxides: Chemical compounds given off by processing and burning gasoline and other fossil fuels. As they decompose, they combine with water to form sulfuric acid.

Supercharging: Use of an air pump to deliver an air-fuel mixture to the engine cylinders at a pressure greater than atmospheric pressure.

Temperature Inversion: A weather pattern in which a layer or "lid" of warm air keeps the cooler air beneath it from rising.

Tetraethyl Lead: A gasoline additive that helps prevent detonation.

Thermal Cracking: A process commonly used to refine oil. The gasoline which is produced has a higher sulfur content than gasoline produced by catalytic cracking.

Thermistor: A semiconductor whose resistance decreases as heat is applied, opposite to the reaction of a normal conductor.

Thermostatic: Refering to a device that automatically responds to temperature changes, in order to activate switches.

Top Dead Center: The exact top of a piston's stroke. Also a specification used when tuning or repairing an engine. Abbreviated: tdc.

Transducer: Devices that change one form of energy into another form of energy. In an ignition system, they may sense a mechanical movement and change it to an electrical signal.

Turbocharger: A supercharging device that uses exhaust gases to turn a turbine that forces extra air into the cylinders.

Two-Stroke Engine: An engine in which a piston completes two stroke to make up one operating cycle.

Vacuum: A pressure less than atmospheric pressure.

Vacuum Lock: A stoppage of fuel flow caused by insufficient air intake to the fuel tank.

Vaporization: Changing a liquid, such as gasoline, into a gaseous state.

Vapor Lock: A condition in which bubbles are formed in a car's fuel system when the fuel gets hot enough to boil. Flow is stopped or restricted as a result.

Venturi: A restriction in an airflow, such as in a carburetor, that speeds the airflow and creates a vacuum.

Venturi Vacuum: Low pressure in the venturi of a carburetor, caused by fast airflow through the venturi.

Volatility: The ability of a liquid to change from a liquid to a vapor.

Volumetric Efficiency: The comparison of the *actual* volume of air-fuel mixture drawn into an engine to the *theoretical maximum* volume that could be drawn in. Written as a percentage.

Index

Air Cleaners:
air ducts, 68
filtering requirements, 65-66, 68
oil bath filter, 68
paper filter, 67
polyurethane filter, 68
replacement of, 67-68
thermostatic controls, 69

Air-Fuel Ratio:
carburetor design, 79-80, 82-84, 87
efficiency of, 27-28, 117, 184
engine requirements, 20, 28-30, 202-203
manifold design, 31-32, 107-108

Airflow Requirements:
determination of, 26
venturi principle, 78
volumetric efficiency, 27

Air Injection:
basic designs, 194-197
Chrysler Corp. systems, 197-198
Ford Motor Co. systems, 197-199
General Motors Corp. systems, 199
principles of, 193-200
pulse systems, 200
second-generation systems, 197

Air Pollution:
components of, 3, 184
history of, 2
legislation, 7
problems of, 6
regulatory agencies, 7
sources of, 38, 184
reduction of, 5, 65, 102, 126, 129, 131, 188, 190, 194, 218

Altitude-Compensating Carburetor, 102-105

American Motors Corporation:
air cleaners, 71-72, 74
catalytic converters, 234
carburetors, 101, 103-104, 136, 150, 161, 173
evaporative emission controls, 44, 46-47
exhaust gas recirculation, 222-223
rollover protection, 38
spark timing controls, 207-208

Antibackfire Valve, 196

Antiknock Valve, 34

Antitoxidant Inhibitors, 34

Atomization of Fuel, 30-31

Atmospheric Pressure, 27, 76-77, 79

Autolite-Motorcraft Carburetors, 103, 136-149

Backfire, 196

Backpressure, 111-112, 221

Bimetal Temperature Sensor, 70

Blowby, 66, 184-185

Boost, 126, 130-131

Bore, 12

Carbon Monoxide, 3-4, 129, 184
control of, 185, 193-194, 230

Carburetors:
altitude-compensating, 102-105
assist devices, 96-102
basic systems, 80-88
filters, 62, 65-75
flooding, 56
icing, 65-66
linkage, 88, 93
placement, 32, 111
types of, 88
venturi designs, 90-91
vacuum in, 78-79

Carter Carburetors, 103-104, 150-160

Catalyst, 230

Catalytic Converter:
design of, 231
care of, 233
emission control, 193, 203, 230
installation, 234, 237
3-way type, 237-238

Catalytic Cracking, 33

Check Valve, 38, 86, 197

Chrysler Corporation:
air cleaners, 72, 74
air injection, 197-198
catalytic converters, 234-236
carburetors, 95-96, 102-103, 150, 161, 173
evaporative emission controls, 44, 48-49
exhaust gas recirculation, 219, 224
rollover protection, 40
spark timing controls, 205, 208-209, 212-213

Clearance Volume, 18

Combination Valve, 43

Combustion, 30, 65, 184, 202, 218

Compression Ratio, 17-18, 126, 129

Cylinder Arrangement, 14

Detonation, 33, 130-131

Diesel Engine, 20-21, 24-25, 118, 128

Displacement, 17-18

Diverter Valve, 196

Draft Tube Ventilation, 184-185

Dynamometer, 29

Electronic Fuel Injection (see: Fuel Injection)

Electric Fuel Pump, 58-60

Emission Controls:
catalytic converters, 230-239
crankcase, 184-190
evaporative, 42-52
exhaust, 193, 218-222
ignition, 203-204, 211
legislation, 7
regulatory agencies, 7

Engines:
diesel, 20-21, 24-25, 118, 128
F-head, 16
four-stroke, 13
horizontally opposed, 15
I-head, 16
inline, 15, 107-108
L-head, 16
normally aspirated, 126, 128
reciprocating, 10, 11
rotary, 22-23
stratified charge, 23-24
two-stroke, 19
V-type, 15, 107-108

Evaporative Emission Control:
American Motors Corp. systems, 44, 46-47

Index

carburetor venting, 43
Chrysler Corp. systems, 46, 48-49
components of, 42-46
Ford Motor Co. systems, 49
General Motors Corp. systems, 51-52
principles of, 43
vapor storage and purging, 44-45

Exhaust Emission Control:
air injection systems, 193-197
catalytic converters, 230-234, 238
engine modifications, 223
exhaust gas recirculation, 218-222
spark timing control, 202-208

Exhaust Gas Recirculation:
American Motors Corp. systems, 222-223
Chrysler Corp. systems, 219, 224
components of, 219
Ford Motor Co. systems, 225-227
General Motors Corp. systems, 219, 228-229
NO_x formation, 218
ported vacuum, 220
principles of, 122

Exhaust Manifold (see: Manifold)

F-Head Engine, 16

Firing Order, 15

Float Valve, 40, 80-81

Flooding, 56-57

Ford Motor Company:
air cleaner, 72-74
air injection, 197-199
catalytic converters, 236-238
carburetors, 91, 96, 98-99, 101, 103, 136, 150, 161, 173
evaporative emission controls, 44, 49
exhaust gas recirculation, 225-227
spark timing controls, 204, 209-211, 214-216

Four-Stroke Engine, 13

Fuel:
additives, 32, 34
atomization of, 30
distribution, 30
gasoline, 26, 32, 37, 233
impurities in, 33
volatility of, 30-32

Fuel Injection:
advantages of, 117
Cadillac system, 122-124
diesel systems, 20, 118
early systems, 117
electronic systems, 118-119, 120-121
mechanical systems, 118, 123
principles of, 117, 124

Fuel Pump:
electrical, 58-60, 119
mechanical, 54-57
operation of, 54, 61, 80

Fuel System:
distribution, 38-41
electrical pumps, 58-60
filters, 61-62
fittings, 42
lines, 38, 40-41
mechanical pumps, 38, 54-57
rollover protection, 38
tanks, 36, 43
tubes, 37
venting requirements, 37

General Motors Corporation:
air cleaners, 74-75
air injection, 199-200
catalytic converters, 234, 238
carburetors, 96, 98-102, 150, 161, 173
evaporative emission controls, 51-52
exhaust gas recirculation, 228-229
rollover protection, 40
spark timing controls, 133, 204-205, 213-214

Gulp Valve, 196

Headers, 113

High-Speed Surge, 222

Holley Carburetors, 104, 161-172

Horizontally Opposed Engine, 15

Hot Spot, 31

Hydrocarbons, 3-4, 32-33, 129, 184
control of, 185, 193-194, 202, 230

I-Head Engine, 16

Ignition Timing, 203-204, 211

Inline Engine, 15, 107-108

Intake Manifold (see: Manifold)

Lead in Gasoline (see: Tetraethyl Lead)

L-Head Engine, 16

Liquid-Vapor Separator Valve, 43, 59

Manifold:
distribution, 30-31, 108
exhaust types, 111-112
heat control, 113
intake types, 30-32, 109, 113
planes, 110
principles of, 31, 107
spread-bore designs, 111
vacuum, 77-78, 81, 114-116

Monolith Converter Construction, 231

Noble Metals, 231

Normally Aspirated Engine, 126, 128

Octane Rating, 33

Oxidation, 230

Oxides of Nitrogen:
as a pollutant, 3-4, 184, 203, 218
control of, 126, 129, 185, 200, 202, 208, 230-231

Particulates, 3, 5

Photochemical Smog, 2, 6

Pintle Valve, 222

Plenum Chamber, 110-111

Pollution (see: Air Pollution)

Poppet Valve, 13

Ported Vacuum, 46, 79

Positive Crankcase Ventilation:
closed systems, 45, 188-191
draft tube ventilation, 184-185
open systems, 185-188
Type 1, 186
Type 2, 186-187
Type 3, 186-187
Type 4, 189

Pressure Differential, 10, 76-77, 184

Pressure-Vacuum Relief Valve, 46

Purge Valve, 45-46

Reciprocating Engine, 10-11

Reduction, 231

Reed Valve, 19

Rochester Carburetors, 104, 173-182

Rotary Engine, 22-23

Scavenging, 113

Sintered Valve Disc, 205

Smog (see: Photochemical Smog)

Spark Timing Control:
American Motors Corp. systems, 207-208
and combustion, 202
and emission control, 203
basic systems, 204-207, 209-211
Chrysler Corp. systems, 205, 208-209, 212-213
electronically controlled timing, 211
Ford Motor Co. systems, 204, 209-211, 214-216
General Motors Corp. systems, 133, 204, 213-214
principles of, 203
speed and transmission controls, 205-207

Stoichiometric Ratio, 28

Stratified Charge Engine, 23-24

Stroke, 12-14

Sulfur Oxides, 5

Supercharging:
advantages of, 126
principles of, 126-127
specific turbocharger designs, 131-133
turbochargers, 127-131
turbocharger controls, 130

Temperature Inversion, 6

Tetraethyl Lead, 34

Thermal Cracking, 33

Thermostatic Control, 69, 114

Turbocharger (see: Supercharging)

Vacuum:
amplifiers, 221
lock, 38, 43
manifold, 77-78
ported, 220
sources of, 10, 79, 84
venturi, 79, 80, 84, 220

Valve Arrangement, 16

Valves:
antibackfire, 96
check, 38, 86, 197
combination, 43, 206
deceleration, 101, 203
diverter, 196
float, 40, 80-81
gulp, 196
liquid-vapor separator, 43, 59
pintle, 222
poppet, 13
pressure-vacuum relief, 46
purge, 46
reed, 19
vacuum delay, 204-205

Vaporization of Fuel, 30-32

Venturi:
Ford's variable venturi, 91
principles of, 78
vacuum, 79-80, 84
variable, 90

Volatility of Fuel, 30-32

Volumetric Efficiency, 26-27

V-Type Engine, 15, 107-108

Answers To Review and NIASE Questions

Chapter 1 Introduction to Fuel Systems and Emission Controls
1.(c) 2.(c) 3.(d) 4.(a) 5.(c) 6.(d)
7.(b) 8.(d) 9.(c) 10.(b)

Chapter 2 Engine Operating Principles
1.(c) 2.(b) 3.(a) 4.(d) 5.(b) 6.(a)
7.(d) 8.(d) 9.(d) 10.(d) 11.(a)
12.(c) 13.(d) 14.(b) 15.(a)

Chapter 3 Engine Air-Fuel Requirements
1.(a) 2.(c) 3.(d) 4.(a) 5.(c) 6.(b)
7.(d) 8.(b) 9.(d) 10.(b) 11.(b)
12.(d)

Chapter 4 Fuel System Overview
1.(c) 2.(b) 3.(a) 4.(c) 5.(d) 6.(c)
7.(a) 8.(d) 9.(c) 10.(d) 11.(b)
12.(b) 13.(a) 14.(c) 15.(d) 16.(b)
17.(c) 18.(a) 19.(d) 20.(d)

Chapter 5 Fuel Pumps and Filters
1.(c) 2.(b) 3.(c) 4.(b) 5.(d) 6.(a)
7.(d) 8.(c) 9.(c) 10.(c) 11.(b)
12.(a) 13.(d) 14.(c) 15.(d)

Chapter 6 Air Cleaners and Filters
1.(d) 2.(b) 3.(a) 4.(c) 5.(b) 6.(a)
7.(c) 8.(a) 9.(c) 10.(b) 11.(d)
12.(a)

Chapter 7 Basic Carburetion
1.(c) 2.(b) 3.(a) 4.(d) 5.(c) 6.(a)
7.(b) 8.(d) 9.(c) 10.(c) 11.(b)
12.(a) 13.(c) 14.(d) 15.(a) 16.(b)
17.(c) 18.(d) 19.(a) 20.(c) 21.(a)
22.(b) 23.(a) 24.(b) 25.(c)

Chapter 8 Intake and Exhaust Manifolds
1.(c) 2.(d) 3.(d) 4.(d) 5.(c) 6.(b)
7.(b)

Chapter 9 Electronic Fuel Injection
1.(b) 2.(c) 3.(d) 4.(d) 5.(b) 6.(d)
7.(b) 8.(c) 9.(b) 10.(a) 11.(c)
12.(b) 13.(c) 14.(a)

Chapter 10 Supercharging and Turbochargers
1.(c) 2.(a) 3.(b) 4.(d) 5.(c) 6.(b)
7.(c) 8.(a) 9.(d) 10.(d) 11.(b)

Chapter 15 Positive Crankcase Ventilation
1.(d) 2.(b) 3.(d) 4.(c) 5.(a) 6.(c)
7.(d) 8.(d) 9.(d) 10.(a) 11.(d)
12.(b) 13.(c) 14.(d)

Chapter 16 Air Injection
1.(b) 2.(a) 3.(c) 4.(d) 5.(a) 6.(b)
7.(b) 8.(d) 9.(c) 10.(b) 11.(a)

Chapter 17 Spark Timing Control
1.(d) 2.(a) 3.(c) 4.(b) 5.(c) 6.(d)
7.(a) 8.(d) 9.(b) 10.(c) 11.(a)
12.(c) 13.(d) 14.(c) 15.(d) 16.(d)
17.(b) 18.(a) 19.(c) 20.(c) 21.(b)
22.(c)

Chapter 18 Exhaust Gas Recirculation
1.(a) 2.(c) 3.(c) 4.(a) 5.(d) 6.(d)
7.(b) 8.(c) 9.(c) 10.(c) 11.(c)
12.(c)

Chapter 19 Catalytic Converters
1.(d) 2.(a) 3.(c) 4.(b) 5.(d) 6.(b)
7.(b) 8.(d) 9.(b) 10.(a) 11.(c)
12.(b) 13.(d) 14.(b) 15.(c)

NIASE Mechanic Certification Sample Test
1.(c) 2.(a) 3.(a) 4.(a) 5.(b) 6.(a)
7.(b) 8.(c) 9.(d) 10.(d) 11.(c)
12.(c) 13.(c) 14.(d) 15.(b) 16.(d)
17.(c) 18.(b) 19.(b) 20.(c) 21.(b)
22.(d) 23.(a) 24.(d) 25.(b) 26.(c)
27.(b) 28.(a) 29.(c) 30.(c) 31.(d)
32.(c) 33.(a) 34.(d) 35.(d) 36.(d)
37.(d) 38.(a) 39.(c)

Acknowledgments

In producing this series of textbooks for automobile mechanics, Chek-Chart has drawn extensively on the technical and editorial knowledge of the nation's carmakers and suppliers. Automotive design is a technical, fast-changing field, and we gratefully acknowledge the help of the following companies in allowing us to present the most up-to-date information and illustrations possible:

Allen Testproducts
American Motors Corporation
Borg-Warner Corporation
Chrysler Corporation
Ford Motor Company
Fram Corporation, A Bendix Company
General Motors Corporation
    AC-Delco Division
    Delco-Remy Division
    Rochester Products Division
    Saginaw Steering Gear Division
    Buick Motor Division
    Cadillac Motor Car Division
    Chevrolet Motor Division
    Oldsmobile Division
    Pontiac Division
Marquette Mfg. Co., a division of
    Applied Power Inc.
Robert Bosch Corporation
South Coast (Calif.) Air Quality
    Management District
Sun Electric Corporation

The authors have made every effort to ensure that the material in this book is as accurate and up-to-date as possible. However, neither Chek-Chart nor Harper & Row nor any related companies can be held responsible for mistakes or omissions, or for changes in procedures or specifications made by the carmakers or suppliers.

This book was reviewed by a selected group of educators from across the nation. Their comments and suggestions were invaluable. These consultants were:

    Larry Bennett, Oakland Community College, Troy, Mich.; Eddie K. Birch, Glendale Community College, Glendale, Ariz.; John Flaherty, Sand Hills Community College, Carthage, N.C.; Clifford H. Ingold, St. Louis Community College, St. Louis, Mo.; Don Nilson, Chabot College, Hayward, Calif.; Jack O'Loughlin, Southern Illinois University, Carbondale, Ill.; George Whitehouse, Vincennes University, Vincennes, Ind.

For Harper & Row, the sponsoring editor is John A. Woods. The production manager is Laura Argento and the designers are James Stockton and Donna Davis.

At Chek-Chart, editorial contributions were made by Jim Ashborn, Sydnie W. Changelon, and William Grinager. Laura Kenyon and Ray Lyons participated in the production of the book, under the direction of Elmer M. Thompson. Original art was produced by Jim Geddes, Gordon Agur, and F.J. Zienty, and coordinated by Gerald McEwan. Paul E. Sanderson is an educational advisor for the series, and the entire project is under the general direction of Robert J. Mahaffay and Robert M. Bleiweiss.

FUEL SYSTEMS AND EMISSION CONTROLS, Classroom Manual and Shop Manual Copyright © 1978 by Chek-Chart, a Division of The H. M. Gousha Company.

All rights reserved. Printed in the United States of America. No part of this publication may be reproduced, stored in a retrieval system, or duplicated in any manner without the prior written consent of Chek-Chart, a Division of the H. M. Gousha Company, P.O. Box 6227, San Jose, CA 95150.

Library of Congress Cataloging and Publication Data:
Chek-Chart, 1978
    Fuel Systems and Emission Controls
    (Canfield Press/Chek-Chart Automotive Series)
Includes index.
1. Classroom Manual.   2. Shop Manual.

ISBN: 0-06-454002-2
Library of Congress Catalog Card No.: 78-448

Fuel Systems and Emission Controls

By Chek-Chart,
a Division of
The H. M. Goushā Company

Ken Layne, CGAM, *Editor*
Kalton C. Lahue, *Contributing Editor*
Gordon Clark, *Managing Editor*

HARPER & ROW, PUBLISHERS, New York
Cambridge, Philadelphia, San Francisco,
London, Mexico City, São Paulo, Sydney